THE IMPULSIVE CLIENT

THEORY, RESEARCH, AND TREATMENT

THE IMPULSIVE CLIENT

CLIENT

THEORY, RESEARCH, AND TREATMENT

Edited by William G. McCown, Judith L. Johnson, and Myrna B. Shure

American Psychological Association, Washington, DC

Published by the
American Psychological Association
750 First Street, NE
Washington, DC 20002

Copies may be ordered from
APA Order Department
P.O. Box 2710
Hyattsville, MD 20784

In the U.K. and Europe, copies may be ordered from
American Psychological Association
3 Henrietta Street
Covent Garden, London
WC2E 8LU, England

Typeset in Goudy by Impressions, A Division of Edwards Brothers, Inc., Madison, WI

Printer: BookCrafters, Inc., Chelsea, MI
Cover designer: Michael David Brown
Technical/production editor: Olin J. Nettles

Library of Congress Cataloging-in-Publication Data

The Impulsive client : theory, research, and treatment / edited by William G.
  McCown, Judith L. Johnson, Myrna B. Shure.
      p.   cm.
    Includes bibliographical references and index.
    ISBN 1-55798-208-2
    1. Impulsive personality. I. McCown, William George. II. Johnson,
Judith. III. Shure, Myrna B.
  RC569.5.I46I47   1993
  616.85′8—dc20   93-11911
                                                            CIP

British Library Cataloguing-in-Publication Data
A CIP record is available from the British Library.

Printed in the United States of America
First edition

This book is dedicated to researchers and clinicians who work with impulsive clients.

CONTENTS

CONTRIBUTORS

Sean Austin, private practice, Columbus Grove, OH

Patricia A. Barnes, Brief Stay Unit, New Orleans Adolescent Hospital

Ernest S. Barratt, Department of Psychiatry and Behavioral Sciences, University of Texas Medical Branch, Galveston

Katherine Bishop, Kenyon College

Michael Bütz, Eastern Montana University

Sheila A. Corrigan, Psychology Service, Veterans Administration Medical Center, New Orleans, LA

Jorge H. Daruna, Division of Child and Adolescent Psychiatry, Tulane University School of Medicine

Philip A. DeSimone, Nathan Kline Institute for Psychiatric Research, Orangeburg, NY

Scott J. Dickman, Department of Psychology, University of Massachusetts–Dartmouth

Eric Y. Drogin, University of Louisville School of Medicine

LaurieAnn Y. Drogin, Jefferson Community College, Louisville, KY

Kathleen L. Eldredge, Stanford University School of Medicine

H. J. Eysenck, Department of Psychology, Institute of Psychiatry, University of London

Sybil B. G. Eysenck, Department of Psychology, Institute of Psychiatry, University of London

Joseph R. Ferrari, Center for Life Sciences, Cazenovia College

Aileen D. Fink, Hahnemann University

Gordon L. Flett, Department of Psychology, York University

Paul L. Hewitt, Department of Psychology, University of Winnipeg

Judith A. Holmes, private practice, Palatine, IL

Judith L. Johnson, Villanova University

Willard L. Johnson, Koochiching Counseling Center, International Falls, MN

Luciano L'Abate, Department of Psychology, Georgia State University; and Cross Keys Counseling Center

Michael R. Lowe, Hahnemann University

Robert M. Malow, Department of Psychiatry, University of Miami Medical Center and Addiction Research and Treatment Center, Highland Park Pavilion/University of Miami Medical Center

Terry Marks-Tarlow, Creativity Research Center of Southern California, La Habra, California

William G. McCown, Nathan Kline Institute for Psychiatric Research and New York University

Ann L. Roedel, Lutheran General Hospital, Niles, IL

Myrna B. Shure, Hahnemann University

Maria E. Touchet, Denison University

Jeffrey A. West, Psychology Service, Veterans Administration Medical Center, New Orleans, LA

Marvin Zuckerman, Department of Psychology, University of Delaware

PREFACE

The primary purpose of this edited volume is to present various theories regarding the etiology and clinical implications of impulsive behaviors. Although this is not a comprehensive anthology, we hope to depict enough diverse theories within one volume to enable readers from dissimilar theoretical perspectives to obtain otherwise inaccessible information. We believe a cross-fertilization of disparate facets of psychology to be critical in maintaining the vibrancy of our discipline through our next centennial. We hope that this volume can provide at least a small dose of "cross-specialization exposure" for a variety of psychologists and other mental health professionals.

Impulsiveness is a vast topic with a large and extensively varied literature. Initially, some of our colleagues pointed out that our attempt to undertake a compilation of chapters with such a diverse construct as impulsivity would be a difficult, if not impossible, task. Others questioned the feasibility of three coeditors with diverse research interests being able to work together at all! Few people (with the possible exception of the tolerant publication staff of APA Books) seemed to understand fully that a book featuring such divergent authors, chapters, and coeditors could serve to celebrate the diversity of our discipline rather than its divisiveness.

Our goal was to maintain the spirit of the scientist–practitioner model by providing a book useful to both clinicians and researchers, with clinical chapters grounded in scientific findings whenever possible. Authors presenting theoretical or research-oriented chapters were encouraged to discuss the practical implications of their research when appropriate. We also wished to highlight the role of impulsiveness in a variety of behaviors where the construct has been historically overlooked. Hence, we have included several chapters involving novel research questions on impulsiveness, such as the chapters by Hewitt and Flett; Ferrari; and Lowe and Eldredge.

Perhaps our most difficult decision as editors regarded areas *not* to include. As alluded to earlier, our first criterion for inclusion was to choose topics we believed to be underemphasized in previous literature on impulsiveness. Our second criterion was to emphasize areas where the role of impulsiveness was not clearly delineated. Consequently, specific chapters regarding several constructs related to impulsiveness (e.g., borderline personality disorder or attention deficit hyperactivity disorder) were not included inasmuch as extensive and comprehensive accounts of these areas are readily available.

Similarly, there were also many potential chapters we wish we could have included, but could not because of constraints of time and space. For example, problems of impulsiveness in the workplace are substantial and may contribute to costly business or industrial errors. HIV disease, major mental illness, medical adherence, and religious and aesthetic values all share a relationship with impulsiveness that warrants further study and explication. Unfortunately, the potential areas of interest far outstripped our resources. Limitations had to be imposed *somewhere* and probably reflect our idiosyncratic biases rather than the importance of a topic. Hence, we ask indulgence from the concerned reader or scholar whose area of interest or expertise has not been included in this book.

ACKNOWLEDGMENTS

Because the domain of this volume is large and necessarily outside of the expertise of any one author, we have made liberal use of outside peer reviewers to assist with anonymous review of many of its chapters. Special thanks—either for atypical encouragement or for participating in the review process—are due to our mentors, colleagues, and friends: Fred Abraham, Marvin Acklin, Bob Abouee, Don Bersoff, Raymond Cattell, Linda Chamberlain, Mark Clark, Mary Ellen Johnson, Al Lagman, Judith Landau-Stanton, Howard Liddle, Lee Matthews, Dan McAdams, Howard Sacks, Alan Sconzert, George Spivack, Crist Stevenson, Patricia Sutker, Gary Thomson, and J. Michael Williams. The first two editors are especially grateful for the commentary provided by Hugh Gannon and Jenny Ornsteen.

Gratitude is also due to the very helpful staff at APA Books, who were quite patient while we experienced numerous delays. Julia Frank-McNeil was immensely helpful in assisting us during the initial process, whereas Mary Lynn Skutley provided the necessary prodding for an edited volume to reach fruition. Special thanks are due to Olin Nettles and Ted Baroody at APA Books, who tirelessly cut through all of our self-imposed obstacles to make sure that this book met its deadlines. Their patience, as well as professionalism, has been greatly appreciated. We look forward to working with all of these people again. Jennifer Posa and Shara Gustman were very kind in providing editorial assistance. Finally, The Nathan Kline Institute and Villanova University were especially considerate in providing resources necessary for completion of this volume.

INTRODUCTION

WILLIAM G. McCOWN, JUDITH L. JOHNSON,
and MYRNA B. SHURE

In the excellent book by Doob (1990), impulsiveness is defined as the absence of reflection between an environmental stimulus and an individual's response. According to this definition, the appropriateness of impulsiveness is largely a function of the demands of the situation at hand. Some stimuli, such as a car suddenly braking on a crowded freeway, may demand immediate behavior without deliberation. Lengthy reflection regarding all of the possible options (e.g., veering, slamming on the brakes, or doing nothing) may prove harmful or even fatal. On the other hand, many situations require careful reflection of all possible responses. A researcher planning an experiment risks uninterpretable results unless he or she thoroughly considers all of the possible methodological confounds and attempts to control for them.

Psychologists have several reasons to be interested in behaviors that are performed impulsively or without reflection. The first reason is theoretical. As Doob (1990) noted, there is always some type of temporal gap between a stimulus and a response. The question of what variables influence this length, how long it is, and the optimal duration for specific situations is both important and interesting to the behavioral scientist. Unfortunately, theorists often overlook the importance of reflectiveness as a behavior that itself warrants systematic study.

Many psychologists are interested in impulsiveness because it may be related to a personality trait (Barratt, 1965). When impulsive behaviors show consistency within the individual, they become of interest to the personality theorist in terms of individual differences in the trait of impulsivity. Furthermore, impulsive behaviors that occur frequently and that are associated with psychological or behavioral dysfunction are of interest to the clinician, because he or she is usually charged with modifying such behaviors in clients seeking treatment for a variety of disorders.

Impulsive behavior is also of interest to psychologists because it is perceived by many to be a rapidly increasing and major social problem

(Brill, 1993). There is a popular belief that many individuals are now less reflective than in previous years or decades and increasingly act without forethought, to the detriment of others. To many lay persons, this tendency partially accounts for spiraling crime rates, increased drug use, and other behavioral problems that appear increasingly common (Public Opinion Online, 1993).

One of the reasons we proposed this edited volume is that we believe, and there is evidence to support the idea, that impulsive behaviors are generally increasing and pose a challenge for society. Even a casual reading of most newspapers suggests that problems relating to an absence of adequate reflection of behavioral consequences are now at an alarming, perhaps even epidemic, proportion. One salient example in the United States is the growing number of deaths among young people (ages 5 to 19) from firearms. Between 1985 and 1990, firearm deaths in this age group rose 64% (Centers for Disease Control, 1993). Although there are undoubtedly many variables operating, it is important to note that more than 70% of these deaths resulted from action that could be described as impulsive or lacking sufficient reflection: ongoing disagreements that get out of control, romantic disagreements that become too heated, or escalated fights over personal property (Saltzman, Mercy, O'Carroll, Rosenberg, & Rhodes, 1992).

As a result, many American citizens are virtual hostages in their own homes, and the quality of life for millions of people, especially in urban areas, suffers incalculably. For example, over 160,000 children have stayed away from school because of fear of handgun or assault weapon violence (U.S. Department of Justice, 1992). Although efforts such as gun control and firearm access restriction will likely stem this tide of fear, aggression, and murder, these solutions alone will not solve the problem of a lack of reflection of behavioral consequences. At best, they can only limit the consequences of impulsiveness by reducing lethality but perhaps not aggression. This is evident in the simultaneously increasing rates of other types of violence. Assault from non-firearm weapons are up 46% in the last decade, despite an increased access to guns by youth (U.S. Department of Justice, 1992).

Impulsivity affects every age group and nearly all aspects of society. Impulsive behaviors are recognized as being a major facet of many childhood behavioral disorders (Koziol, Stout, & Ruben, 1993). Additionally, children are at high risk for being the victims of their parents' or other adults' impulsive behaviors and often have no recourse when a parent or parent figure uses alcohol or illicit drugs impulsively and irresponsibly. The long-term effects of sexual abuse, which often involve impulsive behavior, are still controversial. However, a growing body of research indicates that there are multiple effects of child sexual abuse on both short-term (Conte & Schuerman, 1988) and long-term functioning of the victimized child (Briere & Runtz, 1988). When abuse becomes chronic or severe, it may have permanent psychological consequences, which are only recently being explored (Leifer, Shapiro, Martone, & Kassem, 1991).

Adolescence is commonly viewed as a time when impulsivity may be particularly evident. The popular stereotype is that adolescents think they are invincible and thus behave in a way that may be considered reckless, with little regard for consequences. Yet data suggest that adolescents do not necessarily *think* they are invulnerable. They only *act* as if they are. Indeed, there is a lack of consensus that adolescents underestimate the riskiness of specific behaviors (Quadrel, Fischhoff, & Davis, 1993).

However, clear evidence of adolescent behavioral impulsiveness can be seen in a number of areas. Despite some encouraging trends in certain samples, more adolescents are experimenting with drugs, especially before the age of 15 (Gans & Blyth, 1990). The term "experiment" is probably a misnomer inasmuch as it implies well-conceived plans involving personal risk analysis of the effects of illicit substances. Data suggest that teen drug use is haphazard and is more likely to be influenced by situational factors and availability of the substance. Essentially, such behavior is unplanned, with an absence of deliberation regarding potential consequences.

Impulsiveness is probably most damaging for those with the fewest socioeconomic resources. Minority youths, especially African American young men, are at exceptionally high risk for the effects of impulsive behavior (Hammond & Yung, 1993). Although many popular cultural innovations are attributable to creativity and the absence of rigid constraining reflection in the urban culture (George, 1993), impulsiveness has destructive aspects as well. Young African American urban men are more likely to die of injuries received from a gun wielded by a friend or acquaintance than any other cause. Some of these killings are planned, either as retribution for infractions of the code of the streets or in drug-related disputes. Many more, however, appear "accidental," where tempers get out of hand, with unintended consequences.

Sexual impulsiveness is also at epidemic rates among teenagers. American adolescents are unique among their peers in other developed nations in rates of pregnancy, even when frequency of sexual activity is statistically controlled. More than 87% of these teen pregnancies are unintended, up from 77% a decade ago (Moore, 1992). Furthermore, more than a third of the cases of sexually transmitted diseases appear to have been spread between persons who engage in unplanned sex. With AIDS, where impulsive sexual relations can be fatal, this number may be much higher.

Among adults, problems related to impulsive behaviors are also at unprecedented levels (Brill, 1993). Perhaps in response to issues of lack of control and powerlessness, angry aggression appears to be increasing in disenfranchised groups (Bernard, 1990). Of particular concern is that such aggression is increasingly without focus or occurs seemingly randomly. Domestic violence is also increasing, and although this issue is unclear, spouse abuse clearly has an unplanned, if repetitive, aspect to its occurrence (Langley & Levy, 1977). Criminal activity and subsequent arrests have filled American prisons to capacity, resulting in increasing needs to expand the prisons. Although much of this is related to increased attempts at drug enforcement, there is an alarming increase among prison inmates in the number of crimes that were committed without forethought (Toch, 1992). The assessment of propensity for destructive impulsiveness is often a major concern for the forensic psychologist (Pope, Butcher, & Seelan, 1992).

Impulsiveness has been recognized as a key component of many adult psychiatric disorders (Cloninger, 1987). This is especially true for personality disorders (Costa & McCrae, 1986; Morey, 1993). Impulsive behavior has been linked to substance abuse in general and polydrug abuse in particular (O'Boyle & Barratt, 1993). It has also been implicated in other addictions or para-addictive behaviors, such as excessive shopping, bulimia, and perhaps wrist slashing (Lacey & Evans, 1986; L'Abate, Farrar, & Serritella, 1992). Impulsiveness may also be a key aspect of some types of sexual aggressiveness (McConaghy, 1993) and excessive gambling (Lesieur, 1992).

Psychologists may also be interested in impulsive behaviors because of the inconsistent manner in which society treats their occurrences. The way in which individuals, and also the aggregate social order, address persons who consistently behave impulsively depends in part on how these behaviors are interpreted. Social responses to impulsive behaviors are often contradictory and, consequently, are an interesting topic for social psychologists. On one hand, advertising promotes the message, especially to young people, that unplanned, spontaneous behavior is admirable and fun. Often, this sales pitch is very popular in advertisements for alcohol. It is probably second only to messages glamorizing youth and beauty as the most popular theme for commercials.

On the other hand, society is unclear about the circumstances in which unplanned behavior is socially appropriate. With the possible exception of Mardi Gras and other similar festivals, there are few socially acceptable outlets for impulsiveness. When nonreflective behavior occurs in an inappropriate cultural context, it is often stigmatized, even when it is harmless.

When impulsive behavior hurts the performer or someone else, there is usually a strong social condemnation. The type of response depends on the implicit theory the responder holds regarding the causes of impulsiveness. This response orientation is similar to the one for addictive behaviors, in that there are several implicit belief systems that regulate individual and social reactions (Johnson & McCown, 1992). Perhaps the most popular model of impulsive behavior is the moral model. It assumes that impulsive behavior is evil, either because it hurts others or because it is ultimately self-defeating. Proponents of this tradition believe that societal breakdown of traditional values has resulted in a loosening of impulse control. The solution to controlling problems with impulsive behavior is greater instruction in traditional morality, with harsher sanctions for persons who ignore social constraints. Impulsive behavior from this perspective represents irresponsible behavior, and the solution to its increasing prevalence is greater personal culpability. For those who fail, punishment is seen as a necessary recourse.

Another orientation is the psychotherapeutic paradigm, or, perhaps as it is known in addictions, the disease model. It should be realized that this model does not necessarily imply biological causality and does not inherently make any assumptions regarding the distribution of the traits in question (i.e., dichotomous vs. normally distributed). Instead, the disease model simply suggests that impulsiveness is disordered behavior in need of treatment. Therapies can be as diverse as social learning, cognitive behavioral, behavioral, biological, or psychoanalytic. The assumption, however, is that the impulsive person is disordered or sick, not evil.

A third orientation is that impulsivity is a response to social or personal conditions involving frustration. From a psychological view, support for this is rooted in the classic work of Dollard, Doob, Miller, Mowrer, and Sears (1939). Since the time of that research, many laboratory studies have shown that some animals behave aggressively in response to a variety of stressors, including overcrowding and the failure to receive an expected food reward (Gray, 1991). The frustration–aggression hypothesis assumes that blocking a person's efforts to reach a specific goal results in an aggressive impulse that motivates the person to injure the source of the frustration. Because direct aggression toward the source of frustration is not always possible, aggression may be displaced onto a more neutral target.

Perhaps the dominant perspective endorsed by psychologists is that impulsiveness, aggressiveness, and other nonreflective behaviors represent a deficit in either knowledge or skills (Felson & Tedeschi, 1993). In many ways, the simplest explanation is that impulsive people get into trouble because they do not understand the risks they are taking, or lack specific skills to adapt. This belief has a strong literature and lengthy tradition (e.g., Mischel, 1968). A second and related argument is that people behave impulsively because of the effects of modeling (Bandura, 1986), whereby they observe others doing so and not getting either punished or rewarded.

These social-learning-based explanations have a number of advantages that make them attractive for psychologists. First, they remove the moral stigma associated with impulsive behaviors and the determinacy of the therapeutic model. These explanations are not incompatible with the frustration–aggression model or, for that matter, necessarily incongruent with any of the other orientations. Finally, social learning explanations offer the possibility of interventions that are conceptually straightforward: Provide impulsive persons with better information, role models, and guidance in how to use information, and impulsiveness should be reduced. This strategy has been followed most concisely with adolescents (Baron & Brown, 1991), but has also been used with adults regarding high-risk activities such as smoking and behavior leading to HIV disease.

Still another reason psychologists may be interested in impulsiveness is that so little is truly understood about the phenomenon. Because impulsiveness appears to be a growing problem with wide clinical implications, the reader might think that there would be several clearly articulated and well-known theories regarding its etiology and treatment. However, this is not really the case. Persons interested in research regarding impulsive behavior often have to look harder than they should. Introductory psychology textbooks do not discuss impulsiveness the way that other topics such as aggression, depression, and anxiety are addressed. It is far more common for impulsiveness to be discussed with respect to specific aspects (such as aggression) than in more general terms as a construct.

Clearly, we do not know as much about impulsiveness as we would like, and this should prompt more effort in the psychological and behavioral science community. An exception may be with children: Studies regarding impulsivity and children have been relatively more frequent than in other areas or with other populations, primarily because of the prevalence of attention-deficit hyperactivity disorder (ADHD). In contrast, controlled studies on the reduction of impulsiveness in adults are relatively rare. Even well-controlled studies addressing behavioral patterns specifically linked to impulsivity are uncommon (Powles, 1992). Outcome studies regarding personality disorders are extremely rare and are also fraught with methodological problems (Shea, 1993). Also controversial and lacking in consensus is the treatment with and role of medication in impulsiveness. Although impulsivity clearly has biological correlates (Coccaro, 1989) and perhaps a biological causation, no medication, with the possible exception of stimulants for ADHD, is specifically efficacious in the treatment of impulsive disorders (although fluoxetine for borderline or antisocial personality disorder may be promising; Coccaro, 1993).

One reason that behavioral scientists may have been reticent to study impulsive behaviors and impulsivity is because of the political implications of potential findings. For example, the political implications of the frustration–aggression model are clear: Reduce the stimulus and the response will cease. The social implications of this strategy are not likely to be popular with conservative persons and institutions. Similarly, despite the fact that personality theorists from Adler to the existentialists have accented the importance of personal responsibility for behavior, psychological researchers whose findings suggest that impulsive people are themselves at fault for some or all of their behavior would probably risk the wrath of more liberal colleagues.

Regardless, it is often claimed that violence and other socially destructive impulsive behaviors in America and elsewhere is related to a combination of interactional factors that demand radical social intervention. These include limits on television violence; progressive political policies to eliminate or reduce poverty, racism, sexism, and child abuse; and restrictions on alcohol access and consumption (Coleman, 1985). We do not doubt that these suggestions have their usefulness and would be potentially effective in reducing the effects of impulsive behaviors. Perhaps also effective would be a return to preindustrial lifestyles, adoption of a totalitarian state, or a societal return to uniform religious orthodoxy of the pre-Darwinian era. Desired behavioral endpoints can often be reached from very disparate causal routes (Haynes, 1992) and, in the case of reducing impulsiveness, perhaps from either a liberal or a conservative political pathway. Whether Western democracies can or will adopt political changes to reduce social problems is unclear. In the absence of such change, the psychologist is forced to attempt to work for the common good by better understanding how to change individual behavior in its present context.

ORGANIZATION OF THIS BOOK

The purpose of this book is to increase knowledge and foster research regarding impulsive behavior, in general, and impulsive persons seeking treatment, in particular. Hence, this book was conceptualized and executed in the spirit of the scientist–practitioner model. In this vein, several experts have contributed chapters both reviewing relevant areas of impulsivity and outlining individual programs of research. Additionally, we have sought to include chapters specifically focused on special topics relating to impulsivity as well as intervention concerns. Finally, we hoped to provide researchers and clinicians exposure to a variety of perspectives, especially biopsychological research, which often tends to be unavailable to other subdisciplines of psychology.

This book is divided into three sections. The first section, "Historical Perspective and Etiology of Impulsivity," presents a number of theoretical accounts of why people differ in their absence of reflectiveness. Our strategy here was to include several major theorists whose programs of research have systematically examined this question. There are, unfortunately, a number of other equally well-known researchers who are not represented here. These include such well-known figures as the Blocks, Buss, Cattell, Costa, Gray, Guilford, Kagan, Kendal, McCrae, Millon, Mischel, Plomen, Spivack, Stuss, and many other notable scholars. Unfortunately, space limitations, or in some cases the potential authors' schedules or health, did not allow for the inclusion of contributions by this distinguished list, as much as we might have wished for it. We hope that these authors' works and theories are well represented and apologize to anyone slighted by their omissions.

Chapter 1, by McCown and DeSimone, is an "applied historical" review of impulsiveness as a behavioral disorder. It represents a balance to the view popular in some circles that impulsiveness is only a modern-day problem that did not exist until recent times. This chapter shows that theorists as far back as the ancient Greeks, or even earlier, were concerned about impulsive behaviors and that the possibility of humans behaving impulsively may have been an argument used against the promotion of the spread of democratic governments. The chapter also traces the roots of the 20th century's neglect of the topic of impulsiveness, but is sanguine about future prospects for advancing our knowledge regarding the construct.

In chapter 2, Daruna and Barnes present a summary of knowledge regarding neurodevelopmental factors known to influence impulsiveness. In addition to providing a discussion on precursors of and early influences on impulsivity, these authors review findings relating to central nervous system involvement, including monoamine metabolites and electroencephalogram activity. The role of neural representations in impulsiveness is also addressed. The discussion regarding methodological limitations inherent in psychobiological research is especially illuminating and of interest to any researcher interested in exploring physiological correlates of behavior.

Chapter 3, by Ernest Barratt, discusses one researcher's lifelong career devoted to understanding the causes of impulsive behavior. An excellent example of a systematic program of research, this chapter includes discussions of biological and behavioral correlates of impulsivity and the development of the Barratt Impulsiveness Scale. It is especially interesting to note that the author brings a historical perspective to research into impulsivity, and his own commentary regarding the progression of his research and development of empirical findings is an exemplary model of an investigation conducted in the scientist–practitioner model. Barratt (personal communication, July 1993) has noted that it has often been difficult to obtain funding for research regarding impulsiveness. Now, however, funding agencies are finally beginning to realize the importance of this construct, and Barratt's work is starting to get the attention we feel it richly deserves.

H. J. Eysenck, in chapter 4, presents a critical summary of taxonomic problems with, and reviews research into, the construct of impulsivity. Evidence for the heritability of impulsivity is presented, and its relation to the three-factor model of personality is outlined. The discussion regarding twin studies, heritability, and methods for assessing the relative contributions of genetics and environmental influences to impulsivity is straightforward and excellent and required reading for anyone involved in estimating trait heritability in twin samples. This chapter is particularly valuable for individuals interested in the heritability of personality traits and the manner in which this is established.

Chapter 5, by Zuckerman, argues for a common biological basis for a relation between impulsivity and sensation seeking, a construct previously researched extensively by Zuckerman and his associates (e.g., Zuckerman, 1979). Beginning with a discussion of various scales measuring impulsivity and sensation seeking, the author traces the development of his instrument, the Zuckerman Kuhlman Personality Questionnaire. Impulsivity is discussed with respect to other major factors of personality, and valuable information is provided regarding the psychobiology of impulsivity and sensation-seeking genetics. This chapter should be of particular interest to individuals seeking knowledge on the relation between impulsivity and other personality constructs.

In chapter 6, L'Abate develops his thesis that impulsive behavior is learned in the family according to a linear model of abusive–apathetic and reactive–repetitive parenting styles. L'Abate contributes a relationship perspective for viewing impulsivity and discusses distinctive relational styles in terms of impulsiveness. This chapter provides supporting evidence for a model of impulsivity as learned behavior emanating from a specific familial context, and concludes with therapeutic implications. Therefore, it should be of particular interest to family therapists or others interested in a systemic or interpersonal perspective, or an alternative to the more biological accounts presented in earlier chapters.

In chapter 7, Marks-Tarlow presents a radical view of impulsiveness, namely that it be understood from the perspective of the newly emerging science of chaos theory, or, as it is also known, nonlinear dynamics. Because impulsiveness represents behavior that is, by definition, substantially unpredictable, she disagrees with the notion that the linear model methods of traditional psychological science can encapsulate its essence. Therefore, what clinical and research scales that measure impulsivity are tapping into may not be the construct of most interest to the clinician. Marks-Tarlow further states that impulsiveness is a reasonable response to a chaotic world and that out of impulsiveness the individual eventually "self-organizes," another phenomenon from the newly emerging science of chaos. Her chapter argues for a more hermeneutic attempt at understanding the impulsive person, rather than the psychometric approach supported by most of the authors in this volume.

The title of the next section, "Current Research and Special Populations," is something of a misnomer, inasmuch as chapters also discuss the etiology and treatment of impulsivity, as well as the social implications of impulsive behaviors. On the basis of a number of conversations we had with both clinicians and researchers at the American Psychological Association (APA) conventions in 1989 and 1990, we were convinced that psychologists were interested in a number of topics, including the measure of impulsivity, the application of impulsivity to certain recalcitrant and clinically relevant behavioral patterns, and novel applications of theories regarding impulsivity to other personality variables. Obviously, one book section could not accomplish all of this, and we had to be necessarily selective.

Critics will note what this section does not contain. For example, we do not have chapters on crime, delinquency, impulsive sexual behavior, academic achievement, sexually transmitted diseases, gambling, suicide, or extensive discussions of any of the personality disorders. These and many other topics could have been fruitfully discussed by highlighting the role of impulsive behaviors in their etiology and maintenance through time. Our omission was not based on any excessive importance that we allocate to the topics presented here above other potential areas. Given unlimited resources, or even a second volume, we would certainly expand this section. Our primary aim, however, was to include topic areas that did not have an adequate systematic review in existing literature.

Chapter 8, by Sybil B. G. Eysenck, presents the detailed development of an instrument designed to measure impulsiveness. This chapter, which in some ways is an extension of that of H. J. Eysenck, is valuable for several reasons beyond the empirical studies it highlights and the excellent psychological instrument (the I_7) that it discusses. First, the chapter highlights how impulsiveness fits into a dimensional theory of personality. Second, it documents the process by which a personality trait is continually refined through psychometric findings and the way that this refinement also changes our knowledge about the trait in question. Third, the chapter, perhaps better than any recent work, documents the incredible amount of effort that is involved in establishing the reliability and validity of the measurement of a trait. For students or researchers who have not performed test construction and validation, this chapter is highly recommended

Chapter 9, by Dickman, discusses research efforts to identify information processing and cognitive characteristics associated with impulsivity. Using an information-processing perspective, important individual differences are described, with particular emphasis given to specific cognitive processes associated with high and low impulsives. Dominant theories of impulsivity are examined with respect to information-processing findings. Finally, this chapter concludes with an alternative theory for understanding impulsivity, attentional fixity theory, which explains many of the findings generating from the information-processing literature on impulsivity.

Chapter 10, by Lowe and Eldredge, is perhaps the most detailed chapter in the volume. It examines the role of impulsive behavior in the etiology of a variety of eating disorders. To our knowledge, it is the first publication that systematically addresses this issue, a topic that has both research and clinical implications. This chapter is especially relevant given recent research regarding serotonin and impulsivity (discussed in the chapters by Daruna and Barnes and by Zuckerman) and data that reduction in plasma cholesterol may be associated with a decrease in serotonin, and hence more impulsive behavior (e.g., Krombout, Katan, Nenotti, Keys, & Bluemberg, 1992).

Chapter 11, by Johnson, Mallow, Corrigan, and West, discusses the role of impulsive behavior in a very high-risk population, namely substance abusers. The research from Mallow and his associates' laboratory has largely been dedicated to establishing the role of personality in the addictive process. This chapter goes further, delineating the role of impulsive behaviors in personality disorders associated with substance abuse. Although this chapter is not included in the treatment section discussed later, these authors also present important therapeutic considerations.

Chapter 12 by Hewitt and Flett and chapter 13 by Ferrari are examples of our interest in examining the role of impulsiveness in other personality constructs. We chose the constructs of procrastination and perfectionism because both represent "everyday" behaviors that are well recognized by lay persons. They are also syndromes that in their more extreme forms have clinical dimensions. The literature regarding these two constructs has expanded tremendously in recent years, largely because of the contributions of these authors. We could have easily chosen other personality variables as well: for example, the need for achievement, openness to experience, or aggressiveness. Regardless, the role of impulsive behaviors in manifestations of these personality variables suggests that impulsive action may have a causal role in similar psychological constructs. We hope that the rigorous scholarship of Ferrari and his associates, and of Hewitt, Flett, and the numerous students working in their labs, will be emulated by other behavioral scientists in a variety of applied areas.

The final section of this book is primarily of interest to clinicians, although we are certain that researchers will find fruitful hypotheses to test by examining the chapters' contents.

Chapter 14, by Fink and McCown, is a succinct summary of a very complex literature, namely how to diagnose and treat impulsiveness in children. Another contribution of this chapter is its emphasis on the multiple etiologies of childhood impulse disorders. It also provides a somewhat more psychodynamic orientation regarding etiology than was found in the first section of the volume. Finally, this chapter argues strongly that "monomethods" of assessment of childhood impulsivity are of limited value and should not be standard clinical practice.

Chapter 15, by Holmes, Johnson, and Roedel, could have also been included in the first section because it highlights the etiological role of acquired brain dysfunction in the genesis of impulsivity. This topic is critical for the clinician inasmuch as it highlights the necessity for the health-care establishment to devote resources to the psychosocial adjustment of persons saved by advanced medical procedures. A generation ago, many more people with traumatic brain injury, cerebral vascular occlusions, or severe viral diseases simply died. Today, they are living and may, in many cases, return to a productive lifestyle. However, many such people demonstrate an increase in impulsive behavior when they recover. This is because of problems associated with relatively permanent brain injury. This chapter addresses some of the problems of this population and highlights some potential management strategies.

Topics for the next two chapters were chosen because of the results of a "focus group" we held with a large group of clinicians at the 1989 APA convention in New Orleans. The idea of choosing chapter topics for a scholarly book partially by using a technique made common by the advertising industry may seem unscientific and even a bit crass. We disagree and believe our use of a focus group enabled us to tap areas of genuine interest and to see where there was a perception that the literature was deficit. A recurrent theme among clinicians with whom we spoke was the notion that someone needed to write about how brutally exhausting to the clinician impulsive people can be when in therapy. Common difficulties cited were problems relating to the lack of payment of fees, missed appointments, the tendency of these clients to often present in crisis, and the sheer frustration they cause with the clinician when they fail to follow through on therapeutic recommendations. We feel that chapter 16 by Bütz and Austin addresses these problems head on, with an uncharacteristic frankness not found in most discussions regarding countertransference.

Chapter 17, by Drogin and Drogin, is another chapter that was chosen because of interest from clinical focus group participants. A constant request among the clinicians that we interviewed was for someone to discuss practical legal concerns regarding the treatment of impulsive clients. The sentiment that we found repeatedly expressed was summarized very well by a seasoned Midwestern clinical psychologist: "Everyone knows about the duty to warn," she stated. "What we need is a discussion of how [this legal concept] applies to people who are always at high risk, simply due to their personalities." Because the authors are clinicians first, and legal scholars second, they understand the practical difficulties involved in treating clients who inherently have the capacity to be unpredictable and disruptive.

Chapter 18 was chosen because all three of the coeditors have worked extensively with disenfranchised women and are not happy about the capacity of mental health professionals to provide adequate services to this group. Johnson and Bishop have documented the problems that impulsive women have in obtaining clinical services and the particular societal prejudices against such women. These problems become multiplied when the client is poor, a person of color, older, in ill health, or not physically attractive. The chapter discusses the treatment of such women and advocates a clinical eclecticism rooted in a feminist/empowerment framework. Interestingly enough, it also advocates the use of some of the metaphors of chaos theory, discussed by Marks-Tarlow in chapter 7. The choice of this topic is not meant to imply that there are not special and unique problems encountered when addressing impulsiveness with other populations. For example, useful chapters might have included therapy with impulsive older persons, racial and sexual minorities, medical patients, religious persons, or any number of topics. Unfortunately, we are limited by our space and hope that future work will correct our omissions.

Chapter 19 represents a summary of the work of Shure, Spivack, and their many associates regarding childhood interventions designed to reduce impulsive behaviors. Although more empirical research needs to be conducted in this area, data indicate that regardless of the causes of impulsive behavior in children, we can obtain a substantial and meaningful reduction in its future occurrences by use of interpersonal cognitive problem-solving (ICPS) strategies. We believe that the method of ICPS and similar problem-solving curricula may represent an excellent theoretically derived and empirically supported technique, with potential prevention and treatment efficacy for a variety of populations at risk for nonreflective behavior. Therefore, we feel it is a fitting concluding chapter, not because it synthesizes previous contributions, but because it offers the hope of prevention. However, our concern is that society will not demonstrate sufficient interest in preventing impulsive behaviors to adopt such programs. If this is the case, then we are undoubtedly going to experience greater personal and social hardship in future generations.

REFERENCES

Bandura, A. (1986). *Social foundations of thought and action: A social cognitive theory.* Englewood Cliffs, NJ: Prentice-Hall.

Baron, J., & Brown, R. (Eds.). (1991). *Teaching decision making to adolescents.* Hillsdale, NJ: Erlbaum.

Barratt, E. S. (1965). Factor analysis of some psychometric measures of impulsiveness and anxiety. *Psychological Reports, 16,* 547–554.

Bernard, T. (1990). Angry aggression among the "truly disadvantaged." *Journal of Health Care for the Poor and Underserved, 2,* 175–188.

Briere, J., & Runtz, M. (1988). Symptomatology associated with childhood sexual victimization in a nonclinical adult sample. *Child Abuse and Neglect, 12,* 51–59.

Brill, N. (1993). *America's psychic malignancy: The problem of crime, substance abuse, poverty and welfare-identifying causes with possible remedies.* Springfield, IL: Charles C Thomas.

Centers for Disease Control. (1993). *Quarterly update: Violence and violence prevention.* Atlanta, GA: CDC Office of Public Affairs.

Cloninger, C. (1987). A systematic method for clinical description and classification of personality variants. *Archives of General Psychiatry, 44,* 573–588.

Coccaro, E. F. (1989). Central serotonin in impulsive aggression. *British Journal of Psychiatry, 155,* 52–62.

Coccaro, E. F. (1993). Psychopharmacological studies in patients with personality disorders: Review and perspective. *Journal of Personality Disorders, Suppl.,* 181–192.

Coleman, L. (1985). Perspectives on the medical research of violence. In F. Marsh & J. Katz (Eds.), *Biology, crime and ethics: A study of biological explanations for criminal behavior* (pp. 135–147). Cincinnati, OH: Anderson.

Conte, J. R., & Schuerman, J. R. (1988). The effects of sexual abuse on children: A multidimensional view. In G. Wyatt & G. Powell (Eds.), *Lasting effects of child sexual abuse* (pp. 157–169). Beverly Hills, CA: Sage.

Costa, P., & McCrae, R. (1986). Personality stability and its implications for clinical psychology. *Clinical Psychology Review, 6,* 407–423.

Dollard, J., Doob, L., Miller, N. E., Mowrer, O. H., & Sears, R. (1939). *Frustration and aggression.* New Haven, CT: Yale University Press.

Doob, L. (1990). *Hesitation: Impulsivity and reflection.* New York: Greenwood Press.

Felson, R. B., & Tedeschi, J. T. (Eds.). (1993). *Aggression and violence: Social interactionist perspectives.* Washington, DC: American Psychological Association.

Gans, J., & Blyth, D. (1990). *America's adolescents: How healthy are they?* (AMA Profiles of Adolescent Health series). Chicago: American Medical Association.

George, N. (1993). *Buppies, B boys, baps and bohos: Notes on the Post-soul black culture.* New York: Harper Collins.

Gray, J. (1991). *The psychology of fear and stress* (2nd ed., reprinted with corrections). Cambridge, England: Cambridge University Press.

Hammond, R., & Yung, B. (1993). Psychology's role in the public health response to assaultive violence among young African–American men. *American Psychologist, 48*, 142–154.

Haynes, S. (1992). *Models of causality in psychopathology: Toward dynamic, synthetic and nonlinear models of behavior disorders.* New York: MacMillan.

Johnson, J., & McCown, W. (1992). An overview of substance abuse disorders. In P. Sutker & H. Adams (Eds.), *Comprehensive handbook of psychopathology* (2nd ed.). New York: Plenum Press.

Koziol, L., Stout, C., & Ruben, D. (1993). *Handbook of childhood impulse disorders and ADHD: Theory and practice.* Springfield, IL: Charles C Thomas.

Krombout, D., Katan, M., Nenotti, A., Keys, A., & Bluemberg, B. (1992). Serum cholesterol and rates for suicide, accidents, or violence. *Lancet, 340*, 317.

L'Abate, L., Farrar, J., & Serritella, D. (Eds.). (1992). *Handbook of differential treatments for addictions.* Needham Heights, MA: Longwood & Bacon.

Lacey, J. H., & Evans, C. D. (1986). The impulsivist: A multi-impulsive personality disorder. *British Journal of Addictions, 81*, 641–649.

Langley, R., & Levy, R. (1977). *Wife beating: The silent crisis.* New York: Dutton.

Leifer, M., Shapiro, J., Martone, M., & Kassem, L. (1991). Rorschach assessment of psychological functioning in sexually abused girls. *Journal of Personality Assessment, 56*, 14–28.

Lesieur, H. (1992). *The chase: The compulsive gambler.* Rochester, VT: Schenkman Books.

McConaghy, N. (1993). *Sexual behavior: Problems and management.* New York: Plenum Press.

Mischel, W. (1968). *Personality and assessment.* New York: Wiley.

Moore, K. (1992). *Facts at a glance.* Washington, DC: Childtrends.

Morey, L. (1993). Psychological correlates of personality disorder. *Journal of Personality Disorders, Suppl.*, 149–166.

O'Boyle, M., & Barratt, E. (1993). Impulsivity and DSM–III–R personality disorders. *Personality and Individual Differences, 14*, 609–611.

Pope, K., Butcher, J., & Seelan, J. (1992). *The MMPI, MMPI-2, and MMPI-A in court: A practical guide for expert witnesses and attorneys.* Washington DC: American Psychological Association.

Powles, W. (1992). *Human development and homeostatis: The science of psychiatry.* Madison, CT: International University Press.

Public Opinion Online. (1993, August). *Results of crime and quality of life surveys 1965–1993.* Columbus, OH: Knowledge Index Data Base.

Quadrel, M., Fischhoff, B., & Davis, W. (1993). Adolescent (In)vulnerability. *American Psychologist, 48*, 102–116.

Saltzman, L., Mercy, J., O'Carroll, P., Rosenberg, M., & Rhodes, P. (1992). Weapon involvement and injury outcome in family and intimate assault. *Journal of the American Medical Association, 267*, 3043–3047.

Shea, M. T. (1993). Psychosocial treatment of personality disorders. *Journal of Personality Disorders, Suppl.*, 167–180.

Toch, H. (1992). *Violent men: An inquiry into the psychology of violence.* Washington DC: American Psychological Association.

U.S. Department of Justice. (1992). *Crime statistics.* Washington, DC: U.S. Government Printing Office.

Zuckerman, M. (1979). *Sensation seeking: Beyond the optimal level of arousal* Hillsdale, N.J.: Erlbaum.

I

HISTORICAL PERSPECTIVE AND ETIOLOGY OF IMPULSIVITY

1

IMPULSES, IMPULSIVITY, AND IMPULSIVE BEHAVIORS: A HISTORICAL REVIEW OF A CONTEMPORARY ISSUE

WILLIAM G. MCCOWN and PHILIP A. DeSIMONE

There is little doubt that problems associated with impulsive behaviors are extensive and have far-reaching consequences for the individual and for society. For example, criminal acts performed with little forethought, such as murder and opportunistic rape and theft, are at historically high levels (U.S. Department of Justice, 1992). This alarming tendency is especially true among adolescents (Dryfoos, 1990). Substance abuse, often considered a disorder of impulse control, remains a recalcitrant contemporary issue that is likely to be long-standing. Similarly, AIDS (for which there is no foreseeable cure) is now largely spread in the Western hemisphere through impulsive behavior. Recently, it has been estimated that as much as one half of all psychotherapy resources

Special thanks are due to Christine E. R. Yoder, Princeton Theological Seminary, who provided assistance with this chapter.

are devoted to treating disorders involving impulsive behaviors (McCown, 1993). These include problems of physical and sexual abuse, gambling, childhood impulsiveness disorders, and behavioral problems common to many personality disorders and other psychiatric syndromes (O'Boyle & Barratt, in press).

In addition to social, medical, and public health concerns, the existence of impulsive behavior is of substantial theoretical importance for the behavioral sciences. If most action temporally follows the process of thought, as is asserted by the dominant paradigms in contemporary psychology, how do we explain behavior that occurs—often repeatedly—in the absence of cognitive deliberation? Why do some people persistently engage in undesirable impulsive behavior, action that is clearly not in their own interest? Why is this inappropriate behavior not extinguished because of its ultimately aversive nature? Finally, questions are raised regarding whether psychology can provide useful theory and practice to alter impulsive behavioral patterns. If not, does this suggest fundamental mistakes in our theoretical accounts of behavioral maintenance and change?[1]

BASIC DEFINITIONS

To address these and other relevant questions, scientists and practitioners need a common nomenclature regarding impulsivity. An error clinicians and researchers often make when discussing impulsivity is to confuse it with the construct of *impulses*. This latter term, more in lay use today, has been defined as ephemeral thoughts usually tied to forceful urges (James, 1890). Occasionally, clinicians observe an apparent sparseness of impulses in certain persons, usually associated with advanced illness or neurological conditions. The lack of impulses (*abulia*) may be extreme in degree, to the point of mutism and immobility (Adams & Victor, 1986). More commonly, however, problems with impulses are at the opposite extreme. The popular perception of an "impulsive" person in someone who acts with some degree of frequency on impulsive or unplanned thoughts that others either do not have or have but do not act on (Goslin, 1969).

Whereas *impulse* refers to thoughts, *impulsivity* refers to a constellation of repeated behaviors that are somehow related to these thoughts (Stanford & Barratt, 1992). Despite a seeming consensus on this, a further satisfactory definition of impulsivity has proven elusive, at least

---

[1]As one reviewer of an earlier draft of this chapter stated tersely, "Perhaps the existence of impulsive behavior is an indication of the limits of science to predict and control human beings. . . . In this case our whole paradigm of normal behavior science may be on shaky grounds."

among psychologists and other students of human behavior. Impulsivity has been described by a number of overlapping and sometimes contradictory definitions. These include "human behavior without adequate thought" (L. Smith, 1952), "behavior with no thought whatsoever" (English, 1928), "action of instinct without recourse to ego restraint" (Demont, 1933), and "swift action of mind without forethought or conscious judgment" (Hinslie & Shatzky, 1940). Impulsivity may also mean acting with minimal thought regarding future actions, or acting on thoughts that are not in the individual's or others' best interests (Oxford English Dictionary, 1951). Impulsivity may also possess both positive and negative connotational meanings (see Dickman, chapter 9, this volume). As Eysenck and Eysenck (1985) noted, impulsiveness may connote verve, spunk, and an inclination toward adventure; on the other hand, trait measures of impulsive behavior are associated with a variety of negative conditions, as many of the chapters in this volume will show.

Attempts to describe the construct of impulsivity empirically are at least 40 years old and continue to this day. Grayson and Tolman (1950) used a modal definitions technique with clinicians and found four clusters relevent to this term. These include behavior that occurs (a) without control, inhibition, restraint, or suppression; (b) without thinking, reflection, or consideration; (c) without foresight, adequate planning, or regard for consequences; or (d) with a sense of immediacy and spontaneity. Because definitions vary significantly both in the literature and among scholars, we have asked each chapter author or authors in this volume to provide a definition of impulsiveness or impulsivity applicable to their contribution.

THE CONCEPT OF IMPULSIVITY BEFORE THE 20TH CENTURY: AN ANCIENT LEGACY

The American philosopher and poet George Santayana (1863–1952) noted that those who cannot remember the past are condemned to repeat its failures. Unfortunately, psychologists often disregard this aphorism. Many students are led to believe that contemporary theories of behavior exist in a historical vacuum. They often fail to grasp the significant ties of present research to past threads of understanding developed in the prescientific era. As a result, they too often fail to profit from the rich theorization of previous generations or, as noted, are prone to repeat previous errors. To avoid these mistakes, we believe that an introduction to the concept of impulsiveness should include an account of its historic roots, which extend far back and at least into the time of the ancient Hebrews.

In Genesis 9:18–27, we read that Noah's son Ham impulsively viewed his drunken father's nakedness and that consequently, he and his descendants were cursed by YHWH (one of the three Hebraic names for God in the Old Testament). A few chapters later in Genesis, Lot's wife impulsively turned to look back at the destruction of Sodom. Acting in violation of the YHWH's commandment, she was changed into a pillar of salt. Later in the Old Testament, Elohim (a second name, indicating more merciful aspects of God's character) is not so punitive toward impulsive behavior. Mosaic law differentiated between accidental and impulsive murder, on the one hand, and murder with malice, on the other. "But if he stabbed him suddenly without enmity, or hurled anything on him without lying in wait . . . the congregation shall rescue the manslayer from the hand of the avenger of blood and the congregations shall restore him to his city of refuge" (Numbers 35:22–25). Similar provisions existed in the code of the King of Babylon Hammurabi (c. 1955–1913 B.C.) and, to a lesser extent, in some of the laws of the ancient Egyptians (Baly, 1977).

The ancient Greeks were also interested in impulsive behavior. Homer, Areteaus, and Thoephrastes described in vivid detail varying degrees of behavioral impulsivity and its effects on character. The Greek physician Hippocrates (460–377 B.C.), in what Theodore Millon (1981) has suggested may have been an early description of the borderline personality syndrome, noted that impulsive anger (*kakia,* or emotional harm and evil) and melancholia can coexist in some people, and speculated on its erratic vacillation. Hippocrates also furnished the first formal theories regarding the causes of impulsive behavior. He postulated a fourfold topology of personality, each type associated with specific body fluids. The choleric temperament, associated with impulsivity (*kakia*), arises because of excessive yellow bile. The sanguine temperament, also associated with impulsivity, but of a different type (*aporia,* literally "roadlessness"), occurs because of an excess of blood. Various Greek words approximately translated into the English term "impulse" (i.e., *protennoia,* literally "primal thought"; *metabole,* literally "change" or "transition") also appear in the writings of Epicurus the Cynic (341–270 B.C.) and especially Epictetus (ca. 55–135 A.D.).

It is not surprising that the Greeks were more interested in impulsive behaviors than their historical counterparts. Two traditions in Greek culture battled for intellectual supremacy (Kitto, 1952). The question of the degree of unplanned and irrational behavior that humans were capable of may have played a prominent role in arbitrating between these two philosophies. Democratic thinkers supported the concept that reasonable men (literally only Greek men, and not Greek women) could aptly decide the course of their own society. Others, including Socrates (ca. 470–399 B.C.) believed that people were too irrational to make such choices; benevolent dictatorship was the most humane form of government. Socrates went further and suggested that leaders be drawn only from a special class of persons bred and trained to be athletic, intelligent, and highly emotionally stable. Perhaps having seen the wretched excesses that humans inflicted on each other, Socrates may have been concerned that humans were typically too impulsive to rule one another justly and fairly (Botsford, 1893).

Oddly, there are few references to impulses or impulsivity in the late Hebraic prophets, the Talmud, or very early Christian thinkers. There are exceptions, of course. In the New Testament Gospels (Matt. 21:18–19; Mark 12:12–14), Jesus apparently impulsively curses a fig tree that bears no fruit. Some commentators (Defaul, 1928) have suggested that the Apostle Paul's cryptic "thorn in the flesh" may have been impulsive thoughts or behaviors, especially of a sexual nature, although this is speculative. However, the earliest Christian writers, the pastoral Christian ethicists, are all but silent regarding the nature of impulsive thoughts and behaviors. It was not until Saint Augustine's (354–430 A.D.) *Confessions* (1912), written in the 4th century, that this concept again received attention from a major theologian, and perhaps not until the Reformation that the notion of impulsiveness once again became as much of a concern as it was to the Greeks. The reason for this is not clear. It may be that because early Christianity believed strongly in a corporeal devil, impulsive thoughts and behaviors had an immediate explanation: "The devil was in him." Impulsiveness was no longer a psychological phenomenon, but instead a description of a spiritual condition. With this shift of explanation, the concept became unimportant and irrelevant, as did any theory of psychology that did not place a primary emphasis on Christian theology (Kantor, 1963). What is clear is that between the time of the early Christians and the Reformation, scientific psychology lay dormant, awaiting a resurrection that would not proceed until the scientific method was once again accepted (O'Leary, 1949).

A single English-language word for the noun *impulse* is not documented until the early 16th century. This word was derived from the Latin *impulsus*, the stem of *impellere*, "to move" (Oxford English Dictionary, 1951). The English term *impulsion* appears to have come from the same Latin root, although its more direct stem was the French word *impulsion*, or "influence." The earliest reference to *impulsion* in English was in the 13th century. According to Jennings (1928), an early historian of scientific language, the concept of impulsion was necessary for an Aristotelian account of the universe, popularized by theologians. In the 13th-century interpretation of Aristotelian physics, the behavior of falling bodies was explained by their final-cause nature of "wanting to return to earth." The enigmatic path of the parabolic motion of items projected toward the heavens could be resolved by postulating an irrefutable force that attempted to seize the "will" of each object. The gradual deceleration of upwardly and outwardly projected bodies, clearly evident to logistical mechanics of the 13th century, was accounted for by the fact that the parabolic object initially resisted the impulse to return home until it became too great to ignore. Similarly, an *impulsive cause* (16th-century term) became any cause that was primary, natural, and correct.

Paradoxically, by the 16th century, the term *impulse* was widely used in the English language to describe pejorative thoughts, albeit of a natural course. For example, in the *Principia Theologica* (1554/1932), an anonymous mid-16th century English tract/catechism designed for the slowly growing literate classes, readers were admonished that the devil works by planting evil and powerful "base, common thoughts of impulse and irresistible nature." Among Reformation writers, the moral prescription on impulsive thoughts became more pronounced. The German Reformation leader Martin Luther (1483–1546) counseled the faithful (Henichel, 1911) to "avoid the impulse to [ward] harlotry" and further noted that "drunkenness begins as impulse." In a manner not unlike present-day cognitive behaviorists, Luther cautioned the faithful reader not to "entertain" impulses, lest they "conquer the will."

By the 17th century, the word *impulsive* was being used to describe a persistent *pattern* of dysfunctional behaviors presumably caused by chronic impulsive thoughts. The Scottish theologian John Knox (ca. 1514–1572) spoke (Henichel, 1911) of a "character of Impulse" where "God hath delivered the soul to be dominated by the One our very Saviour died to subdue" (i.e., the Devil). Knox believed strongly in the theological notion of predestination and posited that God allowed those who were not predestined for salvation to be dominated by their impulses to the point where they lost the capacity for free choice. The French theologian and reformer John Calvin (1509–1564) spoke (Henichel, 1911) of those "given over to Impulse in every deed," hastening to indicate that they are "bound over to the Evil one for the path of Sheol" (Hell).

The existence of impulsive thoughts and impulsive behaviors soon became somewhat of a philosophical obstacle for theologians and philosophers of this period (Leckey, 1869). No one was able to resolve the question of how humans could have free will yet at certain times simultaneously be faced with powerful thoughts that seemed to demand action and to strip them of their free will and reason. Satanic influence was always a convenient answer, but one that gradually lost some appeal. Various other novel solutions to this quandary were put forth. The French mathematician and philosopher René Descartes (1596–1650) seemed to have begged the question. His follower Malebranche (1638–1715), a 17th-century Catholic priest, postulated a complete independence of mental thoughts from behaviors. The German philosopher and mathematician Gottfried Wilhelm Leibniz (1646–1716) believed that impulsive thoughts were those that were at odds with the God-ordained pre-existing harmony that abided between matter and thought. The French philosopher and writer Jean-Jacques Rousseau (1712–1778) elevated the virtues of "natural man" acting according to instinct and impulse. He seemed quaintly blind to the occasional horror that people are able to inflict on each other without forethought. The Scottish philosopher and historian David Hume (1711–1776) questioned the degree of human free agency, partly, perhaps, because of the nature of unwanted thoughts and unplanned behaviors.

As orthodox Christianity of the Middle Ages began to lose its grip, it began being replaced by a new way of viewing the world, one that involved observation and reasoning (Kantor, 1963). The first quasi-scientific attempt to systematically describe impulsivity in the post-Roman era was arguably by the French physician and moral philosopher Bonet (1684; c. 1631–1692). Bonet distinguished among impulsive thoughts, obsessive thoughts, impulsive character, and the *folie maniaco-mélancolique,* or erratic and unstable moods that featured impulsive behavior and depression. Like Saint Augustine and other theologically minded individuals, Bonet believed that impulsive or evil cognitions were a part of humanity's lower nature and commonness. However, he distinguished between those of us who are able to cavalierly dismiss such thought and those who "by weakness of mind dwell on such [thoughts] without interruption." These latter individuals were destined to a perpetual moral quandary, because they "know what to do, yet fail to do so." These included the habitually criminal, the inebriant, and "all manners of men who practice moral vice and appear to profit not from instruction." (The philosophical links between Bonet's concepts and biological explanations of impulsiveness, such as those suggested by Zuckerman, chapter 5 in this volume, are noteworthy).

The next 250 years of moral philosophy and early prescientific personality theory would see a periodic oscillation between an emphasis on the process of development of the construct of impulsive thoughts, on the one hand, and interest in the consistency of occurrence of impulsive behaviors in some individuals, on the other. Pinel (1745–1826) fostered further interest in the notion of impulsivity as a pathological process (1801). He was the first to suggest that impulsive personality features may contain a component that is beyond individual volition. Pinel, like many "moral commentators" of his age, including physicians, struggled with the thorny issue of boundaries of individual free will and responsibility versus the fact that some individuals are incapable of "understanding" the consequences of their acts. Pinel noted that certain patients consistently engaged in self-damaging and impulsive acts, despite the fact that their reasoning abilities seemed intact. He also noted that impulsive persons have difficulty completing routine life tasks (a phenomenon discussed in Ferrari, chapter 13, this volume). An even more curious feature was that those suffering from *"la folie raisonnante"* seemed incapable of learning from the detrimental consequences of their impulsive behaviors. Pinel described these cases under the name of *"manie sans delire,"* or insanity without deliriums or without confusion of mind. A number of early psychiatrists made less affirmative contributions toward an empirical theory of behavioral impulsivity. Benjamin Rush (1745–1813), the pioneer of American psychiatry, reintroduced the moral condemnation of impulsive behaviors. J. C. Prichard (1786–1848), an early English "alienist" (i.e., psychiatrist), extended the moral tone voiced by Rush. Prichard (1835) believed that impulsive behavior represented a "defect" in character that deserved social condemnation. Patients afflicted with moral insanity were cursed by overpowering "afflictions and impulses" that compelled them to engage in behaviors they themselves would view as morally evil. This description of the apparent failure of the inhibition of impulses can be applied to a variety of mental illnesses and seems to be nonspecific.

The works of the early physician/phrenologists, especially the German Franz Josef Gall (1758–1828) and his students (i.e., Gall & Spurzheim, 1809), should be mentioned in regard to the development of understanding impulsivity. Phrenology was an honest attempt to construct a science of personality and behavioral functioning based on the structure of the brain, which, it was erroneously believed, correlated with the size and shape of the skull. Gall and other phrenologists attempted an exposition of the relation between the shape of the head and an individual's character. Hence, it is not surprising that they believed impulsivity to be associated with changes in the dimensions of the brain resulting from head trauma. In any case, there is a covariation between the subsequent reduction in the size of the "knot on the head" and a return to normal functioning. What is often overlooked is Gall's insistence (perhaps the rediscovery of an ancient truth of Galen) that head injuries can also cause transitory behavioral changes, characterized by an increase in impulsive behavior and a decrease in general cognitive abilities (Adams & Victor, 1986). More "scientific" neurologists overlooked this fact for another 75 years. (The relation between various neurologic and neuropsychological processes and impulsivity is reviewed by Holmes, Roedel, & Johnson, chapter 15, this volume.)

The German philosopher Immanuel Kant (1724–1804) reinterpreted (1811/1933) the four personality types of Hippocrates and Galen in more contemporary terms, highlighting, as did Galen, the role of impulsive behaviors in the choleric type. Kant was only one of many typological theorists of this century who believed that a specific number of chiefly discrete typologies could categorize human behavior. Other typological theorists include Théodule-Armand Ribot (1839–1916), an early French psychologist who postulated primary dispositions of sensitivity and activity (1890). The emotional type was characterized by vacillation, impulsivity, and lability. Instability was composed of excessive sensitivity and activity. The French theorist Queyart (1829–1902) postulated (1896) the dimensions of emotionality, activity, and mediation. Unlike previous theorists, he believed that typologies could be overlapping. His description of the personality type composed of emotionality and activity bears resemblance to later descriptions of the "multi-impulsive" individual by Lacey and Evans (1986). Perhaps Queyart's most significant contribution—and one overlooked for many years—was his belief that many abnormal features of personality were simply exaggerations of otherwise normal features. (The role of typological versus trait theories is discussed by H. J. Eysenck, Sybil B. G. Eysenck, and Ernest Barratt, chapters 4, 8, and 3 in this volume, respectively.)

It is perhaps fitting that the 19th century closed with the first thoroughly psychological review of the concept of impulses and impulsiveness. Walter Dill Scott (1869–1955), one of the 16 American students for whom Whilhelm Wundt (1832–1920) served as *Erstgutacheter*, or major reader (Benjamin, Durkin, Link, Vestal, & Acord, 1992), completed a nonexperimental dissertation on the psychology of impulses. Dill's dissertation, written in 1900, was primarily a review of previous literature in which he sought to synthesize a scientific definition of the concept. Dill was highly critical of theological and moral implications associated with the idea of impulses. He suggested how the newly established discipline of psychology could make useful contributions to separating scientific fact from moralizing. (This is reflected by Bütz and Austin and by Drogin and Drogin, chapters 16 and 17, respectively, this volume.)

20TH-CENTURY CONTRIBUTIONS TO THE STUDY OF IMPULSIVITY

Ideographic Contributions

In this century, students of human behavior have used two principal methods or metatechniques in researching constructs of personality. The first method, the *ideographic* approach, focuses on individual differences and uses clinical or case studies of individuals. The second method, known as a *nomothetic* approach, empirically studies groups and differences between groups. Much of the contribution of psychiatry toward understanding impulsivity has been ideographic and based on the tradition of careful observation and diagnosis.

The German Ehrlheim Hirt (1849–1922) published one of the first textbooks of psychiatry in 1902, and perhaps the first textbook to use the term *impulsive* as a description of clinically relevant behavior. Titled *Die Temperamente*, Hirt's work applied the doctrine of the four humors to psychiatric cases. Patients who were endowed with an extreme sanguine temperament were noted by Hirt as being characterized by superficial excitability, enthusiasm, and unreliability. Although frequently diagnosed as hysterical, their principal problem, Hirt believed, was a lack of impulse control. The choleric temperament was found in grumbling, angry, vociferous types; it represented a second variant of a personality type unable to control impulses. A combination of choleric and sanguine personality factors produced, Hirt believed, an explosive, angry, "morally insane" individual who was impulsive and disregarded the needs of other people.

Emil Kraeplin (1856–1926) was perhaps the foremost psychiatrist of the first half of the present century, with the exception of Sigmund Freud (1856–1939). The present psychiatric nomenclature reflected in the *DSM–III–R* (American Psychiatric Association, 1987) and in the upcoming *DSM–IV* is essentially a variant of his original systematizing. Kraeplin paid little attention to the study of personality until comparatively late in his career. For the 8th edition of his classic textbook *Psychiatrie: Ein Lehrbuch* (1913), he attempted to discover personality variants that put individuals at high risk for major mental illnesses such as "dementia praecox" and manic–depressive illness. One of these factors was labeled by him as the "cyclothymic" disposition, which Kraeplin believed was a risk factor for manic–depressive illness. Of the four subtypes of this syndrome, two appear to be related to emotional impulsivity. The cyclothymic personality was described as vacillating, arbitrary, impatient, capricious, and *impulsive*. Individuals with an "irascible" personality were described as being ill-controlled, hotheaded, enragable, likely to take umbrage for trivial slights, and also impulsive. Contemporary readers may note that the two subtypes appear to share a common set of attributes, with the "irascible" subtype superimposing a paranoid thought process on "cyclothymic" features.

Perhaps the first theorist to extensively discuss impulsive behavior per se was the German psychiatrist Kahn (1876–1937). Kahn (1928) postulated that the hyperthymic or impulsive person was characterized by excitability, rapidity, and explosiveness, which were related to his or her dimensional positions on polarities of activity, self-orientation, and positive outlook. Similarly, the German psychiatrist Tramer (1888–1952) postulated (1931) an impulsive type: quick to change and quick to discharge tensions. Again, parallels with present theorists in this volume are immediately evident.

The rise of psychoanalysis initially reduced interest in impulsivity as an attribute of personality, focusing instead on the etiology and nature of impulsive thoughts. Clinically, Freud was impressed with the frequency of impulsive thoughts in both normal behavior and in the clinical accounts of "psychoneurotic" patients (Saul, 1947). For Freud, impulses were manifestations of the id that conscious censorship could not suppress. Freud normalized impulses by indicating their universality, their moral neutrality, and their ties to biological processes. In his tripartite theory, Freud conceptualized impulsive behavior as occurring when id impulses break through to the ego, causing the ego to act in a manner not in accordance with mandates of the superego.

Freud (1931/1960) later speculated that a method of character classification might be based on his threefold structure of personality. He postulated the existence of a narcissistic type of personality, totally dominated by ego functioning; a compulsive personality, dominated by the moral restraints of the superego; and an erotic personality, dominated by the drives of the id. Presumably, an impulsive personality would be essentially erotic in nature, dominated by instinctual demands.

Subsequent analytic theorization regarding character typologies had much to say regarding the construct of impulsivity. For example, the Austrian psychoanalyst Wilhelm Reich (1897–1957) conceptualized the "hysterical personality" as containing components of superficiality, impulsiveness, and an inability to sustain endeavors (1933). In this conceptualization, either intense frustration or overindulgence during the psychosexual "phallic stage" may result in poor and impulsive reactions to even minor defeats. Other analytic thinkers such as Abraham, Fenischel, Jones, and Menninger elaborated on these contributions. More recent thinkers in the analytic tradition have made additional, although somewhat speculative, contributions regarding impulsivity, which wait empirical research before they can fully be assessed.

Nomothetic Contributions

Unfortunately, many of the theories of Freud and other analytic writers concerning impulsivity were not subjected to any type of empirical verification. This was not a phenomenon unique to Freudians. The rise of experimental psychology during the last quarter of the 19th century was, in part, a reaction against the antiempirical stance of both moral philosophy and prepsychoanalytic psychiatry (Marx & Hillix, 1963). Most of the prominent, early empirical theorists of personality noted the importance of impulsivity and gave a description of impulsive behaviors in an account of both normal and abnormal personality.

Wundt (1907) was one of the first psychologists to clearly articulate the dimensional nature of traits and to refute a priori the existence of discrete typologies. Wundt's theories of personality were essentially a recapitulation of the Kant/Galen/Hippocrates formulation, with the exception that the four functions were combined to form two orthogonal axes. Wundt simply substituted the four types for an axis of emotionality and an axis of activity. In this manner, the classic "four types" are simply extremes on two dimensional continua. The sanguine "type" is active and nonemotional, whereas the choleric is active and emotional. The phlegmatic is nonactive and nonemotional, whereas the melancholic is nonactive and emotional. Presumably, extreme activity and emotionality would be the cause of temperamental impulsivity.

Some of the most remarkable empirical work in early 20th-century psychology was conducted by the Dutch psychologists Heymans and Wiersma (1909). These researchers performed what amounts to an early method of principal components analysis on the correlation matrix of several personality variables for an extremely large sample of Europeans. They identified three "dimensions" of character that are strikingly contemporary: activity level, emotionality, and susceptibility to external versus internal stimulation. Extreme locations on the continuum produce apparent personality "types." For example, extreme emotionality, activity, and susceptibility to internal rewards produces an "internally motivated and often changing" individual who appears behaviorally impulsive. Related formulations in more contemporary literature include those of Salvatore Maddi and his coworkers (i.e., Maddi, 1968), who postulated 24 different temperaments. High-activation traits with an external orientation were seen as impulsive, whereas low-activation persons with an internal orientation were seen as stable and devoid of flamboyance.

It is unfortunate that most present personality theorists have dismissed the correlation between constitutional factors and personality, outright and with little thought. Although this notion was popularized by many 19th-century authors, the German psychiatrist Ernst Kretschmer (1888–1964) was the best example of a renowned scientist who studied the relationship between body types and behaviors. William Sheldon, his American theoretical descendent, was a psychologist and physician who was concerned with the major components of temperament and their relation to the human physique. Working closely with the gifted measurement theorist S. S. Stevens (Sheldon & Stevens, 1942), Sheldon found significant correlations between basic physique subtypes and temperamental functions. His results were unambiguous in suggesting that there is a close correspondence between temperament as measured by observer ratings and physique as derived from measures taken from photographs. For example, the personality type *somatonia* is most closely related to the muscular build of mesomorphy and tends toward an impulsive and reckless love for adventure, as well as a general disregard for authority or behavioral consequences.

Refinements in the conceptualization of impulsive behaviors awaited the development of more sophisticated mathematical techniques for correlating multiple behaviors across time. Factor analysis is ideally suited for this pursuit (Cattell, 1957). The work of H. J. Eysenck, and later

S. B. G. Eysenck, represents a sophisticated factor analytic approach to the process of understanding personality in general and impulsivity in particular. The Eysencks' body of work, which began shortly after World War II, has long been interested in the role of impulsivity in a dimensional schema of behavioral classification (Eysenck & Eysenck, 1985). Through factor analysis, the Eysencks have identified three orthogonal dimensions of personality: neuroticism versus stability, extroversion versus introversion, and psychoticism versus ego strength. Subsequent work by S. B. G. Eysenck has suggested that impulsivity is related to all three dimensions. Extroverted individuals are likely to behave in a characteristically impulsive way to gain rewards and seek stimulation. This has been labeled as *venturesomeness*. Neurotic individuals behave impulsively because they are changeable. Individuals with elevated scores on the psychotic dimension (not necessarily clinically psychotic individuals) are likely to engage in more sensation-seeking behaviors without reflection, and represent "true impulsiveness." (This and related work is reviewed in the chapters by H. J. Eysenck and Sybil B. G. Eysenck, chapters 4 and 8, this volume, respectively.)

In general, however, nomothetic personality theory suffered under the zeitgeist of classical behaviorism. Operant behaviorism, in particular, with its disinterest in mental processes, was philosophically ill-prepared to investigate individual differences in the notion of action without thinking. Such questions were incompatible with the research agenda of the age, and so powerful was this ideology in American psychology that few psychologists stopped to analyze its weaknesses (L. D. Smith, 1992). More serious nomothetic investigation into the notions of impulsivity awaited a loosening of the behavioral paradigm, which would not occur until the mid 1950s or even later.

POSTBEHAVIORAL INFLUENCES IN THE STUDY OF IMPULSIVITY

Ideographic and nomothetic personality research continues to be popular. However, the postbehavioral influences shaping present-day inquiry into the construct of impulsivity also include at least three newer areas: the biological, the social/cognitive, and the developmental. All such divisions in psychology or any other science are somewhat arbitrary and are probably influenced by the biases of the authors attempting to delineate the groupings. Yet such historical categorization, subjective as it might be, can assist the researcher or clinician in understanding both the previous limitations of the literature and fruitful areas for future collaboration.

Biological/Nomothetic Approaches

The pioneering work of Ernest Barratt (1963, 1967) was highly influential in providing a link between learning theory and biological mechanisms associated with individual differences in impulsivity. Barratt's work (discussed by Barratt, chapter 3, this volume) initially represented a neobehavioral attempt to link impulsiveness to learning theory popular at the time. It quickly progressed toward a more biological orientation, an admirable attempt considering that behavioristic theorization was still dominant.

The work of Arnold Buss and Robert Plomin (e.g., Buss & Plomin, 1975) is discussed extensively in this volume by Sybil B. G. Eysenck (chapter 8) and by H. J. Eysenck (chapter 4). These researchers have defined impulsivity as a primary dimension of personality characterized by a tendency to respond quickly rather than to inhibit one's responsiveness. Its opposite is labeled *deliberateness*. The work of Buss and Plomin is most notable because it combines not only factor analytic methods but also findings from multivariate genetic research and from developmental psychology.

Cognitive–Behavioral and Social Learning Approaches

True to their roots in operant learning, cognitive–behavioral and social learning theorists have been characteristically less interested in individual differences than in the general process of changing behaviors, including impulsivity. However, during the early 1950s and the 1960s, (then) behaviorists such as Albert Bandura and Walter Mischel developed theoretical accounts regarding how individuals learn to suppress impulsive behaviors. Mischel's work was one of the bulwarks that weakened the behaviorists' grip on the science of human performance. Mischel attempted to tackle the thorny question (from a classical behaviorist perspective, at least) of how self-imposed delay of gratification was possible (Mischel, 1961). Individuals who display a generalized failure to delay gratification are often labeled by others as "impulsive." However, Mischel's research suggested that delay of reward is actually a complex web composed of previous experiences, attention, anticipation of reward, and anticipation associated with postponing receipt of reward (Mischel, 1966).

From Mischel's and many other social learning theorists' perspectives, such expectations regarding rewards do not necessarily generalize from situation to situation. This implies that whatever consistency in personality is claimed either is learned or is simply in the mind of the observer, the latter being a very popular belief in the 1970s. Although the denial of the consistency of personality is somewhat of a "straw man position" (Mischel, 1986), resolution of the person–situation debate that occurred during the 1980s makes more extreme personological theories empirically untenable. Most contemporary personality theorists now believe that much of human behavior is consistent, at least in many situations. A frequent weakness of social learning and cognitive behavioral theorists during the last 20 years has been their tendency to ignore the consistency of personality traits, including impulsiveness.

However, despite these shortcomings, social learning theory is almost unquestionably the dominant perspective in contemporary psychology. Among clinicians, cognitive behaviorism has become almost an orthodox theoretical position, nearly as unquestioned as behaviorism was 50 years ago. A major reason is pragmatic: Cognitive–behavioral techniques are often highly effective and comparatively easy to learn. Not surprisingly, given the severity of problems associated with impulsive behaviors, a number of very powerful clinical interventions have been developed and tested by behaviorists, and especially by cognitive behaviorists. (These are reviewed briefly by Fink & McCown, chapter 14, this volume. A detailed cognitive/informational processing account of impulsivity is discussed by Dickman, chapter 9, this volume.)

Familial or other systems theories—which are often closely aligned with social learning accounts (McCown & Johnson, 1993)—suggest that impulsive behavior is a learned reaction against specific family patterns. In this volume, L'Abate presents an interesting and clinically relevant account of the role of family patterns in the development of impulsive behaviors (chapter 6). Marks-Tarlow (chapter 7) presents an account of impulsiveness using the newly developed and rapidly evolving paradigm of chaos theory. Her account incorporates both systemic and individual factors, and although somewhat speculative at present, it is highly suggestive of the possibilities of a genuinely new strategy of understanding human behavior.

Developmental Theories

Developmental psychologists have long noted that children tend to be more impulsive than adults. However, during the 18th and 19th centuries, this knowledge was anecdotal. Perhaps more than anyone else, the work of Jerome Kagan and his associates (e.g., Kagan, Rossman, Day, Albert, & Phillips, 1964) "legitimized" the study of impulsivity, making it a respectable and occasionally popular topic in the empirical developmental literature. Kagan's longitudinal work (e.g., Kagan, Lapidus, & Moore, 1979) has suggested that impulsivity is an enduring behavioral trait and that evidence of it can be ascertained early in the life of the infant. Kagan's work has further indicated the manner in which developmental theories have now become more intertwined with biological facets. (This fruitful trend toward theoretical integration is discussed by Daruna & Barnes, chapter 2, this volume, and is also reflected in many of the other chapters in this volume, including those of Zuckerman and the Eysencks.)

The work of Shure (1986) and of Spivack and Shure (1989) illustrates the growing hybrid that exists between developmental and social learning theories. Shure's work has shown that it is possible to change the process of thinking in children of high-risk, inner-city families so that the results produce less impulsive behaviors and less antisocial outcomes. Shure and Spivack's work has emphasized both developmental and cognitive behavioral interventions designed to teach children how to think, not what to think. Similar work is being pursued by a number of other investigators, regarding teaching adolescents and adults how to think less impulsively. These successful programs have in common an emphasis on the relevancy of problem-solving training by making sure that the didactic material is appropriate to the developmental level of the student or client receiving practice or instruction. Researchers now also realize the importance of culturally competent instructors accompanied by culturally sensitive presentations of key concepts (Spivack & Shure, 1989). (Chapter 19 in this volume by Touchet, Shure, & McCown indicates the potential utility of a problem-solving approach to reduce impulsive behaviors.)

CONCLUSION: IMPLICATIONS OF THE STUDY OF IMPULSIVENESS

Impulsiveness is clearly a common phenomenon that not unoccasionally has serious health and psychological implications. Probably since the time of the ancient Hebrews, and certainly since the ancient Greeks, impulsiveness has been recognized as a major component of psychopathology. There is a large "prescientific" literature regarding the personality trait that we now label impulsivity, suggesting that problems with impulsive behavior are not a new phenomenon but have existed since antiquity. Interest began to surface when theological conceptions began to loosen and individuals, rather than the devil, were seen as responsible for human behavior. However, theologians and philosophers felt a discomfort with the notion of impulsiveness, recognizing, perhaps as the Greeks did, that it necessarily alters the image of humankind.

During the first half of this century, ideographic and a somewhat atheoretical nomothetic inquiry dominated psychological discussion regarding the causes of impulsive behavior. During the second half of this century, research has expanded substantially, partially as a result of the demise of the dominant paradigm of orthodox behaviorism. Presently, there is some evidence that disparate theoretical approaches may be converging and that researchers of a variety of subdisciplinary orientations may be benefiting from and influencing each other's explorations and conceptualizations.

As we learn more about impulsivity, we may find ourselves in an uncomfortable position, somewhat like the Greeks and the Renaissance thinkers. With advances in psychology and other behavioral sciences, our growing knowledge may well lead to a variety of reconceptualizations about human behavior. When recognized to be primarily a function of nature, circumstance, and nurture, impulsive behaviors can be freed from negative stigmas and will be seen as a concern for scientists, not theologians. People with problems of impulsive control will be treated, psychologically and perhaps also biologically, rather than judged. This shift may be upsetting to traditionalists, and we would not be surprised if future progress generates a reciprocal backlash. Whether society will be comfortable enough to allow scientific progress to continue is not at all assured.

As we learn more about impulsiveness we will also learn more about its prevention. This much-needed knowledge will have vast social implications. For example, as we discover more about environmental contingencies that mediate negative impulsive behaviors, we will probably be able to apply these results for purposes of prevention. In fact, a very good case can be made that we have sufficient knowledge to implement effective prevention technology today. And yet it is all too rare to see appropriate application of this psychological knowledge, even among groups where prevention is desperately needed. Greek and Renaissance thinkers struggled with the concept of what the existence of impulsive behavior says about human nature. A more relevant problem today may be what the refusal of society to try out scientifically sound strategies to prevent psychological dysfunction says about our collective moral nature.

REFERENCES

Adams, R., & Victor, M. (1986). *Principles of neurology* (3rd ed.). New York: McGraw-Hill.

American Psychiatric Association. (1987). *Diagnostic and statistical manual of mental disorders* (3rd ed., rev.). Washington, DC: Author.

Augustine. (1912). *Confessions* (W. Watts, Trans.). London: Heinemann.

Baly, D. (1977). *God, man, and history in the testament.* New York: Random House.

Barratt, E. S. (1963). Behavioral variability related to stimulation of the cat's amygdala. *Journal of the American Medical Association, 186,* 773–775.

Barratt, E. S. (1967). The effects of Thiazesim LSD-25, and bilateral lesions of the amygdalae in the release of a suppressed response. *Recent Advances in Biological Psychiatry, 9,* 229–240.

Benjamin, L. T., Jr., Durkin, M., Link, M., Vestal, M., & Acord, J. (1992). Wundt's American doctoral students. *American Psychologist, 47,* 123–131.

Bonet, T. (1684). *Sepulchretum* [The cemetary]. Paris.

Botsford, G. (1893). *The Athenian constitution.* New York: Everyman Library.

Buss, A., & Plomin, R. (1975). *A temperamental theory of personality development.* New York: Wiley.

Cattell, R. B. (1957). *Personality and motivation structure and measurement.* New York: World.

Defaul, R. (1928). *The early Christians.* London: Hodder & Stoughton.

Demont, L. (1933). *A concise dictionary of psychiatry and medical psychology.* Philadelphia: Lippincott.

Dryfoos, J. (1990). *Adolescents at risk: Prevalence and prevention.* New York: Oxford University Press.

English, H. (1928). *A student's dictionary of psychological terms.* Yellow Springs, OH: Antioch Press.

Eysenck, H. J., & Eysenck, M. (1985). *Personality and individual differences.* New York: Plenum Press.

Freud, S. (1960). Libidinal types. In *Collected papers* (J. Strachey, Ed. & Trans.; Vol. 21, pp. 216–236). London: Hogarth. (Original work published 1931)

Gall, F., & Spurzheim, J. (1809). *Recherches sur le système nerveux* [Research on the nervous system]. Paris: Schoell.

Goslin, D. (Ed.). (1969). *Handbook of socialization theory and research.* Chicago: Rand McNally.

Grayson, H., & Tolman, R. (1950). A semantic study of concepts of clinical psychologists and psychiatrists. *Journal of Abnormal Psychology, 45,* 216–231.

Henichel, R. (1911). *Fathers of the reformation.* London: Gorman.

Heymans, G., & Wiersma, E. (1909). Beitrage zur speziellen Psychologie auf Grund einer Massenuntersuchung. [Implications of the psychology of personal types to lifework]. *Zeitschrift für Psychologie, 51,* 22–128.

Hinslie, L., & Shatzky, J. (1940). *Psychiatric dictionary.* New York: Oxford University Press.

Hirt, E. (1902). *Die Temperamente* [The temperaments]. Leipzig, Germany: Barth.

James W. (1890). *The principles of psychology.* (Vol. 1). New York: Holt.

Jennings, A. (1928). *A brief history of scientific language.* London: Lewis.

Kagan, J., Lapidus, D., & Moore, M. (1979). Infant antecedents of cognitive functioning: A longitudinal study. *Annual Progress in Child Psychiatry and Child Development,* 46–77.

Kagan, J., Rossman, B. L., Day, D., Albert, J., & Phillips, W. (1964). Information processing in the child: Significance of analytic and reflective attitudes. *Psychological Monographs, 78* (1, Whole No. 578).

Kahn, E. (1928). *Psychopathischen Personlichkeiten* [Psychopathic personality]. Berlin: Springer-Verlag.

Kant, I. (1933). *Collected works* (Vol. 2). Oxford, England: Oxford University Press. (Original work published 1811)

Kantor, J. R. (1963). *The scientific evolution of psychology.* (Vol. 1). Granville, OH: Principia Press.

Kitto, H. D. (1952). The Greeks. Harmonsworth, Middlesex, England: Penguin Books.

Kraeplin, E. (1913). *Psychiatrie: Ein Lehrbuch* [Psychiatry: A textbook]. Leipzig, Germany: Barth.

Lacey, J., & Evans, D. (1986). The impulsivist: A multi-impulsive personality disorder. *British Journal of Addictions, 83,* 391–393.

Leckey, W. (1869). *History of European morals from Augustus to Charlemagne.* London: Routledge.

Maddi, S. (1968). *Personality theories.* Homewood, IL: Dorsey.

Marx, M., & Hillix, W. (1963). *Systems and theories in psychology.* New York: McGraw-Hill.

McCown, W. (1993, August). *The ideodynamics of impulsive families.* Paper presented at the 101st Annual Convention of the American Psychological Association, Toronto, Canada.

McCown, W., & Johnson, J. (1993). *The treatment resistant family: A consultation/crisis intervention treatment strategy.* New York: Haworth.

Millon, T. (1981). *Disorders of personality: DSM III: Axis II.* New York: Wiley.

Mischel, W. (1961). Delay of gratification, need for achievement, and acquiescence in another culture. *Journal of Abnormal and Social Psychology, 62,* 543–552.

Mischel, W. (1966). Theory and research on the antecendents of self-imposed delay of reward. In B. A. Maher (Ed.), *Progress in experimental personality research* (Vol. 3, pp. 85–132).

Mischel, W. (1986). *Introduction to personality* (4th ed.). New York: Holt, Rinehart & Winston.

O'Boyle, M., & Barratt, E. S. (in press). Impulsivity and DSM-III-R personality disorders. *Personality and Individual Differences.*

O'Leary, P. L. (1949). *How Greek science passed to the Arabs.* London: Routledge.

Oxford English dictionary. (1951). Oxford, England: Oxford University Press.

Pinel, P. (1801). *Traité médico-philosophique sur l' aliénation mentale* [medico-philosophical treatise on mental alienation]. Paris: Richard, Caille et Ravier.

Prichard, J. (1835). *A treatise on insanity.* London: Sherwood, Gilbert, & Piper.

Principia theologica. (1932). New York: Hanfer. (Original work published 1554)

Queyart, F. (1896). *Les caractères et l' éducation morale* [Personality traits and moral education]. Paris: Alcan.

Reich, W. (1933). *Charakteranalyse* [Character analysis]. Leipzig, Germany: Sexpol Verlag.

Ribot, T. (1890). *Psychologie des sentiments* [Psychology of emotions]. Paris: Delahaye.

Saul, L. (1947). Emotional maturity. Philadelphia: Lippencott.

Sheldon, W., & Stevens, S. (1942). *Varieties of temperament: Psychology of constitutional differences: New York: Harper.*

Shure, M. (1986). Interpersonal problem-solving: A cognitive approach to behaviour. In R. Hinde, A. Perret-Clermon, & J. Stevenson-Hinde (Eds.), *Social relationships and cognitive development* (pp. 191–207). Oxford, England: Oxford University Press.

Smith, L. (1952). *A dictionary of psychiatry for the layman.* London: Maxwell.

Smith, L. D. (1992). On prediction and control: B. F. Skinner and the technological ideal of science. *American Psychologist, 47,* 216–223.

Spivack, G., & Shure, M. B. (1989). Interpersonal Cognitive Problem Solving (ICPS): A competence-building primary presentation program. *Prevention in Human Services, 6,* 151–178.

Stanford, M. S., & Barratt, E. S. (1992). Impulsivity and the multi-impulsive personality disorder. *Personality and Individual Differences, 13,* 831–834.

Tramer, M. (1931). Psychopathic personalities. *Schweizer medizinische Wochenschrift, 217,* 272–322.

U.S. Department of Justice. (1992). *Crime statistics.* Washington, DC: Author.

Wundt, W. (1907). *Volkerpsychologie* [Folk psychology]. Leipzig, Germany: Engelmann.

2

A NEURODEVELOPMENTAL VIEW OF IMPULSIVITY

JORGE H. DARUNA and PATRICIA A. BARNES

The term *impulsivity* is usually reserved for maladaptive behavior (Barratt & Patton, 1983). The behavioral universe thought to reflect impulsivity encompasses actions that appear poorly conceived, prematurely expressed, unduly risky, or inappropriate to the situation and that often result in undesirable consequences. When such actions have positive outcomes, they tend to be seen not as signs of impulsivity, but as indicative of boldness, quickness, spontaneity, courageousness, or unconventionality (for further discussion, see Dickman, chapter 9, this volume).

Impulsive behavior is more typical of young individuals (Kopp, 1982). Differences in impulsivity can be reliably observed among toddlers under 2 years of age (Power & Chapieski, 1986; Silverman & Ragusa, 1990). It is also usually possible, on the basis of emerging motor competence, to differentiate between more and less impulsive infants (Rothbart, 1988). Thus, impulsivity may be regarded as a quality of behavior that is evident to differing degrees throughout the life span and that if extreme relative to one's peers becomes disabling and should be viewed as a clinical disorder. It is the goal of this chapter to present a framework for understanding impulsivity in neurodevelopmental terms with the hope that it will facilitate discourse among clinicians with different theoretical orientations.

MEASUREMENT OF IMPULSIVITY

Methods for the measurement of impulsivity are varied and depend on the age of the subject. These are discussed in greater depth by Fink and McCown (chapter 14, this volume); however, a brief review here may be helpful. Adult measures of impulsivity (e.g., Eysenck, 1983) rely on the self-assessment of behavior through the use of questionnaire items (e.g., *Are you usually carefree? Would you do almost anything for a dare?*). On the other hand, impulsivity in children is most frequently measured using rating scales completed by adults (e.g., Conners, 1982) or more directly by testing them with tasks that require response inhibition. For example, one of the more popular tasks is the Matching Familiar Figures Test (MFFT; Kagan, Rossman, Day, Albert, & Phillips, 1964). Successful performance on this test necessitates a delay in response while carefully comparing several similar stimuli to a target stimulus in order to determine which one is identical to the target. Another frequently used task demands that the child refrain from approaching some desirable object (Mischel, 1958; Sears, Rau, & Alpert, 1965). Other tasks include those that require the child to make slow motor responses (e.g., draw a line as slowly as possible) (Maccoby, Dowley, Hagan, & Degerman, 1965), to carefully plan and execute motor responses to get through a maze of lines rapidly and without errors (Porteus, 1968), or to inhibit a response to a simple go signal when a simple no-go signal is also presented (Schachar & Logan, 1990).

It should be noted that correlations among the various measures of impulsivity are generally low (Gaddis & Martin, 1989; Olson, 1989), suggesting that impulsivity can be situation or task specific.

IMPULSIVITY: PRECURSORS AND EARLY INFLUENCES

Evidence from studies of identical twins separated at birth suggests that the tendency to behave impulsively has a significant genetic component (Zuckerman, 1991; see also Zuckerman, chapter 5, this volume.) However, this cannot be taken as solely supporting a genetic influence because brain organization at birth is not simply a manifestation of genes (Edelman, 1988). Brain organization also reflects epigenetic influences affected by the manner in which environmental factors impinge upon the mother and alter the intrauterine environment (see Figure 1).Therefore,

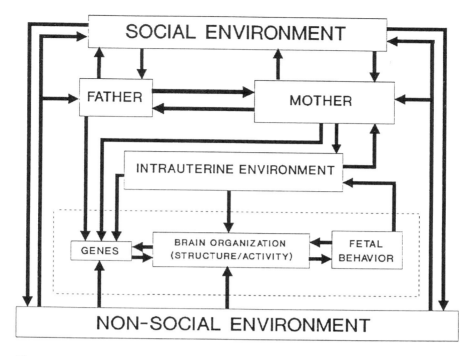

Figure 1. Prenatal brain organization is subject to multiple influences. "Social environment" encompasses current and past interpersonal experiences impinging upon the mother and would be subjectively expressed in the form of the emotional state and associated bodily changes. "Non-social environment" refers to all other influences that affect the mother (e.g., nutrition, exposure to pathogens, toxins, drugs, radiation, etc.). Note that social and nonsocial environments influence each other.

until more direct measures of genetic influence become available, behavioral similarity between twins raised apart is probably best conceptualized as reflecting the interaction of genes with aspects of the prenatal environment.

Early behavior has been found in some studies to be predictive of later impulsivity. One study (Carey, Fox, & McDevitt, 1977) found evidence that a difficult temperament at 6 months of age predicted impulsive performance on the MFFT during the school years. Our research group has also observed that more impulsive children are described retrospectively as having been more difficult infants (Daruna & Rau, 1989). However, other studies (e.g., Olson, Bates, & Bayles, 1990) have shown that measures obtained in the first year of life are frequently not significantly predictive of later impulsivity. This seems reasonable given that the first year of life is a period of rapid development, high plasticity, and restricted behavioral capability, which may mask the condition of the brain. Thus, one may encounter infants in early life who appear headed for impulsivity according to some index (e.g., difficult temperament) but who will never exhibit impulsive behavior because (a) the original presentation reflected only a transient lack of synchrony in the development of relevant neural systems or (b) the receiving social environment had characteristics capable of attenuating the bias toward impulsive behavior. Similarly, infants who do not appear headed for impulsivity may eventually show impulsive behavior because (a) they possess subtle neural malformations (Vitiello, Stoff, Atkins, & Mahoney, 1990) that manifest only later as the behavioral repertoire develops or (b) the receiving social environment has characteristics that strongly select for and reinforce impulsive behavior.

With respect to the influence of the social environment, it is noteworthy that certain features of early social interactions with caregivers appear to be important as predictors of impulsivity in a variety of tasks. Low-impulsive children tend to have mothers who are less punitive and engage in more frequent positive verbal interactions with their children (Olson et al., 1990; Power & Chapieski, 1986). These aspects of the mother–child relationship increase the prediction of impulse control above that possible using measures of cognitive level or behavioral compliance. Also, mothers who encourage independence have less impulsive children, and maternal intrusiveness or the tendency to overcontrol appears to be associated with high impulsivity (Silverman & Ragusa, 1990).

MODIFICATION OF IMPULSIVITY

Efforts to modify impulsivity have generally emphasized cognitive behavioral training or pharmacological intervention. Specific strategies for behavioral self-control and pharmacological treatments that increase the release of monoamine neurotransmitters in the brain have been demonstrated to be somewhat effective in the management of impulsive behavior.

Training programs have taken many forms (e.g., Kendall & Braswell, 1985; Nelson & Behler, 1989; Santostefano, 1978; Spivack & Shure, 1989). Typically, these approaches expose the child to specific tasks and exercises in an effort to develop the child's ability to take into account the relevant dimensions of a situation and to actively regulate motor responses. Some training programs (Spivack & Shure, 1989; Touchet, Shure, & McCown, chapter 19, this volume) emphasize the verbal analysis of social situations and are specifically designed to facilitate the multidimensional representation of such situations and of the options for responding adaptively by weighing the consequences. In general, cognitive/behavioral interventions are efficacious (Routh & Mesibov, 1980), although in some instances the effects are transient or fail to generalize to real-life situations.

Pharmacological treatments have also been useful in modifying impulsivity in children. Stimulant drugs such as methylphenidate, dexedrine, and pemoline have been especially helpful (Wilens & Biederman, 1992). In addition, monoamine oxidase inhibitors (Zametkin, Rapoport, Murphy, Linnoila, & Esmond, 1985) and tricyclic compounds (Greenhill, 1992) have been found to benefit impulsive children. These drugs are all capable of increasing monoamine levels at synapses.

The efficacy of these types of interventions appears consistent with the idea that impulsivity may ultimately reflect inadequately detailed cognitive representation of situations traceable, at least partially, to low arousal. This is because low arousal has been theorized to limit the available capacity for neural processing (Kahneman, 1973) and monoamine release has been associated with increased arousal (Pape & McCormick, 1989). It is tempting to speculate that cognitive/behavioral interventions may work by making the individual more cognitively efficient (i.e., capable of making optimal use of available representational capacity), whereas pharmacological interventions may directly increase arousal and thereby augment representational capacity.

BIOLOGICAL DIFFERENCES AND IMPULSIVITY

Recent human studies of monoamine metabolites and the brain's spontaneous electrical activity (see Zuckerman, 1991, for a comprehensive review) in high- and low-impulsive individuals provide additional support for the view that impulsivity may reflect low arousal and its effect on neural representational capacity.

Monoamine Metabolism

Much of the research on the function of monoaminergic systems in humans has been conducted by measuring metabolite levels in plasma, cerebrospinal fluid (CSF), or urine and by quantification of enzyme activity or monoamine-reuptake membrane channels in platelets. Unfortunately, these measures are clearly indirect and entirely useless for gauging regional neural activity. They require a number of assumptions to permit even weak inferences regarding the function of monoaminergic systems in the brain. For instance, it must be assumed that metabolite levels are indicative of the overall neural activity of specific neurotransmitter systems. However, it is possible that differences in metabolite levels reflect factors such as metabolite efflux or transport out of the central nervous system, rather than neurotransmitter utilization in synaptic transmission. Similarly, platelet enzyme activity or the number of monoamine reuptake channels on platelet membranes must be assumed to be reflective of the situation at synapses, even though at the synapses, but not in platelets, enzyme activity and membrane characteristics are subject to regulation by neuronal activity.

These are very serious limitations. In fact, there is already some direct evidence that monoamine oxidase (MAO) activity in platelets is not correlated with MAO activity in the brain (Young, Laros, Sharbrough, & Weinshilboum, 1986). Nonetheless, researchers have forged ahead and observed some differences in indicators of monoamine activity as a function of impulsivity. Platelet MAO activity has been found to be lower in impulsive adults, especially those prone to exhibit aggression, antisocial tendencies, or thrill-seeking behaviors (Zuckerman, 1991). Studies of 3-methoxy-4-hydroxyphenylglycol (MHPG), 5-hydroxyindoleacetic acid (5-HIAA), and homovanillic acid (HVA), the major breakdown products for norepinephrine (NE), serotonin (5-HT), and dopamine (DA), respectively, present a complex picture. For instance, in one study, the relationship of MHPG to personality measures depended on what type of bodily fluid (i.e., CSF, plasma, urine) was used to estimate its level (Ballenger et al., 1983). There was only a trend for lower plasma MHPG in impulsive individuals. CSF 5-HIAA and HVA also tended to be lower in impulsive adults and those lacking ego strength or a sense of self-efficacy. Others (Schalling, Ashberg, & Edmoor, 1984, cited in Zuckerman, 1991) have found that CSF 5-HIAA is lower in more impulsive individuals, but the extent of the correlation depends on group characteristics and is significant only in nondepressed patients and not in depressed patients or nonpatient volunteers. CSF HVA and MHPG levels showed a similar trend, being lower in the more impulsive nondepressed patients. These studies show some consistency in that metabolite levels tended to be lower in more impulsive individuals and higher in more socially competent or anxious subjects. This is particularly the case for 5-HIAA levels.

Studies of children and adolescents have been only partially consistent with the picture emerging from adult studies. For instance, even though impulse control tends to increase with age, MAO activity and the levels of 5-HIAA and HVA all decrease with age (Kruesi et al., 1990; Stoff et al., 1989). Moreover, in contrast with the adult picture, at least one study has found MAO activity to be higher in more impulsive youngsters (Stoff et al., 1989). However, the results for 5-HIAA levels are more congruent with the results of adult studies. At a given age, low 5-HIAA levels tend to be associated with aggressiveness; higher 5-HIAA levels are seen in the more socially competent youngsters. It has also been observed that imipramine binding to platelets, an index of the number of 5-HT reuptake sites, is lower in children exhibiting impulsive aggression (Birmaher et al., 1990). These studies jointly indicate that children with less impulse control, particularly with respect to aggressive behavior, have lower 5-HIAA levels, fewer 5-HT reuptake sites, and higher MAO activity. This pattern of results defies straightforward interpretation in that one can make a good case for either a greater or a lesser 5-HT availability at postsynaptic sites as underlying impulsivity.

Despite the limitations of the studies and the ambiguities presented by the data, it seems fair to conclude that the available evidence in humans hints at the involvement of monoaminergic nonspecific ascending projections in the regulation of impulse control. The more consistent findings involve the serotonergic system. However, the available evidence is simply not sufficient to draw unambiguous conclusions regarding the nature of serotonergic postsynaptic effects in impulsive individuals. Research with animals (e.g., Soubrie, 1986) has suggested that continued efforts in this area are justified, particularly as noninvasive imaging techniques become more refined to permit measurement of specific neurochemical entities in regions of the living brain.

Spontaneous Electrical Activity of the Brain

The electroencephalogram (EEG) is sensitive to the arousal state of the subject. The amplitude of brain waves as a function of their frequency (i.e., the EEG's power spectrum) reflects changes in the state of the subject from deep sleep to attentive anticipation. These changes recorded at the scalp are ultimately indicative of the frequency with which neuronal groups within the neocortex (probably clusters of minicolumns) oscillate from maximum to minimum excitability within a given period of time and the degree of synchrony in these oscillations across areas of the neocortex (Daruna, 1986; Nunez, 1981).

The dynamics of regional cortical activity are influenced by signals arriving via cortico-cortical, thalamo-cortical, and nonspecific pathways such as the monoaminergic projections originating within the brainstem (Lopes da Silva, 1991). It is especially noteworthy, given the possible involvement of monoamine neuron activity in the regulation of impulse control, that increased activity of the noradrenergic and serotonergic nonspecific projections facilitates thalamocortical transmission (Pape & McCormick, 1989) and contributes to EEG desynchronization, a measure of increased cortical arousal.

Studies of the EEG power spectrum in adults (e.g., O'Gorman & Lloyd, 1987) have tended to show that individuals low on impulsivity exhibit lower power (i.e., have more desynchronized activity) within the alpha band (8–13 Hz) of the EEG. We have also observed (Daruna & Rau, 1989) that young children (4 to 7 years of age) rated by their parents as low in impulsivity have significantly lower power within the alpha band (see Table 1).

EEG coherence, a different measure of the degree of coupling in activity across cortical regions (Thatcher, 1992), has not been sufficiently

TABLE 1
Mean (SEM) Age and EEG Alpha Power in Young Children (4 to 7
Years) Differing in Impulsivity on the Conners Scale

| | Impulsivity | | | |
| | Low (n = 10) | | Medium (n = 12) | |
Measure	$\bar{X}$	SEM	$\bar{X}$	SEM
Conners Scale				
Score (*SD* units)	−1.25	0.10	−0.15	0.13**
Age (years)	5.5	0.32	5.7	0.37
Alpha (8–12 Hz) power (μV^2)*				
Left Frontal (F3)	17.5	2.5	33.5	5.0**
Right Frontal (F4)	18.7	3.0	33.9	4.3***
Left Parietal (P3)	22.1	5.0	49.2	5.0**
Right Parietal (P4)	20.6	5.3	58.2	12.7**

*Eyes-open, linked-ears reference.   **$p < .02$.

investigated in relation to impulsivity. However, preliminary evidence suggests that the frontal areas of the brain have higher coherence (i.e., are more synchronized) in the more impulsive subjects (Thatcher, personal communication, April, 1992).

These findings appear consistent with the interpretation that low-impulsive subjects are more aroused, as indexed by less synchronized brain activity. Therefore, they may be capable of representing situations in more detail, hence acting on the basis of more information than do impulsive persons.

CONCEPTUALIZING IMPULSIVITY IN NEURODEVELOPMENTAL TERMS

As one examines the various descriptions of impulsive behavior, it is evident that a common feature of such behavior is the failure to take into account sufficient relevant aspects of the situation confronting the individual, or characteristics of the individual pertinent to the situation, as part of the prelude to action. Moreover, it appears as if, frequently, possible negative or temporally distant consequences have not been adequately incorporated in the decision-making process. The failure to take into account relevant information may ultimately reflect a reduced capability to consider multiple factors simultaneously or in close temporal proximity. This can occur because representational capacity is either limited or has low dimensionality, such that a single representation takes up a significant portion of the available capacity. One can also argue that representational capacity of high dimensionality could give rise to behavior that appears impulsive, because at any instant, a subset of the representations (not necessarily the most relevant subset) could be sufficient to trigger action that would seem poorly planned.

In the brain, representational capacity can be viewed as the number of processing units (e.g., small clusters of cortical neurons arranged in minicolumns) optimally ready to respond at a given moment. The dimensionality of representational capacity refers to the extent to which neural networks incorporating such clusters are able to respond independently of each other and thus simultaneously code for different aspects of a situation. Representational capacity and its dimensionality are dependent on the ability of cortical minicolumns to be active in a desynchronized manner. This is partially a function of the reciprocal interconnectivity among cortical, thalamic, and brainstem neuronal groups, which is largely established prenatally (Rakic, 1988) and only partially under genetic guidance (Edelman, 1988). For instance, the monoaminergic nonspecific projections develop very early during gestation (Verney, Zecevic, Nikolic, Alvarez, & Berger, 1991) and show adult-like characteristics by birth or shortly thereafter (Nobin & Bjorklund, 1973). Subsequent stimulation of these neural networks, in the form of postnatal experience, further refines the microstructure of the brain within the constraints imposed by what has already transpired. Stimulation exerts its influence by mechanisms involving changes in the configuration of neuronal membrane constituents, leading to ionic fluxes that initiate alterations in the neuron's macromolecular composition and ultimately affect its morphology. These changes are expressed in the proliferation, expansion, or disappearance of synapses; the thickening, myelination, or elimination of axon branches; and the growth or death of neurons. In essence, patterns of stimulation can, within limits, shift the structure of neural networks either toward or away from the architecture needed for optimal representational capacity.

Impulsive behavior may result not only from factors affecting capacity or dimensionality, but also from representational imbalances, that is, biases away from taking into account past negative experiences or possible future outcomes. Such imbalances may ultimately be traceable to the functional asymmetries that appear to characterize the left and right hemispheres (Nass & Gazzaniga, 1987). For instance, there is growing evidence that the hemispheres show specialization with respect to the regulation of approach and avoidance behaviors and the associated positive and negative emotional states (Fox & Bell, 1990; Heilman & Bowers, 1990). It is also evident that within each hemisphere the frontal regions are more specialized than other regions with respect to behavioral inhibition (Miller, 1992). Representational imbalances leading to impulsive action can arise as a result of a variety of structural features, such as those secondary to tissue damage, or those reflective of the degree to which the nonspecific projections have spatially asymmetric distributions and thus differentially determine capacity or dimensionality in brain regions specialized for processing different types of information.

Still another possible basis for impulsivity does not implicate representational factors like capacity, dimensionality, or imbalance, but is more directly due to threshold factors (i.e., the ease with which specific representations can elicit motor acts) or timing factors (i.e., the temporal overlap of signals emanating from the various representations to regulate action). These features can independently serve to differentially weigh representations in the computation leading to action. The regulation of motor output regions by other specific regions can be affected by a wide range of factors, including the length and diameter of axons projecting to motor areas from the other regions, the level of myelination of the axons, the precise point of synaptic contact on the postsynaptic neurons, and other characteristics such as the quantity of transmitter released, postsynaptic receptor density, and the preexisting instantaneous level of membrane polarization. The nonspecific projections could also play a role at this level by affecting activation thresholds at key points in a manner that would influence the timing of signals converging on systems regulating motor output.

Given the preceding considerations, it seems unlikely that impulsive behavior is reflective of any single aspect of brain organization. Instead, a composite dynamic state of brain organization, reflective of capacity, dimensionality, and asymmetry at the representational level and of thresholds or timing factors at the level of motor control, would seem minimally necessary to adequately model impulsivity. Nonetheless, it seems reasonable to expect that structural and functional characteristics of the monoaminergic pathways would be relevant to the occurrence of impulsive behavior, because they influence dynamic aspects of brain organization.

The possible involvement of monoaminergic pathways underscores the importance of very early experience in the regulation of impulsivity. This is because monoaminergic pathways develop quite early; thus, nongenetic factors that affect their organization must operate during prenatal life. Moreover, experience during infancy would be very influential in the integration of these early developing projections into complex neural networks in a manner that could either promote or neutralize any original bias toward impulsivity. Therefore, any association between the activity of systems like the monoaminergic projections and behavioral tendencies like impulsivity must also reflect the contribution of experiential factors, at least in the sense that these factors have served to stabilize the structure of neural networks in a manner that permits monoaminergic activity to be a major determinant of impulsivity.

SUMMARY AND FURTHER SPECULATIONS

The term *impulsivity* denotes behavior that is not adequately matched to the situation or that is prematurely executed and has maladaptive consequences. Impulsive individuals are those who exhibit such behavior across time and in a variety of contexts. Impulsive behavior is a manifestation of dynamic features of brain organization that determine (a) neural representational capacity, dimensionality, and symmetry; (b) the activation thresholds of motor systems; and/or (c) the temporal convergence of neural signals regulating motor output.

The dynamic aspects of brain organization underlying impulsivity are ultimately traceable to microstructural features of neural circuitry, which are largely established prenatally and cannot be entirely genetic in origin. Therefore, prenatal environmental factors, not yet understood, can make a contribution to an organization of the brain in utero that could predispose individuals toward impulsivity. Early behavioral manifestations of impulsivity have been difficult to establish, but some evidence has suggested that a difficult temperament may be a precursor to impulsivity. Also, early experiences with punitive and overcontrolling parents have been associated with later impulsivity.

Behavioral interventions that diminish impulsivity typically involve training individuals to generate more elaborate representations of specific situations. Pharmacological treatments that increase monoamine availability at synapses also appear to reduce impulsivity. Studies of monoamine metabolism and the electrical activity of the brain in high- and low-impulsive individuals have provided evidence that these groups differ in cortical arousal, which has been in turn theorized to determine neural representational capacity.

In light of current knowledge, it seems reasonable to view brain organization at birth as biased with respect to the potential for impulsive behavior in later years. Genetic factors, although undeniably contributing, are probably not as significant as they seem at first glance. Maternal experiences during pregnancy and management of infant behavior should have a profound effect on the structure of the brain and the potential for impulsive action.

The tendency toward impulsivity may appear in some infants as a difficult temperament. If such a behavioral tendency leads to understimulation (to keep the baby from becoming fussy) or harsh handling of the infant by caregivers (because of frustration), the consequence may be a molding of brain organization in a direction that further reinforces the bias toward impulsivity. As the child gets older and becomes more mobile, impulsivity begins to manifest more directly and may become further entrenched if it is controlled aggressively (Lewis, 1992).

The foregoing speculations suggest that impulsivity (i.e., "get it while you can") may be a successful adaptation to impoverished, dangerous, or hostile environments. It is naturally available in the genome and gains increasing phenotypic expression because of selection forces applied by postconception environments, beginning in the form of prenatal epigenetic effects. A role for prenatal influences should be expected because if the mother lives in an impoverished and dangerous setting, it would seem adaptive for her internal milieu to exert pressure on fetal brain organization in a direction predisposing to behavioral tendencies, such as impulsivity, that may be more successful in that environment. It is also critical to recognize that once such a tendency emerges, for whatever reason, it will act as an organizer of the environment. In other words, impulsivity can provoke the deprivation and aggression that it is best suited to endure, and thus lead to its own stabilization.

In light of the preceding discussion, we would like to close by emphasizing the need to focus attention on events during the prenatal period and the early months of life as crucial to the development of traits like impulsivity. Moreover, it is essential that the clinician and the parent be aware that "natural reactions" to impulsive behavior (e.g., punishment or overcontrol) may be counterproductive. What impulsive children may need is sustained help to recognize their propensity to act on the basis of partial information, and interventions designed to help them sustain arousal and adequately represent situations as a prelude to action. It is important not to lose sight of the fact that interventions, be they cognitive, behavioral, pharmacological, or educational, are *all* biologically mediated. Research efforts must give high priority to the question of how to best combine treatments in specific cases to yield maximum benefit.

REFERENCES

Ballenger, J. C., Post, R. M., Jimerson, D. C., Lake, C. R., Murphy, D. L., Zuckerman, M., & Cronin, C. (1983). Biochemical correlates of personality traits in normals: An exploratory study. *Personality and Individual Differences, 4,* 615–625.

Barratt, E. S., & Patton, J. H. (1983). Impulsivity: Cognitive, behavioral, and psychophysiological correlates. In M. Zuckerman (Ed.), *Biological bases of sensation seeking, impulsivity, and anxiety* (pp. 77–116). Hillsdale, NJ: Erlbaum.

Birmaher, B., Stanley, M., Greenhill, L., Twomey, J., Gavridescu, A., & Rabinovich, H. (1990). Platelet imipramine binding in children and adolescents with impulsive behavior. *Journal of the American Academy of Child and Adolescent Psychiatry, 29,* 914–918.

Carey, W. B., Fox, M., & McDevitt, S. C. (1977). Temperament as a factor in early school adjustment. *Pediatrics, 60,* 621–624.

Conners, C. K. (1982). Parent and teacher rating forms for the assessment of hyperkinesis in children. In P. A. Keller & L. G. Ritt (Eds.), *Innovations in clinical practice: A source book* (Vol. 1). Sarasota, FL: Professional Resource Exchange.

Daruna, J. H. (1986). Mental development, response speed, and neural activity. In M. G. Wade (Ed.), *Motor skill acquisition of the mentally handicapped: Issues in research and training* (pp. 189–212). Amsterdam: Elsevier-North-Holland.

Daruna, J. H., & Rau, A. E. (1989). *Low-impulsivity in young children: Temperamental precursors, cognitive function, and brain activity.* Unpublished manuscript.

Edelman, G. M. (1988). *Topobiology: An introduction to molecular embryology.* New York: Basic Books.

Eysenck, H. J. (1983). A biometrical–genetical analysis of impulsive and sensation seeking behavior. In M. Zuckerman (Ed.), *Biological bases of sensation seeking, impulsivity, and anxiety* (pp. 1–27). Hillsdale, NJ: Erlbaum.

Fox, N. A., & Bell, M. A. (1990). Electrophysiological indices of frontal lobe development: Relations to cognitive and affective behavior in human infants over the first year of life. *Annals of the New York Academy of Sciences, 608,* 677–698.

Gaddis, L. R., & Martin, R. P. (1989). Relationship among measures of impulsivity for preschoolers. *Journal of Psychoeducational Assessment, 7,* 284–295.

Greenhill, L. L. (1992). Pharmacologic treatment of attention deficit hyperactivity disorder. *Psychiatric Clinics of North America, 15,* 1–27.

Heilman, K. M., & Bowers, D. (1990). Neuropsychological studies of emotional changes induced by right and left hemispheric lesions. In N. L. Stein, B. Leventhal, & T. Trabasso (Eds.), *Psychological and biological approaches to emotion* (pp. 97–113). Hillsdale, NJ: Erlbaum.

Kagan, J., Rossman, B. L., Day, D., Albert, J., & Phillips, W. (1964). Information processing in the child: Significance of analytic and reflective attitudes. *Psychological Monographs, 78*(1, Whole No. 578).

Kahneman, D. (1973). *Attention and effort.* Englewood Cliffs, NJ: Prentice-Hall.

Kendall, P. C., & Braswell, L. (1985). *Cognitive–behavioral therapy for impulsive children.* New York: Guilford Press.

Kopp, C. B. (1982). Antecedents of self-regulation: A developmental perspective. *Developmental Psychology, 18,* 199–214.

Kruesi, M. J. P., Rapoport, J. L., Hamburger, S., Hibbs, A., Potter, W. Z., Lenane, M., & Brown, G. L. (1990). Cerebrospinal fluid, monoamine metabolites, aggression, and impulsivity in disruptive behavior disorders of children and adolescents. *Archives of General Psychiatry, 47,* 419–426.

Lewis, D. O. (1992). From abuse to violence: Psychophysiological consequences of maltreatment. *Journal of the American Academy of Child and Adolescent Psychiatry, 31,* 383–391.

Lopes da Silva, F. (1991). Neural mechanisms underlying brain waves: From neural membranes to networks. *Electroencephalography and Clinical Neurophysiology, 79,* 81–93.

Maccoby, E. E., Dowley, E. M., Hagan, J. W., & Degerman, R. (1965). Activity level and intellectual functioning in normal preschool children. *Child Development, 44,* 274–279.

Miller, L. A. (1992). Impulsivity, risk-taking, and the ability to synthesize fragmented information after frontal lobectomy. *Neuropsychologia, 30,* 69–79.

Mischel, W. (1958). Preference for delayed reinforcement: An experimental study of a cultural observation. *Journal of Abnormal and Social Psychology, 66,* 57–61.

Nass, R. D., & Gazzaniga, M. S. (1987). Cerebral lateralization and specialization in human central nervous system. In V. B. Mountcastle, F. Plum, & S. R. Geiger (Eds.), *Handbook of physiology: The nervous system* (Vol. 5, Pt. 2, pp. 701–761). Bethesda, MD: American Physiological Society.

Nelson, W. M., III, & Behler, J. J. (1989). Cognitive impulsivity training: The effects of peer teaching. *Journal of Behavior Therapy and Experimental Psychiatry, 20,* 303–309.

Nobin, A., & Bjorklund, A. (1973). Topography of the monoamine neuron systems in the human brain as revealed in fetuses. *Acta Physiologica Scandinavica, Suppl. 388,* 1–40.

Nunez, P. L. (1981). *Electric fields of the brain: The neurophysics of EEG.* New York: Oxford University Press.

O'Gorman, J. G., & Lloyd, J. (1987). Extraversion, impulsiveness, and EEG alpha activity. *Personality and Individual Differences, 8,* 169–174.

Olson, S. L. (1989). Assessment of impulsivity in preschoolers: Cross-measure convergences, longitudinal stability, and relevance to social competence. *Journal of Clinical Child Psychology, 18,* 176–183.

Olson, S. L, Bates, J. E., & Bayles, K. (1990). Early antecedents of childhood impulsivity: The role of parent–child interaction, cognitive competence, and temperament. *Journal of Abnormal Child Psychology, 18,* 317–334.

Pape, H. C., & McCormick, D. A. (1989). Noradrenaline and serotonin selectively modulate thalamic burst firing by enhancing a hyperpolarization-activated cation current. *Nature, 340,* 715–718.

Porteus, S. D. (1968). New applications of the Porteus Maze Tests. *Perceptual and Motor Skills, 26,* 787–798.

Power, T. G., & Chapieski, M. L. (1986). Child rearing and impulse control in toddlers: A naturalistic investigation. *Developmental Psychology, 22,* 271–275.

Rakic, P. (1988). Specification of cerebral cortical areas. *Science, 241,* 170–176.

Rothbart, M. K. (1988). Temperament and the development of inhibited approach. *Child Development, 59,* 1241–1250.

Routh, D. K., & Mesibov, G. B. (1980). Psychological and environmental intervention: Toward social competence. In H. E. Rie & E. D. Rie (Eds.), *Handbook of minimal brain dysfunctions: A critical view* (pp. 618–644). New York: Wiley.

Santostefano, S. (1978). *A biodevelopmental approach to clinical child psychology: Cognitive controls and cognitive control therapy.* New York: Wiley.

Schachar, R., & Logan, G. D. (1990). Impulsivity and inhibitory control in normal development and childhood psychopathology. *Developmental Psychology, 26,* 710–720.

Schalling, D., Ashberg, M., & Edmoor, G. (1984). *Personality and CSF monoamine metabolites.* Unpublished manuscript.

Sears, R., Rau, L., & Alpert, R. (1965). *Identification and child rearing.* Stanford, CA: Stanford University Press.

Silverman, I. W., & Ragusa, D. M. (1990). Child and maternal correlates of impulse control in 24-month-old children. *Genetic, Social, and General Psychology Monographs, 116,* 435–473.

Soubrie, P. (1986). Reconciling the role of central serotonin neurons in animal and human behavior. *Behavioral and Brain Sciences, 9,* 319–364.

Spivack, G., & Shure, M. B. (1989). Interpersonal Cognitive Problem Solving (ICPS): A competence-building primary prevention program. *Prevention in Human Services, 6,* 151–178.

Stoff, D. M., Friedman, E., Pollock, L., Vitiello, B., Kendall, P. C., & Bridger, W. H. (1989). Elevated platelet MAO is related to impulsivity in disruptive behavior disorders. *Journal of the American Academy of Child and Adolescent Psychiatry, 28,* 754–760.

Thatcher, R. W. (1992). Cyclic cortical reorganization during early childhood. *Brain and Cognition, 20,* 24–50.

Verney, C., Zecevic, N., Nikolic, B., Alvarez, C., & Berger, B. (1991). Early evidence of catecholaminergic cell groups in 5- and 6-week old human embryos using tyrosine hydroxylase and dopamine-beta-hydroxylase immunocytochemistry. *Neuroscience Letters, 131,* 121–124.

Vitiello, B., Stoff, D., Atkins, M., & Mahoney, A. (1990). Soft neurological signs and impulsivity in children. *Developmental and Behavioral Pediatrics, 11,* 112–115.

Wilens, T. E., & Biederman, J. (1992). The stimulants. *Psychiatric Clinics of North America, 15,* 191–222.

Young, W. F., Laros, E. R., Sharbrough, F. W., & Weinshilboum, R. M. (1986). Human monoamine oxidase: Lack of brain and platelet correlation. *Archives of General Psychiatry, 43,* 664–669.

Zametkin, A., Rapoport, J. L., Murphy, D. L., Linnoila, M., & Esmond, D. (1985). Treatment of hyperactive children with monoamine oxidase inhibitors: I. Clinical Efficacy. *Archives of General Psychiatry, 42,* 962–966.

Zuckerman, M. (1991). *Psychobiology of personality.* Cambridge, England: Cambridge University Press.

3

IMPULSIVITY: INTEGRATING COGNITIVE, BEHAVIORAL, BIOLOGICAL, AND ENVIRONMENTAL DATA

ERNEST S. BARRATT

This chapter will briefly review my impulsivity research and include selected "unpublished thoughts" that led to different directions of research for various time periods. Describing a research program in retrospect often makes the sequence of experiments and related hypotheses appear more rational than they really were at the time. The research program discussed in this chapter had a fortuitous beginning, a "searching for answers" midlife, and a conclusion that is still in progress. At this point, I believe it is important to define personality traits such as impulsivity and impulsiveness in a broad context.

The main point I would like to make is that personality traits should be defined by other than self-report or rating scale procedures (the psychometric approach). It has traditionally been assumed that traits defined by the psychometric approach reflect a "true" nosology of "natural" events. Biological, behavioral, and cognitive correlates are then sought to better understand these psychometrically defined traits. Rarely are psychometric measures revised on the basis of these "correlates." In contrast, one could just as readily start with biological measures and look for psychometric correlates. It is my position that a convergence of data among a wide range of measurements should be used to define personality traits, including impulsiveness. I use a general systems model of personality that is primarily of heuristic value in defining personality traits. Hopefully, the following overview of my research will provide a basis for understanding why I consider this broad, interdisciplinary view of personality theory to be important.

This chapter will be divided into three parts. I will first discuss my early research on impulsiveness, especially its relationship to anxiety. Then, the more formal search for biological and behavioral correlates of impulsivity, as well as revisions of the Barratt Impulsiveness Scale (BIS), will be presented. Lastly, I will describe the consideration of impulsiveness in a general systems model of personality; this will include a brief overview of my current research on impulsive aggression.

IN THE BEGINNING

At mid-century (late 1940s and early 1950s), anxiety had been related to the drive level and habit strength constructs in the Hull (1943) and Spence (1956) learning theory paradigm. In this context, Taylor (1953) developed a scale of manifest anxiety, based on habit strength and drive, that was related to performance on a wide variety of laboratory tasks, including verbal learning (Taylor, 1958), complex perceptual–motor tasks (Shepherd & Abbey, 1958), and serial learning (Taylor & Spence, 1952).

During approximately this same period, several self-report personality scales were developed on the basis of factor analyses of item pools (e.g., Guilford & Zimmerman, 1949; Thurstone, 1953). In reviewing the intercorrelation matrices for these personality scales, I noted that there was a cluster of items that suggested an impulsiveness trait (acting without thinking) that had a relatively low correlation with a cluster of anxiety items. Relating the psychometric research to the learning theory research, I wondered if impulsiveness was related to what Hull had termed "oscillatory inhibition" (O) in his 1943 theory. Spence (1956) discussed "O" as "intraindividual" variability of performance in perceptual–motor or learning tasks.

Several anxiety scales were available at this time in addition to the Taylor Manifest Anxiety Scale (e.g., Cattell, 1957), but these were part of omnibus personality scales. However, the various inventories that related to "impulsiveness" as I wanted to measure it within the Hullian/

Spence model did not appear broad enough to measure a predisposition toward intraindividual variability of behavior. I therefore rewrote items from selected scales and added items to arrive at my first Impulsiveness Scale (Barratt, 1959). As part of teaching a clinical psychology laboratory, I would have students complete the impulsiveness and anxiety scales, and relate their scores to performances on various laboratory tasks. The results of these instructional laboratory experiments not only confirmed the published results of the relationship of anxiety to performance on selected laboratory tasks, but demonstrated that impulsiveness was related to performance as well.

Because the impulsiveness and anxiety measures were orthogonal, I often selected subjects for experiments who were at the extremes of the four quadrants of a scatter diagram, and compared their performance on selected laboratory tasks (Barratt, 1959, 1963b, 1967b). During the 1960s, I pursued two general lines of research based on my early results: (a) continued laboratory studies of the relationship of impulsiveness to both biological (e.g., central nervous system and autonomic nervous system [ANS] measures) and performance measures and (b) continued revision of the BIS to arrive at a more homogeneous item pool that did not overlap with anxiety. Thus, early revisions of the BIS involved many unpublished studies of not only the interrelationships of impulsiveness items (and part–whole analyses) but also the relation between impulsiveness and anxiety items.

I conducted several experiments based on the original idea that impulsiveness might be a measure of Hull's O and relate to intraindividual variability of performance. I confirmed the hypothesis that the BIS was significantly related to intraindividual variability of performance on perceptual–motor tasks and also to intraindividual variability of ANS measures (Barratt, 1963b). I also noted that high-impulsiveness subjects had problems with "planning ahead" on laboratory tasks like the Porteus Maze Test (Barratt, 1967b). Furthermore, on the Porteus mazes, tracings for high-impulsive subjects were "wiggly" and suggestive of a fine tremor. The relationship of impulsiveness to the accuracy of fine perceptual motor performance was later confirmed in a study of impulsivity related to paced tapping (Barratt, Patton, Olsson, & Zuker, 1981). (Hull had speculated about a neural basis for 0 and for the intraindividual variability of performance.)

On a wide range of laboratory tasks among groups selected on the basis of both impulsiveness and anxiety, the high-impulsive, low-anxiety subjects were the most inefficient of the four groups (Barratt, 1967b). I also found that the response set appeared to be related to effects of impulsiveness on intraindividual variability of performance, with high-impulsive subjects having difficulties in changing response set. On the basis of these data, I began to speculate about brain relationships.

In a search for brain correlates of these behaviors, an infrahuman research program paralleled my human-level research from 1963 to about 1976 (selected papers include Barratt, 1963a, 1967a; Barratt, Creson, & Russell, 1968; Francois, Barratt, & Harris, 1970; McDiarmid & Barratt, 1969). I conducted brain–behavioral experiments using acute and chronic preparations with implanted electrodes in monkeys and cats. I did both single-cell and macroelectrode stimulation and recordings at selected brain sites. This research was summarized in my 1972 review paper and will not be reviewed in depth here. It was this research, however, that set the stage for my current research on the effects of Dilantin (phenytoin) and other anticonvulsants on impulsive aggressive behavior. To appreciate the importance of that research in shaping my current impulsiveness research at the human level, a summary is appropriate. I noted the following:

> With regard to the acquisition of these behavioral predispositions, we suggest that during early childhood, individuals become conditioned to respond to a wide variety of stimuli with patterns of responses or behavioral predispositions which have survival value for the individual (or group). The more frequently used response patterns become readily available behavioral predispositions or "personality traits." The conditioning that results in the formation of these patterns is probably a very subtle (unconscious) type of conditioning, similar to that which Miller (1969) and his colleagues have demonstrated for many visceral and glandular processes. These behavioral predispositions are related to changes in neuronal structures and function; different behavioral predispositions involve different neural circuits or systems which interact with each other. More inclusive behavioral predispositions such as impulsiveness and anxiety probably involve fairly elaborate neural–endocrine systems. We propose that impulsiveness (impulse control) involves a neural system which includes, primarily, interrelationships among the orbitofrontal cortex, selected limbic system nuclei (especially the basolateral amygdala), basal ganglia nuclei, and the cerebellum. Anxiety involves essentially the hypothalamic–hypophyseal axis, the ascending reticular activating system, and the orbitofrontal cortex; the effects of the reticular system on cortical functioning are especially important here.
>
> Our rationale for proposing these two systems relates to the characteristics which we have observed to date in individuals with varying levels of anxiety and impulsiveness. Let us briefly summarize these characteristics. The impulsive person typically: acts without thinking, acts on the spur of the moment, is restless when required to sit still, likes to take chances, is happy-go-lucky, has difficulty in concentrating, and is a doer and not a thinker. The anxious person typically: is aware of bodily functions in stressful situations (blushing, hands sweating, nausea, heart beating), worries, feels tense, feels more sensitive than other people, has difficulty sleeping because of fears, is

not calm, and angers easily. Essentially, impulsiveness involves the control of the expression of thoughts and actions while anxiety involves feelings about external or internal stimuli. We suggest that anxiety is a feeling. Without going into detail, only a few observations will be made to tie together, then our "neuropsychologizing" with the behavioral and feeling characteristics of impulsiveness and anxiety. [We (Barratt, Pritchard, Faulk, & Brandt, 1987) have demonstrated event-related cortical potential differences related to anxiety and impulsiveness.]

The forebrain, especially the orbitofrontal cortex, has been related in many studies to inhibitory control (neurophysiological) of other cortical and subcortical nuclei and to changes in motor (behavioral) control (Schlag & Scheibel, 1967). Both limbic system nuclei (e.g., the basolateral amygdala) and the reticular activating system have been functionally related to orbital frontal activity. Thus, our two separate systems could have a combined effect in the orbital frontal cortex, although this is obviously not the only point at which a combined effect could occur.

The amygdala has been suggested to be related to approach–avoidance behavior (Goddard, 1964). Our findings of the functional connections between the basolateral amygdala and the putamen and globus pallidus suggest a possible route by which the amygdala could be involved in "monitoring" of motor performance. Further, the functional relationships which we found between the amygdala and hippocampus and the olive and cerebellum could also be related to the inhibitory role of the cerebellum relative to the neocortex. Dilantin, which suppresses the activity of the cerebellar cortex, has been reported to be effective in controlling impulsiveness in patients (Turner, 1967). The uncal area is known to be a locus for generating seizure activity (Gloor, 1960), which certainly represents an involuntary change in level of motor control. Last, our findings that lithium appears to be related to the functioning of the orbital frontal cortex coupled with the therapeutic effects of lithium in manic patients suggests the possible involvement of this area in impulsive behavior. The inefficiency of high impulsive subjects (especially the HILA subjects) on motor tasks could relate to this system.

The feelings which characterize anxiety could be related to: (1) the hypothalamic–hypophyseal control of endocrine functions which result in ANS changes and other somatic changes (nausea) felt by the anxious person; many recent research findings suggest this possibility (e.g., Levi, 1969); (2) the cognitive awareness of tenseness resulting from nonspecific reticular control of cortical activity; the orbitofrontal cortex could be a key area for the effects of the reticular activating system relative to feelings of tenseness. (Barratt, 1972, pp. 219–220)

The discussion quoted above set the stage for my next series of experiments. I planned to further develop the BIS in a new direction partly on the basis of our laboratory results. I would include items that related to motor responses, planning ahead, and changing a cognitive "set." In my clinical experiences, I also started to appreciate that my patients were "impulsive" in these three ways.

The BIS-6 contained six subscales: motor control, intraindividual variability, impulsive interests, risk taking, interpersonal relationships, and impulse control. The intercorrelations of these subscales were .48 or less ($N = 149$). The BIS-5 total score correlated .82 with the BIS-6 total score. I had related the BIS-5 to other selected measures of impulsiveness and anxiety and had confirmed, in general, that impulsiveness and anxiety were separate domains of personality, as previously noted. In contrast with the BIS-5 and the BIS-6, the BIS-10 was later developed on the basis of my overall research results and contained only three subscales: motor, cognitive, and planning ahead. The BIS-10 will be described later in more depth.

One final laboratory experiment of importance during this early period was the search for psychophysiological correlates of eyelid conditioning among subjects selected for anxiety and impulsiveness. The main results indicated that high-impulsiveness subjects conditioned better to a "positive" stimulus than to a "negative" stimulus. The results further indicated that the high-impulsive, less anxious subjects were less aroused (less activated) than the other subjects, as evidenced by electroencephalogram spectral analyses.

THE SEARCH FOR THE VALIDITY AND INTEGRATION OF DATA

I extended my research to applications of the construct of impulsiveness as measured by the BIS. In several clinical and field studies, I related impulsiveness, anxiety, and other selected personality constructs to different everyday life situations. Most of these data were not formally published, but the results were described in reports to the sponsoring agencies. I also observed "impulse control" among my patients.

In a 3-year study of medical students (Barratt & White, 1969), we divided the students at the beginning of their freshman year into four groups on the basis of anxiety and impulsiveness scores. We interviewed them periodically, as well as their professors, spouses, and peers who knew them well. The high-impulsive, high-anxiety students appeared to have social problems but also were good "con artists"; their professors often ranked them as top students. In contrast, they were described as "troublemakers" at fraternity parties, or, if married, they had marital problems. The high-impulsive, high-anxiety subjects all sought some counseling during their first 3 years of medical school; their problems were generally serious psychiatric problems. The high-anxiety, low-impulsive students had typical anxiety problems, including being fearful, worrying, or having free-floating anxiety. The low-impulsive, low-anxiety students appeared less highly motivated and not as interested, for example, in social mobility. They reported few psychiatric problems and appeared to take life problems in stride. It appeared that impulsiveness as measured by the BIS had some practical clinical value. In analyzing these data, the need to consider impulsiveness and anxiety in a broader "personality theory" context became obvious.

In an extensive study of interview sensitization techniques (1969–1975), I studied the personality, ability, and developmental characteristics of people who could assess others well versus those who were poor assessors (Barratt, 1975). In one of the early experiments in this project, three groups of subjects in different settings (Jewish medical fraternity, non-Jewish medical fraternity, and Catholic brothers and priests) who lived together and interacted on a daily basis were asked to rate each other for the following: best friend, most stable, person they could discuss personal problems with, best liked in house, most likely to help solve group problems, most socially polished, most intellectually honest, and most trustworthy. Impulsiveness did not relate to any of the peer ratings for the members of the two fraternities. For the Catholic priests, impulsiveness was negatively related to being stable and trustworthy. Mean impulsiveness scores were not significantly different across groups. What was responsible for these results?

Social environment appeared partially associated with impulsiveness in peer ratings. Other personality traits were consistently related across all three groups to selected peer ratings. Social extroversion, for example, was related positively to being perceived as most stable by the fraternity members and negatively related to stability by the priests. Again, I realized the need to look at the relationship of impulsiveness to personality in a broader context. Parenthetically, one of the results of this research project was that persons who were able to assess impulsiveness best in others had above average levels of impulsiveness themselves.

In a study of juvenile delinquency (1976–1978), I compared juvenile delinquents, "normal" controls, and juvenile–psychiatric inpatients on a wide range of measures (Barratt, 1978). I obtained 52 neuropsychological and personality measures, as well as visual-evoked potentials to measure augmenting/reducing. There were 187 adolescent controls, 210 juvenile delinquents, and 156 adolescent psychiatry inpatients in the study. Impulsivity as measured by the BIS was higher for the delinquents than for the other two groups, but only among the older adolescents. Another measure of impulsiveness, the Delay Avoidance subscale on the Brown Holtzman Study Habits Inventory (Brown & Holtzman, 1967), was higher at all age levels for the delinquents compared with the other two groups. Also, the Kagan Matching Familiar Figures Test (Kagan, Pearson, & Welch, 1966) impulsiveness score was significantly higher for the delinquents at all ages compared with the other two groups.

In a follow-up study (Barratt, 1981), selected results of the juvenile delinquency study were confirmed. This follow-up study also extended the cognitive psychophysiological research. The results were consistent with earlier findings that high-impulsive persons were visual-evoked-potential augmenters. This indicated that cognitive measures were related to impulsiveness, and suggested further differences in the "gating" of sensory information related to impulsiveness. This study also demonstrated that high-impulsive subjects have a fast cognitive tempo and underestimate time judgments. Similar results were obtained in paced tapping (Barratt et al., 1981). The time perception measures were related to the augmenting/reducing psychophysiological measures. Here, then, there was an interrelationship among measures of brain activity, cognition, and impulsiveness. The augmenting/reducing correlation with impulsiveness has been replicated several times (Barratt et al., 1987).

In a study of time zone anchors related to impulsiveness, we found that high-impulsive subjects were more "present" oriented than "future" or "past" oriented (Barratt, Adams, & White, 1977). Using a modification of Cottle's (1976) time zone measure, we found that high-impulsive subjects were more willing to predict the future but, as noted, were more present oriented.

As noted previously, high-impulsive subjects generally tend to be inefficient in performing perceptual–motor tasks (Barratt, 1985b). In reaction time experiments, the high-impulsive subjects responded more slowly if there was a warning signal interval in the range of 600 to 800 ms (depending on task demands); otherwise, they responded more quickly. Why would the high-impulsive subjects respond more slowly with a warning signal in selected reaction time tasks? I tentatively concluded that these results were related to "changing a mental set" and frontal lobe functions.

In the mid-1970s, I realized that I had accumulated a wide range of data characterizing impulsiveness but did not have a basis for synthesizing the data. If I was going to use the laboratory and field data to better measure impulsiveness, what direction should be taken? Where and how could I look most meaningfully for the convergence of data points that would assist in revising the BIS and in developing other measurement techniques of impulsiveness?

TOWARD A GENERAL SYSTEMS THEORY MODEL OF PERSONALITY

My colleagues and I had used a wide range of conceptual models at varying times in our research (Barratt & Patton, 1983). These models differed in complexity and usefulness but were not inclusive enough to enable an overall, cohesive view of our data. We had used mathematical latency models, a risk-taking and limbic-system model, a model of intraindividual variability, and an operant model of response control. These were helpful with circumscribed data (for example, reaction times related to the mathematical latency models), but not with overall data sets from the many experiments. At this point we reviewed our data again and asked, "What should a personality theory model do and what questions do we want to answer?" Essentially, we wondered how impulsiveness plays a role in describing individuals.

We had viewed impulsiveness as a first-order personality trait. We had theorized that it was part of an "action-oriented" cluster of personality traits that in a broad sense included, for example, extroversion, sensation seeking, and hypomania (Barratt & Patton, 1983). We also theorized that anxiety was part of a cluster of "feeling" traits that would, for example, include dysthymia or sadness, anger, and levels of depression. We could accept impulsiveness as part of extroversion, as the Eysencks had proposed early on (see H. J. Eysenck, chapter 4, and S. B. G. Eysenck, chapter 8, this volume). But this reasoning addressed the place of impulsiveness only within the *structure* of personality. There were also *process* issues. If a neuroanatomist observes the brain, he or she will describe structures and pathways. If a physiologist views the brain, he or she will describe processes. How can structure and function be interrelated in personality research? Personality trait theories that result from multivariate analyses of self-report or rating scales are usually hierarchial in nature.

The current debate over the number of "super" personality factors (Costa & McCrae, 1992; Eysenck 1991, 1992; Zuckerman, 1992) is aimed at understanding personality structure. As noted earlier, it is assumed in trait theory personality research that these factors represent a valid view of people from a naturalistic viewpoint. Biological or cognitive correlates are often sought to better understand the nature of the traits, but these traits as defined by the multivariate psychometric process are basic. But how do the biological and cognitive correlates relate to defining impulsiveness? Are these biological and cognitive measures actually *correlates* of impulsiveness, or do they in part help *define* impulsiveness as much as the self-report measures do? It was here that we realized we should take a different approach.

Fiske (1971) outlined the steps in arriving at the measurement of concepts including personality concepts. He started with natural observations (Figure 1). The natural observations that one makes will ultimately determine the concepts and measurements that are used to define personality. For example, a sociologist would have a different set of observations than a neuroscientist when selecting "natural" events that will ultimately define a person. So what should our model include?

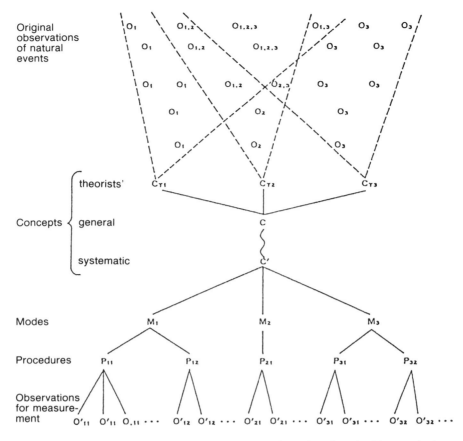

Figure 1. Links between concepts and observations (reprinted with permission from Fiske, 1971).

In my search for a model, I was influenced by several authors. First, Lazare's (1973) description of four hidden conceptual models (medical, psychological, behavioral, and social) that are used by psychiatrists in making clinical decisions was influential. The medical model is primarily based on a biological view of people; it is a "disease"-based model that leads primarily to somatic treatments. In contrast, the psychological model is primarily a psychoanalytic approach to understanding people, with an emphasis on cognitive structure and functions; psychotherapy is the chief treatment modality. The behavioral model emphasizes learned behaviors within a reward paradigm, and the fact that behaviors are modified by the scheduling of stimuli. Finally, the social model emphasizes the social environment, interpersonal relationships, and group affiliations in a person's life space. Lazare noted that "we lack a comprehensive set of personal 'laws' that include the models described here" (p. 348). What was important to me was that these four models really paralleled four different and basic categories of concepts that are used to describe people: biological, cognitive, environmental, and behavioral.

It appeared to me that these four categories were the minimum number of categories that could be used to classify all descriptions of people starting with natural events. Most personality theories have emphasized one or two of these categories, but relatively few have recognized the need to systematize all four categories of data. In keeping with the age-old mind–body and nature–nurture issues, a number of "unifying" theories have been proposed to resolve these and other conflicts (e.g., Goodman, 1991; Kluckhon & Murray, 1949; Murphy, 1947; Stats, 1991; Weiss, 1973).

As noted, I concluded a priori that there were four basic and minimal ways to classify natural observations and that a model of people should build on these four categories. My next goal, then, was to find a way to systematize the four categories of basic concepts outlined by Lazare. Systems take many forms and have many inherent formal properties (Klir, 1972; Weinberg, 1975) and have been used to model everything about human beings from cells (Rabitz, 1989) to society (Kuhn, 1974). Weiss (1973) defined a system as follows:

> Pragmatically defined, a system is a circumscribed complex of relatively bounded phenomena, which within these bounds, retains a relatively stationary pattern of structure in space or of sequential configuration in time in spite of a high degree of variability in the details of distribution and interrelations among its constituent units of lower order. Not only does the system maintain its configuration and integral operation in an essentially constant environment, but it responds to alterations of the environment by an adaptive redirection of its componential processes in such a manner as to counter the external change in the direction of optimum preservation of its systemic integrity. (p. 40)

Weiss's definition included properties that can describe a person as a system and that I thought were essential to developing a personality model. This is in contrast with a hierarchial system. The bounded phenomenon in Weiss's system would be the person in his or her life space (Lewin, 1935). Personality would have a structure that is relatively constant over time but with interactions within the structure. It would thus have adaptive properties that allowed it to cope with changing environments. It would, in addition, be susceptible to breakdowns (e.g., psychopathological states).

I was also influenced, in my choice of a system to interrelate the four categories of data and concepts, by Ashby's (1960) negative feedback model, which he developed to explain the role of the brain in adaptive behavior. He also had four classes of concepts in his model, including the brain, behavior, motivation, and the environment. I changed the brain

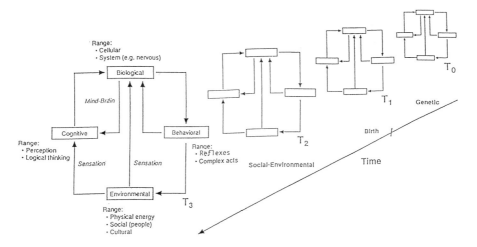

Figure 2. General systems model used as a basis for defining personality states and traits. Each panel (T₀–T₃–Tₙ) is a cross-section at a given time period in a person's life. These structural relationships change over time. Each of the four basic categories of concepts and measurements interrelate in varying degrees to define components of personality and ability (reprinted with permission from Barratt, 1991).

category to biological and motivation to cognition. The resulting model (Figure 2) provided the basic framework for a state–trait model of personality with four basic categories of concepts.

This model has been described briefly elsewhere (Barratt, 1985a, 1991; Barratt & Patton, 1983) and will not be discussed in depth here. Selected attributes of the model will be reviewed here to provide a basis for better understanding my concept of impulsiveness as a personality trait.

The model (Figure 2) suggests that there is a "structure" to one's personality that can be viewed over time (T1, T2, etc.). The time dimension is important and relates to what Weiss (1973) called the "sequential configuration." This necessitates measuring structure at varying time durations to describe processes. In line with Buss and Plomin's (1975) temperament theory of personality development, I proposed that what I call "traits" and what Buss and Plomin called "temperaments" have a genetic component and are manifest in childhood. These traits provide the structure for the personality theory and describe the main patterns (or profiles) of the interrelationships of measurements among and within the four categories. For example, there may be five major ways of describing the structure on the basis of factor analyses of self-report measures (the "Big Five"; Eysenck, 1991). The structure can be a hierarchy imposed on the system as a whole, but the elements of the structure (e.g., traits) are understood in terms of measurements within the four underlying components of the system. Changes within various components of the system are necessary for adaptation to occur. These changes represent the functioning or processes within the system and help maintain stability. The environment is included as part of the model in this closed negative feedback system. Most general systems models of personality theory are feedback systems with "open" input and output channels to the environment. The assumption about environment here is similar to that expressed by Lewin (1935) in his concept of life space—the environment is circumscribed at any given moment in time by the limits of one's perception of it. Thus, it is not an "open" but a "closed" system.

Each of the four categories of concepts can be defined by subsystems, and related models have been developed elsewhere (e.g., Kuhn, 1974). Those will not be pursued at this point but will be alluded to later when impulsiveness and aggression are defined.

The idea of the structure of the overall model being represented by personality traits needs to be elaborated on. As noted, I think that traits should be defined by measurements from all four categories. This is not unique to my system. For example, in the third revised edition of the *Diagnostic and Statistical Manual of Mental Disorders* (*DSM–III–R*), personality is defined as "deeply ingrained patterns of *behavior*, which include the way one relates to, perceives, and *thinks* about the *environment* and oneself" (American Psychiatric Association, 1987, p. 403, emphasis added). This definition includes three of my four categories. It is further noted in the *DSM–III–R* that prominent aspects of personality structure that endure over time are "traits" and can be measured (p. 403). Likewise, less enduring but still measurable characteristics of the personality are called "states." However, there are no universally accepted state and trait descriptions of personality structure, as previously noted. The time window in our model for viewing structure and function is related to the state–trait distinctions. (No one has developed a reliable "state" measure of impulsiveness.) There are many other personality descriptors that are not considered in detail here: for example, the overlap in the definitions of mood, state, trait, temperament, and emotions.

In the model (Figure 2), there are two subsystem feedback loops. One involves the biological–cognitive components. As noted, this represents the mind–body controversy, which in my opinion still exists. Cognition (mind) is represented in my model as a separate category of concepts that are always inferential. This does not mean that there are separate physical systems for the brain (biological) and cognitive functions. It does mean that measurements have been made and taxonomies proposed for thought processes (Hunt & Agnoli, 1991; Simon, 1992) that cannot be directly related to the brain at this time. Because I had proposed a "cognitive impulsiveness" subtrait (Barratt, 1985b), it was important that the category of cognition be included in my personality model.

This model is not a personality model in the typical sense, as noted earlier. It is primarily heuristic. What does it add to our understanding of impulsiveness? First, it emphasizes the need to integrate data from the four basic or minimal categories of "natural observations" as outlined in Fiske's (1971) diagram. My data clearly fall into those four categories, as discussed elsewhere (Barratt, 1975, 1985a, 1991, in press). The model also suggests that measures of impulsiveness like the BIS may have limitations in defining impulsiveness from a naturalistic viewpoint. The role of cognition in impulsiveness is a good example of this.

I have noted that a fast cognitive tempo and lack of planning ahead are characteristic of high-impulsive subjects. This is shown, in part, by their underestimation of time intervals. It was also evident in the cognitive tempo of many of my patients that high levels of impulsiveness relate to "speed" of thought (Barratt, 1985a). In the BIS-10, I included a cognitive subscale of impulsiveness on an a priori basis. However, as discussed elsewhere (Barratt, in press), I and others (Luengo, Carrillo-de-la-Peña, & Otero, 1991) have not been able to identify a cognitive subfactor in analyses of self-report measures. In a current factor analytic study (Barratt, Patton, & Stanford, 1993) of the BIS items as answered by 374 college students, I did not isolate a cognitive factor. Is cognitive impulsiveness important clinically? Different types of patients have different cognitive tempos. Melges (1982) noted that in manic disorders, the rate of mental activity is accelerated. High-impulsive subjects have fast cognitive tempos like manic patients (I found lithium to be related to intraindividual variability of performance in my infrahuman research; Barratt, 1972). In my recent study of college students' self-report responses, the item *I have racing thoughts* did load on a factor with "motor impulsiveness" items (e.g., *I act on the spur of the moment*). It is possible that people cannot reliably assess their own thought processes with regard to cognitive tempo. However, from a clinical and laboratory viewpoint, I think the subfactor of cognitive impulsiveness exists and is an important feature of impulsiveness. How can it be measured?

I am currently designing a computer-based laboratory for measuring impulsiveness by self-report, performance, and cognitive psychophysiological measures. I have shown in my current research on impulsive aggression that impulsiveness relates to frontal–parietal cortical activity during the performance of an oddball task with "deviant stimuli." I plan to extend these results, along with the results of my earlier research, to capitalize on the convergence of data from the four basic classes of our systems model to arrive at a more comprehensive method for measuring impulsiveness.

This brings me to my current research interests. I am studying impulsive aggression and have demonstrated that people with a tendency to "lose their temper easily" (have a "short fuse") possess a high level of both impulsiveness and anger (Barratt, 1991). We have also shown that people who commit a wide range of different types of criminal acts have high levels of impulsiveness (Stanford & Barratt, 1992).

I speculated earlier about the relationship of impulsiveness, as measured by the BIS, to Eysenck's factors. In a recent study (O'Boyle & Barratt, 1993), among substance abusers, the BIS correlated .66 with Eysencks's psychoticism subscale of the Eysenck Personality Questionnaire (EPQ), and .08 with the EPQ E subscale. This is consistent with Eysenck's interpretation of impulsiveness within a hierarchial personality structure. Consistent with the Stanford and Barratt (1992) results, it was also found in this study that higher impulsive subjects were dependent on a larger number of different types of drugs. An important question is why impulsiveness appears to be related to circumscribed motivations rather than to have a general motivational effect. Perhaps I will get the answers in my new laboratory approach.

In summary, in my model there is a convergence of data to indicate that impulsiveness is a first-order personality trait that has a partially genetic basis. Impulsiveness is also related to serotonin levels in the brain and affects cognitive functioning primarily through frontal lobe executive functions (and S2 serotonin receptors), which, in turn, affect parietal lobe sensory integration functions. Impulsivity also relates to inefficient performance on selected perceptual–motor tasks and increases intraindividual variability of behavior. Finally, impulsiveness possesses two psychometrically determined subtraits (an idea–motor trait and a "planning ahead" trait) and has a cognitive component that is difficult to measure with self-report scales (Barratt, 1991).

REFERENCES

American Psychiatric Association. (1987). *Diagnostic and statistical manual of mental disorders* (3rd ed., rev.). Washington, DC: Author.

Ashby, W. (1960). *Design for a brain.* New York: Wiley.

Barratt, E. S. (1959). Anxiety and impulsiveness related to psychomotor efficiency. *Perceptual and Motor Skills, 9,* 191–198.

Barratt, E. S. (1963a). Behavioral variability related to stimulation of the cat's amygdala. *Journal of the American Medical Association, 186,* 773–775.

Barratt, E. S. (1963b). Intraindividual variability of performance: ANS and psychometric correlates. *Texas Reports Biology and Medicine, 21,* 496–504.

Barratt, E. S. (1967a). The effects of Thiazesim LSD-25, and bilateral lesions of the amygdalae in the release of a suppressed response. *Recent Advances in Biological Psychiatry, 9,* 229–240.

Barratt, E. S. (1967b). Perceptual motor performance related to impulsiveness and anxiety. *Perceptual and Motor Skills, 25,* 485–492.

Barratt, E. S. (1972). Anxiety and impulsiveness: Toward a neuropsychological model. In C. D. Spielberger (Ed.), *Anxiety: Current trends in theory and research* (Vol. 1, pp. 195–222). New York: Academic Press.

Barratt, E. (1975). *Interview sensitization techniques: Assessing behavioral predispositions from live (and video-taped) interviews* (Final Report, Contract No. 62-0138-76R). Washington, DC: Central Intelligence Agency.

Barratt, E. S. (1978). *Multivariate diagnostic assessment of juvenile delinquents* (Final Report, Grant No. EA-78-COl-4900). Austin, TX: Criminal Justice Division, Office of the Governor.

Barratt, E. S. (1981). Time perception, cortical evoked potentials, and impulsiveness among three groups of adolescents. In J. R. Hays & K. S. Solway (Eds.), *Violence and the violent individual* (pp. 87–96). New York: SP Medical and Scientific Books.

Barratt, E. S. (1985a). Impulsiveness defined within a systems model of personality. In C. D. Spielberger & J. M. Butcher (Eds.), *Advances in personality assessment* (Vol. 5, pp. 113–132). Hillsdale, NJ: Erlbaum.

Barratt, E. S. (1985b). Impulsiveness subtraits: Arousal and information processing. In J. Spence & C. Izard (Eds.), *Motivation, emotion and personality* (pp. 137–146). Amsterdam: Elsevier.

Barratt, E. S. (1991). Measuring and predicting aggression within the context of a personality theory. *Journal of Neuropsychiatry, 3,* 535–539.

Barratt, E. S. (in press). Impulsiveness and aggression. In J. Monohan & H. Steadman (Eds.), *Violence and mental disorder.* Chicago: University of Chicago Press.

Barratt, E. S., Adams, P. M., & White, J. H. (1977). *Time perception related to cortical augmenting/reducing in adolescence* (Annual report). Washington, DC: Office of Naval Research.

Barratt, E. S., Creson, D. L., & Russell, G. (1968). The effects of lithium salts on brain activity in the cat. *American Journal of Psychiatry. 125,* 530–536.

Barratt, E. S., & Patton, J. H. (1983). Impulsivity: Cognitive, behavioral, and psychophysiological correlates. In M. Zuckerman (Ed.), *Biological bases of sensation seeking, impulsivity, and anxiety.* Hillsdale, NJ: Erlbaum.

Barratt, E. S., Patton, J., Olsson, N. G., & Zuker, G. (1981). Impulsivity and paced tapping. *Journal of Motor Behavior, 13,* 286–300.

Barratt, E., Patton, J., & Stanford, M. (1993). *Construct validity of the Barratt Impulsiveness Scale.* Unpublished manuscript.

Barratt, E. S., Pritchard, W. S., Faulk, D. M., & Brandt, M. E. (1987). The relationship between impulsiveness subtraits, trait anxiety, and visual N100 augmenting/reducing: A topographic analysis. *Personality and Individual Differences, 8,* 43–51.

Barratt, E., & White, R. (1969). Impulsiveness and anxiety related to medical students' performance and attitudes. *Journal of Medical Education, 44,* 604–607.

Brown, W. F., & Holtzman, W. H. (1967). *Survey of study habits and attitudes* [Manual]. New York: Psychological Corporation.

Buss, A., & Plomin, R. A. (1975). *Temperament theory of personality development*. London: Wiley-Interscience.

Cattell, R. (1957). *Handbook for the IPAT anxiety scale*. Champaign, IL: Institute for Personality and Ability Testing.

Costa, P. T., & McCrae, R. R. (1992). Four ways five factors are basic. *Personality and Individual Differences, 13*, 653–665.

Cottle, T. (1976). *Perceiving time*. New York: Wiley.

Eysenck, H. J. (1991). Dimensions of personality: 16, 5, or 3? Criteria for a taxonomic paradigm. *Personality and Individual Differences, 12*, 773–790.

Eysenck, H. J. (1992). Four ways five factors are not basic. *Personality and Individual Differences, 13*, 667–673.

Fiske, D. (1971). *Measuring the concepts of personality*. Chicago: Aldine.

Francois, G. R., Barratt, E. S., & Harris, C. (1970). Assessing the spontaneous cage behavior of the squirrel monkey. *Primatology, 11*, 89–92.

Gloor, P. (1960). Amygdala. In J. Field, H. W. Magoun, & V. E. Hall (Eds.), *Handbook of neurophysiology* (Vol. 2, pp. 1395–1420). Washington, DC: American Physiological Society.

Goddard, G. V. (1964). Functions of the amygdalae. *Psychological Bulletin, 62*, 89–109.

Goodman, A. (1991). Organic unity theory: The mind–body problem revisited. *American Journal of Psychiatry, 148*, 553–563.

Guilford, J. P., & Zimmerman, W. S. (1949). *The Guilford–Zimmerman Temperament Survey* [Manual]. Beverly Hills, CA: Sheridan Supply Company.

Hull, C. L. (1943). *Principles of behavior*. New York: Appleton-Century.

Hunt, E., & Agnoli, F. (1991). The Whorfian hypothesis: A cognitive psychology perspective. *Psychological Review, 98*, 377–389.

Kagan, J., Pearson, L., & Welch, L. (1966). Conceptual impulsivity and inductive reasoning. *Child Development, 37*, 583–594.

Klir, G. (1972). *Trends in general systems theory*. New York: Wiley Interscience.

Kluckhon, C., & Murray, H. (1949). *Personality in nature, society, and culture*. New York: Knopf.

Kuhn, A. (1974). *The logic of social systems*. London: Jossey-Bass.

Lazare, A. (1973). Hidden conceptual models in clinical psychiatry. *New England Journal of Medicine, 288*, 345–350.

Levi, L. (1969). Neuroendocrinology of anxiety. *British Journal of Psychiatry*, Special Publication No. 3.

Lewin, K. A. (1935). *Dynamic theory of personality*. New York: McGraw-Hill.

Luengo, M., Carrillo-de-la-Peña, M., & Otero, J. (1991). The components of impulsiveness: A comparison of the 1.7 Impulsiveness Questionnaire and the Barratt Impulsiveness Scale. *Personality and Individual Differences, 12*, 656–667.

McDiarmid, C. G., & Barratt, E. S. (1969). Techniques for psychophysiological research with squirrel monkeys. In *Proceedings of the Second International Congress of Primatology* (Vol. 1, pp. 246–253). New York: Karger/Basel.

Melges, F. T. (1982). *Time and the inner future (a temporal approach to psychiatric disorders)*. New York: Wiley.

Miller, N. E. (1969). Learning of visceral and glandular responses. *Science, 163*, 434–445.

Murphy, G. (1947). *Personality: A biosocial approach to origins and structure*. New York: Harper.

O'Boyle, M., & Barratt, E. S. (1993). Impulsivity and DSM–III–R personality disorders. *Personality and Individual Differences, 14*, 609–611.

Rabitz, H. (1989). Systems analysis at the molecular level. *Science, 246*, 221–226.

Schlag, J., & Scheibel, A. (1967). Forebrain in inhibitory mechanisms [Special issue]. *Brain Research, 6*.

Shepherd, A. H., & Abbey, D. S. (1958). Manifest anxiety and performance on a complex perceptual–motor task. *Perceptual and Motor Skills, 8*, 327–330.

Simon, H. (1992). What is an explanation of behavior? *Psychological Science, 3*, 150–161.

Spence, K. W. (1956). *Behavior theory and conditioning*. New Haven, CT: Yale University Press.

Stanford, M. S., & Barratt, E. S. (1992). Impulsivity and the multi-impulsive personality disorder. *Personality and Individual Differences, 13*, 831–834.

Stats, A. (1991). Unified positivism and unification psychology. *American Psychologist, 46*, 899–912.

Taylor, J. A. (1953). A personality scale of manifest anxiety. *Journal of Abnormal and Social Psychology, 48*, 285–290.

Taylor, J. A. (1958). The effects of anxiety level and psychological stress on verbal learning. *Journal of Abnormal and Social Psychology, 57*, 55–60.

Taylor, J. A., & Spence, K. W. (1952). The relationship of anxiety level to performance in serial learning. *Journal of Experimental Psychology, 44*, 61–64.

Thurstone, L. L. (1953). *Examiner manual for the Thurstone Temperament Schedule*. Chicago: Science Research Associates.

Turner, W. J. (1967). The usefulness of diphenylhydantion in treatment of non-epileptic emotional disorders. *International Journal of Neuropsychiatry, Suppl. 2*, S8–S20.

Weinberg, G. (1975). *An introduction to general systems thinking*. New York: Wiley.

Weiss, P. (1973). *The science of life: The living system—A system for living*. Mt. Kisko, NY: Futura.

Zuckerman, M. (1992). What is a basic factor and which factors are basic? Turtles all the way down. *Personality and Individual Differences, 13*, 675–681.

4

THE NATURE OF IMPULSIVITY

H. J. EYSENCK

TAXONOMIC PROBLEMS OF IMPULSIVITY

Personality, and the behaviors that constitute personality, are best viewed in terms of traits such as impulsivity, sociability, persistence, and activity. These traits, in turn, are correlated to form higher-order factors or dimensions, such as Extraversion (E), Neuroticism (N), or Psychoticism (P) (Eysenck & Eysenck, 1985). Thus, we are dealing with a hierarchical system (Eysenck, 1947) possessing a large number of traits but only a small number of dimensions (Eysenck, 1991). It is sometimes argued that such a view, supported through factor analytic studies, fails to explain, as well as name, the variables involved (e.g., Wickland, 1990). Although such a criticism may be justly applied to some poorly executed studies, it does not do justice to the integration of psychometrics and experimental study that characterizes much modern research. An example of the experimental and theoretical work on impulsivity, serving to explain the underlying causes of many of the correlational discoveries, is the book by Reznick (1989) titled *Perspectives on Behavioral Inhibition*.

Another historically important criticism of the trait approach is that of Mischel (1968), who cast doubt on the explanatory value of personality

57

traits and suggested that situational variables carried much more of the explanatory variance. These points have been answered more than adequately (Magnusson, 1981), and it is currently widely recognized that both traits and situations determine individual conduct.

It is of course not sufficient to postulate a personality trait; we must demonstrate its phychometric existence. One of the psychometrically least satisfactory inventories, the Minnesota Multiphasic Personality Inventory (MMPI), illustrates this need. Factor analytic studies have shown that each of the MMPI scales is an amalgam of psychometrically heterogeneous items that are poorly correlated and do not define any kind of universal trait or type. Items often correlate better with traits other than the one they are supposed to measure, and the factors that emerge from analyses bear no relation to those postulated (see Eysenck & Eysenck, 1985, for a review of factorial analyses).

Another objection often made by critics not familiar with the method of factor analysis is that we only get out what we put in, so that the analysis result is a tautology. This would only be true if we knew beforehand what we put into a questionnaire, but of course that is not true, as the case of the MMPI clearly demonstrates. Consider another example, the case of social shyness. J. P. Guilford & Guilford (1936) postulated the existence of a factor of this nature, obviously on the basis of the theory that all of the items used would define one single general factor. In contrast, Eysenck (1956) argued that there might be two kinds of social shyness, with different etiologies. Introverts are shy because they do not want to be with other people; neurotics are shy because they are afraid of other people. Factor analysis demonstrated the existence of two independent factors in the S (Sociability) scale, one correlating highly with E, the other with N. Clearly, we may have a theory about what we put into a questionnaire, but we must check up on that theory by conducting proper correlational studies.

The need for such correlational studies is particularly imperative in relation to impulsivity, because the term has been used in many different contexts. The most impressive study to demonstrate this need is a recent paper by Dickman (1991), who reported three studies showing that two minimally correlated types of impulsivity could be distinguished. *Dysfunctional* impulsivity is defined as the tendency to act with less forethought than do most people with equal ability, and this tendency is a source of difficulty. *Functional* impulsivity, on the other hand, is the tendency to act with relatively little forethought when such a style is optimal. "The present work indicates that these two tendencies are not highly correlated and that they bear different relations both to other personality traits and to the manner in which certain basic cognitive processes are executed" (p. 95). Typical questions illustrating functional and dysfunctional types of impulsivity are "I am good at taking advantage of unexpected opportunities, where you have to do something immediately or lose your chance" (functional) and "Often, I don't spend enough time thinking over a situation before I act" (dysfunctional).

Dickman (1991) pointed out that most current scales are concerned with dysfunctional impulsivity, which is perhaps not surprising in view of the preoccupation of clinical psychologists with problems of living. Typical correlations reported by Dickman for the "narrow impulsivity" scale (see S. B. G. Eysenck, chapter 8, this volume) were .73 with dysfunctional impulsivity and .20 with functional impulsivity. This suggests (when reliabilities are taken into account) that the dysfunctional impulsivity scale generally measures the same thing as the "narrow impulsivity" scale, whereas the functional impulsivity scale has only a slight relation to "narrow impulsivity." Other interesting correlations show disparate correlations of functional versus dysfunctional impulsivity with venturesomeness ($r = .44$; $r = .20$), orderliness ($r = .224$; $r = -.46$), and cognitive structure ($r = .28$; $r = -.51$).

The factor analytic approach to impulsivity is illustrated in the chapter by S. B. G. Eysenck and is solely concerned with dysfunctional impulsivity. It might be better to name functional impulsivity "spontaneity" or some similar word to indicate its separateness from dysfunctional impulsivity; having two independent traits with the same name is embarrassing and leads to complications. Certainly, the dictionary definition of the adjective *impulsive* suggests such an interpretation. To be impulsive is to be "characterized by actions based on sudden desires, whims, or inclinations" (Collins, 1988, p. 504). This definition, implying emotional involvement, corresponds badly with the definition of functional impulsivity. A secondary definition of impulsive is "spontaneous"; thus, common speech recognizes both meanings.

Regardless, it is certainly true that most, if not all, previous studies of impulsivity (e.g., Barratt, 1965; Gerbing, Ahadi, & Patton, 1987; J. S. Guilford, Zimmerman, & Guilford, 1976; Schalling, Edman, & Asberg, 1983) have interpreted the term in the dysfunctional sense, and it will be so interpreted in this chapter. Impulsivity, as will be shown in chapter 8, does break down into subfactors that show only moderate correlations. These correlations are approximately the same size as those among the four components of Zuckerman's (1983) sensation-seeking scales, which are Disinhibition (DIS), Thrill and Adventure Seeking (TAS), Experience Seeking (ES), and Boredom Susceptibility (BS). The fact that in both cases the trait (impulsivity or sensation seeking) breaks down into subfactors shows that the original hierarchical model may not be quite correct. There are traits at Level 1 and dimensions or higher-order factors at Level 2, but traits such as impulsivity and sensation seeking clearly break down into subtraits, so that we need at least three levels: subtraits, traits, and dimensions.

TABLE 1
Intercorrelations Among Four Impulsivity Subfactors and Four Sensation-Seeking Subfactors, and Correlations Among These and P and E

	Impulsivity				Sensation-seeking				P	E
Subfactor	IMP, N	Risk-taking	Non-planning	Liveliness	DIS	TAS	ES	BS	.34	.36
IMP, N		.41	.36	.30	.31	.19	.15	.27	.28	.29
Risk-taking	.36		.37	.24	.38	.46	.21	.31	.33	.17
Nonplanning	.40	.43		.23	.23	.23	.31	.20	.03	.42
Liveliness	.18	.21	.13		.17	.19	.14	.12		
DIS	.17	.37	.20	.11		.26	.42	.44	.30	.29
TAS	−.02	.42	.16	.08	.23		.39	.18	.09	.16
ES	.05	.27	.35	.02	.36	.27		.23	.30	.12
BS	.17	.33	.29	.12	.41	.04	.22		.29	.10
P	.24	.34	.31	.00	.30	.09	.30	.29		
E	.20	.27	.07	.34	.29	.16	.12	.10		

Note. Above the diagonal = female subjects. Below the diagonal = male subjects. IMP = Impulsiveness; N = Narrow; DIS = Disinhibition; TAS = Thrill and Adventure Seeking; ES = Experience Seeking; BS = Boredom Susceptibility; P = Psychoticism; E = Extroversion. Adapted by permission from Eysenck (1983, p. 19).

In actuality, the situation is even more complicated, as is shown in Table 1, which gives correlations among the four subfactors of impulsivity, the four subfactors of sensation seeking, and the dimensions of personality P (psychoticism) and E (extroversion). The four subfactors for sensation seeking (SS) have already been introduced; those for impulsiveness (IMP) are Narrow Impulsivity (N), Risk Taking, Nonplanning, and Liveliness. Their nature is obvious from their titles, and they will be discussed in more detail in chapter 8. All of the eight subfactors lie in the P+/E+ quadrant, and all eight are intercorrelated. Obviously, there is much overlap. TAS has little to do with P; liveliness correlates only with E; narrow impulsivity and risk taking, but also DIS, correlate with both P and E; BS mainly correlates with P, as does nonplanning. Clearly, both concepts cover a fairly similar area. Furthermore, some IMP subfactors correlate as highly with SS subfactors as with other impulsivity subfactors (e.g., risk taking), as do some SS subfactors with impulsivity (e.g., DIS). There are some obvious problems in the taxonomy of personality traits. Some clarification has been brought by the factor analytic studies summarized by S. B. G. Eysenck (chapter 8, this volume), but the need for conceptual clarification and psychometric analysis is clear.

Measurements of impulsivity would be expected to have some relation to everyday impulsive behavior, particularly of an abnormal kind (e.g., hyperactive children, impulsive fire setting, drug addiction, impulsive criminal acts). There is ample evidence for such correlations in the literature (Eysenck & Gudjonsson, 1989; Kennedy & Grubin, 1990; Lacey & Evans, 1990; Stanford & Barratt, in press). This shows that "impulsivity" is more than a psychometric curiosity.

The Heritability of Impulsivity

Identification of a trait through psychometric proceedings is only the first step in nailing down a concept, and the second step involves the quest for information concerning the amount of relative variance contributed by genetic and environmental factors. Once these questions have been answered it can be decided whether causal questions can best be addressed through reference to biological factors, social factors, or both. In this section, I will consider the heritability of impulsivity (Eaves, Eysenck & Martin, 1989; Eysenck, 1983). (Both of these sources contain information concerning the methodology adopted here; for the most part, I have relied on the analysis of twin data.)

There is an outmoded "classical" approach to the estimation of heritability, using correlations between relatives, in particular twins, and culminating in the estimation of various ratios describing the relative importance of genetic and environmental influences on trait variation. This approach has led to ratios such as the H of Holzinger (1929), the E of Neel and Schull (1954), and the HR of Nichols (1965). It essentially concentrates on heritability and makes estimates of heritability on the basis of so many unverified assumptions that their genetic meaning must remain doubtful.

In contrast, the biometrical–genetical approach was initiated by Fisher (1918), extended and applied by Mather and Jinks (1974), and applied to the analysis of human behavior (Jinks & Fulker, 1970). Essentially, this approach goes beyond simple estimates of heritability to an assessment of the kinds of genetic action and mating systems operating in the population ("genetic architecture"). From this point of view, therefore, estimates of heritability are of relatively limited importance; what is of interest is partitioning the total phenotypic variance into a number of genetic, environmental, and interactional factors and determining their relative contribution to the total variance.

The most fundamental formula in this connection analyzes the phenotypic (P) variance into the sum of genotypic (G) and environmental (E) variance: $V(P) = V(G) + V(E)$. In other words, the total phenotypic variance is equal to the sum of the genotypic and environmental variance. When the analysis is based on twins, the analysis of variance partitions total trait variation into two sources: that between pairs and that within. To the extent that pairs resemble each other, the mean square between (B) will be greater than that within (W), the ratio $(B-W/B+W)$ being a measure of this resemblance known as the *intraclass correlation*. It is to these mean squares that we equate our genetic and environmental components $V(G)$ and $V(E)$.

However, because people are typically raised in families, the environmental part of the model must be elaborated, replacing E with two components. One component reflects the effects of the home background together with shared or common experiences; the other reflects experiences that typically differ for children despite the fact that they are reared together. These two sources of variance are commonly referred to as *between-family environmental variance*, or *common environment* (CE), and *within-family environmental variance*, or *specific environment* (SE). Thus, the total phenotypic variance is now equal to $V(G) + V(CE) + V(SE)$.

Heritability itself is best understood in terms of the components of variance that enter into it. The genetic variance can be divided into four main components: $V(A)$, which is the additive genetic variance; $V(D)$, which is the nonadditive genetic variance due to dominance at the same gene loci; $V(EP)$, which is the nonadditive genetic variance due to interaction among different gene loci, called *epistasis*; and $V(AM)$, which is the genetic variance due to assortative mating (i.e., the increment to the total variance attributable to the degree of genetic resemblance between mates on the traits in question). This gives us a formula for the genetic variance of $V(G) = V(A) + V(D) + V(EP) + V(AM)$.

The nongenetic variance is composed of the following factors: $V(E)$, which is the additive environmental variance that is independent of the genotype; $V(GE)$, which is variance due to the interaction (i.e., nonadditive effects) of genotypes and environments; $Cov(GE)$, which is the covariance of genotypes and environments, also sometimes known as *correlated environments*; and $V(e)$, the error variance due to the unreliability of measurements. $V(e)$ is traditionally mixed in with the environmental variance, although, strictly speaking, the formula should be rewritten to exclude this and deal with the genetic and environmental variance of the "true" score alone.

It can be seen that the general formula for phenotypic variance embraces a large number of terms and that a numerical estimate of the importance of these various terms is of much greater interest and importance than a simple estimate of the heritability. In fact, there are two heritabilities: the so-called *narrow heritability*, which is the proportion of additive genetic variance, and heritability in the broad sense (i.e., the proportion of the total phenotypic variance that the genetic variance is). There are many other complexities attending the concept of heritability. Thus, some estimates of heritability in the broad sense include $Cov(GE)$ in the numerator; this is done on the assumption either that the covariance is due to the genotype or that a particular method of estimating does not permit separation of $V(G)$ and $Cov(GE)$.

When relating empirical data from twin studies, familial studies, adoption studies, or inbreeding depression studies to the model, one finds

TABLE 2
Contributions of Genetic and Environmental Factors to the Total Variance of Four Impulsive Factors, for Male and Female Subjects Separately

Subfactor	Genetic		Environmental			
			Common		Specific	
	Male	Female	Male	Female	Male	Female
IMP, N	.40	.37	.20	.24	.40	.39
Risk-taking	.17	.36	.27	.20	.55	.44
Nonplanning	.38	.36	.16	.20	.46	.44
Liveliness	.15	.16	.30	.24	.55	.60

Note. Error variation has not been deducted from the contributions of specific environmental factors. IMP = Impulsiveness; N = Narrow. Adapted by permission from Eysenck (1983, p. 14).

that the data do not usually allow an estimation of all of the variables in the equation. Some of the variables can be directly estimated; for example, assortative mating can be estimated by measuring the correlation between spouses on a given trait. Other variables in the equation may be estimated directly from social events, which in effect constitutes a genetic experiment. Thus, it is possible to estimate the importance of dominance in inbreeding depression (e.g., the lower scores obtained from the offspring of married relatives such as cousins). Frequently, however, the procedure is to conduct an analysis starting with a very simple assumption, such as that all of the phenotypic variances are due to between-family environmental factors; when this is found to give a very poor fit to the data, a second variable is brought in such as the additive genetic variance. If the combination still gives a poor fit, one may go on to introduce the within-family environmental variance, and so forth. In this way, we can show that certain variables that would be very difficult to test directly (e.g., interaction effects) are not required because a simple analysis adequately represents the empirical data.

A much more detailed introduction to the methods is given in Eaves et al. (1989); here we must consider only the main findings from our studies. The first study used 588 pairs of twins, of whom 75 were unlike-sex pairs, the rest being divided between male and female monozygotic and dizygotic twins of like sex (Eysenck, 1983). All were given a 52-item impulsivity questionnaire; 40 of the questions were selected as best representing the four component factors of impulsivity (see S. B. G Eysenck, chapter 8, this volume). Results are reported separately for male and female subjects (Table 2).

The results need to be corrected for measurement error; they would otherwise lead to exaggerated values for specific environmental variance. When this is done, heritability for IMP.N (narrow impulsivity) is .57 and

TABLE 3
Contributions of Genetic and Environmental Factors to the Total Variance of Four Sensation-Seeking Factors, for Male and Female Subjects Separately

| | Genetic | | Environmental | | | |
| | | | Common | | Specific | |
Subfactor	Male	Female	Male	Female	Male	Female
DIS	.51	.41	.49	.59	.00	.00
TAS	.45	.44	.02	.02	.53	.54
ES	.58	.57	.03	.03	.39	.40
BS	.41	.34	.11	.13	.48	.53

Note. Error variation has not been deducted from the contributions of specific environmental factors. DIS = Disinhibition; TAS = Thrill and Adventure Seeking; ES = Experience Seeking; BS = Boredom Susceptibility. Adapted by permission from Eysenck (1983, p. 18).

.60 for male and female subjects, respectively, and the heritability of different aspects of impulsivity is generally in excess of 50%. Common environmental variance, as usual, was less than specific environmental variance (Eaves et al., 1989).

The results for sensation seeking, in a similar experiment using 422 pairs of twins, were similar (Fulker, Eysenck, & Zuckerman, 1980). Table 3 shows the results. Again correction for unreliability would elevate all heritabilities beyond the .50 level, as in the case of impulsivity.

One may conclude that genetic factors account for most of the variance in impulsive behavior and that the environmental variance is largely specific (within-family) rather than common (between-family). These findings are not unique to impulsivity (and sensation seeking), but generally apply to most, if not all, personality traits and dimensions (Eaves et al., 1989).

The Biological Basis of Impulsivity

The preceding section clearly indicates the importance of genetic factors in impulsivity, and thus the search for causal factors is most likely to be successful if directed toward biological factors. Of course, environmental factors may also contribute to impulsivity, but the evidence firmly negates many current theories that implicate the family, parenting styles, and similar childhood influences. Such factors would generate between-family environmental variance, and as we have seen, this is largely absent from our equation. Within-family (specific) variance is important, but difficult to study; it is composed of singular events happening to one child but not the others in a given family, and hence presents awkward problems to the researcher. Research in this area is not impossible, but there is a dearth of results in the literature, such that we know next to nothing about the environmental events that might be included in this category.

Eysenck's (1967) original theory of extraversion would seem to be particularly applicable to impulsivity. According to this theory, impulsivity should be regarded as due to low cortical arousal, which in turn would be related to poor functioning of the reticular activating system (Strelau & Eysenck, 1987). Arousal activates the cortex, which inhibits the activity of lower clusters; if arousal is lowered, inhibition is removed, allowing impulsive behaviors to occur with greater freedom. Both P and E are characterized by low cortical arousal, and as we have seen, impulsivity is correlated with both P and E. Stimulant drugs increase arousal, and depressant drugs reduce it, each affecting impulsive behavior accordingly (Eysenck, 1963). The best example of such action is the successful administration of stimulant drugs to hyperactive children, whose behavioral impulsivity is significantly lowered as a consequence (Eysenck & Gudjonsson, 1989). The well-established correlation between impulsivity and crime, and the involvement of alcohol (a depressant drug) in such criminal activity, further illustrates the far-reaching consequences of this theory (Eysenck & Gudjonsson, 1989).

It would be wrong to give the impression that the cortical arousal theory is universally accepted, or that it covers all of the data. However, there has been more experimental work supporting this theory than any other theory in this field. Indeed, it is doubtful if there is any other theory in existence that could claim similar explanatory status. However, there are many anomalies crying out for experimental enquiry (Strelau & Eysenck, 1987), and there are a number of factual findings that may be difficult to fit into any particular theory (Zuckerman, 1991).

One important clue to the nature of impulsivity is the fact that it is typically increased by brain damage and neurosurgery. In particular, both damage to the right hemisphere frontal areas within the frontal lobes and damage to the orbito-frontal area (one of the main sites of entry from the cortex to the limbic system) appear to be intimately involved in the disinhibition of behavior. Together with the apparent reduction of anxiety, disinhibition of behavior is the consequence of such damage. Right temporal lobe epilepsy has also been linked to impulsive behavior. These results are compatible with the cortical arousal theory, but there are alternative possibilities (Zuckerman, 1991).

Another line of research has emphasized the fact that low levels of serontonergic activity are associated with impulsivity (Soubrie, 1986). Dopamine has also been associated with impulsive behavior (Gray, 1987). Related to the studies of neurotransmitters are investigations of platelet monoamine-oxidase (MAO), an enzyme present in all tissues, with the highest brain concentration in the thalamus. Generally, low levels of

MAO are characteristic of high P, high E, and impulsiveness (Zuckerman, Ballenger, & Post, 1984), with impulsivity also found in low-MAO animals. (The most recent evidence was reviewed by Zuckerman, 1991.) The serotonin metabolite 5-HIAA has been similarly implicated, as has homovanillic acid. Cortisol has shown a negative relationship with impulsivity. These and other similar relations were discussed in much greater detail than would be possible here by Zuckerman (1991).

There are interesting psychophysiological correlates of impulsivity. Electroencephalogram studies by O'Gorman and Lloyd (1987) found high impulsives to be less aroused and showing more power in the alpha band (less arousal), particularly for narrow impulsivity. Similarly, impulsivity was found to be related to augmentation of the visual-evoked potential (i.e., an increase in amplitude accompanying increased stimulus intensity) (Barratt, Pritchard, Faulk, & Brandt, 1987), a finding also observable in cats (Lukas & Siegel, 1977) (see Barratt, chapter 3, this volume, for further discussion of this point). For nonimpulsives, there is decreasing amplitude with increasing stimulus intensity.

The examples given here of the relation between impulsivity and various biological factors do not by any means exhaust the rich literature that has grown around the general concepts of extraversion, psychoticism, arousal, and impulsivity. The textbook *Psychobiology of Personality* by Zuckerman (1991) is a rich mine of information on these topics and a storehouse of information on the relationships involved. I can do no more here than draw attention to this wide and rapidly increasing field. It promises to put the causal network underlying the biological factors fostering individual differences in impulsivity on a firmer foundation.

The emphasis here on biological factors should not be misinterpreted to mean that environmental influence is ineffectual or cannot modify impulsive behavior. Genetic factors, acting through the physiological and hormonal systems of the body, are powerful, but not completely powerful. Modification of emotional and muscular reactions is possible; there is a large literature on the growth of inhibitory potential in animals demonstrating that possibility, which also exists in humans (Honig & Staddon, 1977). Thus, nature–nurture debates are never either–or but instead concern the relative contributions of each.

Impulsive behavior, clinically manifested, can be controlled to a considerable extent by means of behavior therapy (Rimm & Masters, 1974; Redd, Porterfield, & Andersen, 1979; Suedfeld, 1980). Levels of impulsivity can be changed through behavior therapy or the use of stimulant and depressant drugs; biology is important, but not decisive. It just so happens that we know more about the biological roots of impulsive behavior than about environmental ones. It is time we redressed that imbalance.

SUMMARY

Impulsive behavior is composed of many correlated types of impulsivity, most of them dysfunctional, although at least one functional (socially useful) type has been recognized. Factor analysis has revealed that these types of dysfunctional impulsivity are related to the three major dimensions of personality of Psychoticism, Neuroticism, and Extraversion. Impulsivity overlaps to a considerable extent with Sensation Seeking, which also breaks down into a number of subfactors.

Genetic studies have demonstrated a strong hereditary basis for impulsivity, accounting for over 50% of the total phenotypic variance. Environmental effects are largely within-family (i.e., specific to each child in a family) and are minimally related to common environments, or between-family variance. This makes the empirical study of causal influences much more difficult but brings the study of impulsiveness into line with the great majority of other traits of personality that have been studied by geneticists (Eaves et al., 1989).

Genetic determination suggests a search for physiological, neurological, and hormonal sources of impulsive behavior, and many of these have been discovered. The major theory in the field, associating impulsivity with low cortical arousal, has found some support, but many anomalies remain. A good deal of detailed research remains to be done before we can be said to have an adequate understanding of the nature of impulsivity. In the meantime, we do know enough information to change a person's (or an animal's) level of impulsivity by either chemical or behavioral methods. But clearly we have come some way toward an understanding of the concept.

REFERENCES

Barratt, E. S. (1965). Factor analysis of some psychometric measures of impulsiveness and anxiety. *Psychological Reports, 16*, 547–554.

Barratt, E. S., Pritchard, W. S., Faulk, D. M., & Brandt, M. E. (1987). The relationship between impulsiveness sub-traits, trait anxiety, and normal N100-augmenting-reducing: A typographic analysis. *Personality and Individual Differences, 8*, 43–51.

Collins. (1988). *Dictionary and thesaurus*. London: Author.

Dickman, S. J. (1991). Functional and dysfunctional impulsivity: Personality and cognitive correlates. *Journal of Personality and Social Psychology, 58*, 95–102.

Eaves, L., Eysenck, H. J., & Martin, N. (1989). *Genes, culture and personality: An empirical approach*. New York: Academic Press.

Eysenck, H. J. (1947). *Dimensions of personality*. London: Routledge & Kegan Paul.

Eysenck, H. J. (1956). The questionnaire measurement of neuroticism and extraversion. *Rivista di Psicologia, 50,* 113–140.

Eysenck, H. J. (Ed.). (1963). *Experiments with drugs*. Elmsford, NY: Pergamon Press.

Eysenck, H. J. (1967). *The biological basis of personality*. Springfield, IL: Charles C Thomas.

Eysenck, H. J. (1983). A biometrical–genetical analysis of impulsive and sensation-seeking behavior. In M. Zuckerman (Ed.), *Biological basis of sensation seeking, impulsivity and anxiety* (pp. 1–27). Hillsdale, NJ: Erlbaum.

Eysenck, H. J. (1991). Dimensions of personality: 16, 5 or 3? Criteria for a taxonomic paradigm. *Personality and Individual Differences, 8,* 773–790.

Eysenck, H. J., & Eysenck, M. W. (1985). *Personality and individual differences: A natural science approach*. New York: Plenum Press.

Eysenck, H. J., & Gudjonsson, G. (1989). *The causes and cures of criminality*. New York: Plenum Press.

Fisher, R. A. (1918). The correlation between relatives on the supposition of Mendelian inheritance. *Transactions of the Royal Society* (Edinburgh), *52,* 399–433.

Fulker, D. W., Eysenck, H. J., & Zuckerman, M. (1980). A genetic and environmental analysis of sensation-seeking. *Journal of Research in Personality, 14,* 261–281.

Gerbing, D., Ahadi, S., & Patton, J. (1987). Toward a conceptionalization of impulsivity: Components across the behavioral and self-report domain. *Multivariate Behavioral Research, 22,* 11–22.

Gray, J. (1987). The neuropsychology of emotional personality. In S. M. Stakl, S. D. Iverson, & E. C. Goodman (Eds.), *Cognitive neurochemistry* (pp. 171–190). Oxford, England: Oxford University Press.

Guilford, J. P., & Guilford, R. B. (1936). Personality factors S, E and M, and their measurement. *Journal of Psychology, 2,* 109–127.

Guilford, J. S., Zimmerman, W. S., & Guilford, J. P. (1976). *The Guilford–Zimmerman Temperament Survey handbook*. San Diego, CA: EdITS.

Holzinger, K. J. (1929). The relative effect of nature and nurture on twin differences. *Journal of Educational Psychology, 20,* 245–248.

Honig, W. K., & Staddon, W. R. (1977). *Handbook of operant behavior*. Englewood Cliffs, NJ: Prentice-Hall.

Jinks, J. L., & Fulker, D. W. (1970). Comparison of the biometrical genetical, MAVA, and classical approaches to the analysis of human behavior. *Psychological Bulletin, 73,* 311–349.

Kennedy, H. G., & Grubin, D. H. (1990). Hot-headed or impulsive? *British Journal of Addiction, 85,* 639-643.

Lacey, J. H., & Evans, C. D. (1990). The impulsivist: A multi-impulsive personality disorder. *British Journal of Addiction, 81,* 641–649.

Lukas, J. H., & Siegel, J. (1977). Cortical mechanisms that augment or reduce evoked potentials in cats. *Science, 196,* 73–75.

Magnusson, D. (Ed.). (1981). *Toward a psychology of situations: An interactional perspective.* Hillsdale, NJ: Erlbaum.

Mather, K., & Jinks, J. L. (1974). *Biometrical genetics.* London: Chapman & Hall.

Mischel, W. (1968). *Personality and assessment.* New York: Wiley.

Neel, J. V., & Schull, W. J. (1954). *Human heredity.* Chicago: University of Chicago Press.

Nichols, R. C. (1965). The National Merit Twin Study. In S. G. Vandenberg (Ed.), *Methods and goals in human behavior genetics* (pp. 89–114). New York: Academic Press.

O'Gorman, J. G., & Lloyd, J. (1987). Extraversion, impulsiveness, and EEG alpha activity. *Personality and Individual Differences, 8,* 169–174.

Redd, W. H., Porterfield, A. L., & Andersen, M. L. (1979). *Behavior modification.* New York: Random House.

Reznick, J. S. (Ed.). (1989). *Perspectives on behavioral inhibition.* Chicago: University of Chicago Press.

Rimm, D. C., & Masters, J. C. (1974). *Behaviour therapy.* New York: Academic Press.

Schalling, D., Edman, G., & Asberg, M. (1983). Impulsive cognitive style and inability to tolerate boredom in psychophysiological studies of temperamental vulnerability. In M. Zuckerman (Ed.), *Biological basis of sensation-seeking, impulsivity, and anxiety* (pp. 123–145). Hillsdale, NJ: Erlbaum.

Soubrie, P. (1986). Reconciling the role of central serotonin neurons in animal and human behavior. *Behavioral and Brain Sciences, 9,* 319–364.

Stanford, M. S., & Barratt, E. S. (in press). Impulsivity and the multi-impulsive disorder. *Personality and Individual Differences.*

Strelau, J., & Eysenck, H. J. (1987). *Personality dimensions and arousal.* New York: Plenum Press.

Suedfeld, P. (1980). *Restricted environmental stimulation.* New York: Wiley.

Wickland, R. A. (1990). *Zero-variable theories and the psychology of the explainer.* New York: Springer-Verlag.

Zuckerman, M. (Ed.). (1983). *Biological basis of sensation-seeking, impulsivity, and anxiety.* Hillsdale, NJ: Erlbaum.

Zuckerman, M. (1991). *Psychobiology of personality.* Cambridge, England: Cambridge University Press.

Zuckerman, M., Ballenger, J. C., & Post, R. M. (1984). The neurobiology of some dimensions of personality. *International Review of Neurobiology, 25,* 391–436.

5

SENSATION SEEKING AND IMPULSIVITY: A MARRIAGE OF TRAITS MADE IN BIOLOGY?

MARVIN ZUCKERMAN

The traits of impulsivity and sensation seeking are conceptually and empirically wedded. The reasons for this relationship may lie in their common biological bases and ultimately in their genetic determinants. Although they are related, there are also distinctions between these two traits. This is particularly apparent in the subtrait of sensation seeking called Thrill and Adventure Seeking. The purpose of this chapter is to highlight similarities and differences in this domain, as well as to present a biological account of impulsiveness and sensation seeking.

Thrill and adventure seeking often involves physical risk taking where behaving impulsively can be fatal. However, in the disinhibitory or experience-seeking forms of sensation seeking, impulsivity may lead to social embarrassment or even legal problems, but in these expressions it rarely leads to immediate death. For example, mountain climbers are generally high sensation seekers (Zuckerman, 1983) and, in fact, show one of the biological markers for sensation seeking, low levels of monoamine oxidase (Fowler, von Knorring, & Oreland, 1980). Regardless of how they may behave at sea level, mountain climbers are not impulsive in their approach to the activities involved in climbing. Their route up the mountain is carefully planned, weather conditions are taken into account, and gear is checked and rechecked before the climb. Indeed, impulsive or careless climbers are not welcome because of their potential to endanger both their own lives and the lives of others. Although risk taking is intrinsically involved in mountain climbing, it is not the major reason for, or sole point of, climbing. Instead, the experience is the reward, and incidental risk is something to be minimized through training, skill, and planning. However, the low sensation seeker would observe that there is no risk at all when admiring the mountain from a safe distance at sea level.

The most recent definition of sensation seeking is "the seeking of intense, complex, and novel experiences and the *willingness* to take risks for the sake of such experiences" (Zuckerman, in press, emphasis added). There are some forms of sensation seeking in which the risk is more intrinsic to the reward of the experience. For example, compulsive gamblers only enjoy gambling when their wagers, potential wins, and risks of losing are high, because only extreme risk can produce the high levels of arousal they seem to enjoy (Anderson & Brown, 1984). Risk taking for its own sake is found in impulsive sensation seekers who act with little or no cognitive appraisal of risk. At the opposite extreme are obsessives who think too much about the risk and the possibilities of negative outcomes—as Hamlet muses in his famous soliloquy (*The Tragedie of Hamlet, Prince of Denmark*, III, i, 84–85), "And thus the native hue of resolution is sicklied o'er with the pale cast of thought."

RELATIONSHIPS BETWEEN SENSATION SEEKING AND IMPULSIVITY

Table 1 shows the relationships between various impulsivity scales and the General Sensation Seeking Scale (Zuckerman, 1979). Many of these relationships are moderate to high, particularly those with the Impulsivity, Harmavoidance, and Cognitive Structure subscales of Jackson's Personality Research Form (Jackson, 1974). Harmavoidance is particularly related to the Sensation Seeking (SS) subscale of Thrill and Adventure Seeking, because the risk avoided is physical risk of bodily harm. Jackson's Impulsivity scale, which correlates about equally with all of the SS subscales, describes a person who "tends to act on the 'spur of the moment' and without deliberation; gives vent readily to feelings and wishes, speaks freely; may be volatile in emotional expression" (Jackson, 1974, p. 7). Cognitive Structure, on which low sensation seekers tend

TABLE 1
Correlations Between Sensation Seeking and Impulsivity Scales and the Sensation Seeking Scale IV: General Scale (Zuckerman, 1979)

Impulsivity scales	Men	Women
Harmavoidance (PRF)	−.81	−.72
Impulsivity (PRF)	.51	.42
Cognitive Structure (PRF)	−.64	−.45
Impulsivity N (Eysencks')	.20	.30
Risk Taking (Eysencks')	.57	.57
Nonplanning (Eysencks')	.35	.34

Note. All correlations significant, $p < .01$. Correlations with PRF (Personality Research Form; Jackson, 1974) are from Zuckerman (1974), 82 male and 71 female students. Correlations with the Eysencks' scales are from S. B. G. Eysenck & Eysenck (1977), 254 male and 625 female twins.

to score high, describes a person who "does not like ambiguity or uncertainty in information; wants all questions answered completely; desires to make decisions based upon definite knowledge rather than guesses or probabilities" (Jackson, 1974, p. 6).

In an earlier version of the S. B. G. Eysenck and Eysenck Impulsivity scales (1977), the broad construct of impulsivity was subdivided into subscales measuring Narrow Impulsivity (similar to Jackson's definition above), Risk Taking (opposite of Harmavoidance), and Nonplanning (opposite of Cognitive Structure). Sensation seeking is moderately correlated with Risk Taking and significantly, but less strongly, correlated with Nonplanning and Narrow Impulsivity.

Personality scales often correlate to the extent that there are similarities in the content of their items. As S. B. G. Eysenck describes in chapter 8 of this volume, sensation seeking in the form of venturesomeness has been incorporated into the Eysencks' construct of impulsivity (S. B. G. Eysenck & Eysenck, 1978) along with narrow impulsivity (fast decision making without forethought). Naturally, venturesomeness correlates highly with sensation seeking scales (Zuckerman, in press), but this tells us little about the intrinsic relationship between the two traits.

Gerbing, Ahadi, and Patton (1987) factor analyzed the items in a number of putative impulsivity scales, including Barratt & Patton's (1983), S. B. G. Eysenck and Eysenck's (1977), H. J. Eysenck and M. W. Eysenck's (1985), and Buss and Plomin's (1975) scales. They also included the Sensation Seeking Scale items from Form V (Zuckerman, 1979; Zuckerman, Eysenck, & Eysenck, 1978). Twelve factors were found, the first two of which seem to be Impulsivity in the narrow sense (acting on the spur of the moment) and Thrill Seeking, or the enjoyment of risk taking. The observed scale correlations and the correlations of these factors with other factors are given in Table 2. Also shown in this table are the correlations with some performance definitions of impul-

TABLE 2
Correlations Between Thrill Seeking and Other Factor Scales Derived from Factor Analyses of Impulsivity Scales

Component factor scales	Scale *r*s	Factor *r*s
Impulsivity (n)	.52	.59
Quick Decisions	.35	.45
Avoids Planning	.35	.39
Energetic	.22	.26
Happy-Go-Lucky	.27	.30
Impulsive Purchaser	.25	.30
Unreflective	.23	.28
Distractible	.12	.12
Restless	.19	.19
Impatient	.07*	.16
Avoids Complexity	−.20	−.23
Simple Reaction Time	−.15	−.14
Time Perception	−.01*	−.02*
Matching Familiar Figures Test	−.12*	−.16

Note. All correlations significant (*p* < .05) except those indicated with an asterisk. Adapted with permission from Gerbing, Ahadi, & Patton, 1987, p. 371.

sivity, including simple reaction time, time perception, and combined error and latency scores on the Matching Familiar Figures Test (MFFT; Kagan, 1966).

The thrill (sensation) seeking and narrow impulsivity scales and factors correlate highly with each other (*r* = .52 and *r* = .59, respectively). Both scales are also moderately correlated with quick decision making and avoidance of planning factors, and correlated significantly but less highly with energetic, happy-go-lucky, impulsive buying, unreflectiveness, and restlessness factors. Avoidance of complexity was negatively correlated with thrill seeking, but uncorrelated with impulsivity. Thrill seeking, but not impulsivity, predicted fast reaction times and latencies on the MFFT. Despite these last two differences, the sensation seeking and impulsivity (narrow) factors showed a remarkable similarity in their correlations with other factors, as well as a high degree of relatedness between themselves.

IMPULSIVITY AND SENSATION SEEKING WITHIN THE MAJOR PERSONALITY DIMENSIONS

Two recent studies included a number of impulsivity and sensation seeking scales in factor analyses designed to define the major factors in basic personality or temperament measures (Zuckerman, Kuhlman, & Camac, 1988; Zuckerman, Kuhlman, Thornquist, & Kiers, 1991). In both studies, measures of these two traits combined with measures of social-

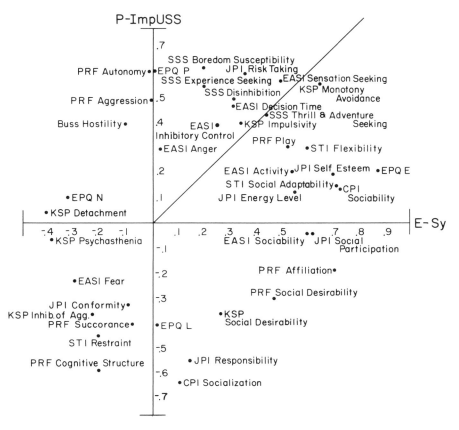

Figure 1. Loadings of personality scales plotted on the psychopathy–impulsive unsocialized sensation seeking (P-ImpUSS) and sociability (E-Sy) factor dimensions. (From "Personality in the Third Dimension: A Psychobiological Approach," by M. Zuckerman, 1989, *Personality and Individual Differences, 10,* p. 396. Copyright 1989 by Pergamon Press. Reprinted by permission.)

ization in a dimension of personality called *Impulsive Unsocialized Sensation Seeking* (ImpUSS). Additionally, both studies suggested H. J. Eysenck and Eysenck's (1976) Psychoticism (P) scale to be the best marker for this dimension.

The dimension is illustrated in the factor plots between P-ImpUSS and Sociability factors from the two studies in Figures 1 and 2. In the first study (Zuckerman et al., 1988), oblique rotations were used, so Figure 1 shows the correlations between factors. As shown in Figure 1, all four of the subscales from the SSS (Zuckerman et al., 1978) as well as Buss and Plomin's (1975) EASI impulsivity scales for fast decision time and lack of inhibitory control, fell into the ImpUSS dimension also marked by the Eysenck's P scale and an Autonomy scale at the positive pole. Socialization, responsibility, and cognitive structure scales fell at the op-

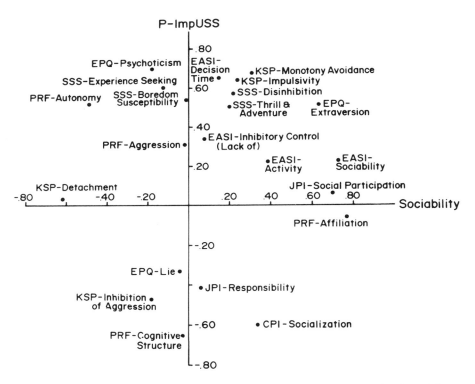

Figure 2. Loadings of personality scales plotted on the psychopathy–impulsive unsocialized sensation seeking (P-ImpUSS) and sociability factor dimensions.

posite end of the dimension. Some scales, like the Schalling and Edman (1987) Karolinska Scales of Personality (KSP) Impulsivity and Monotony Avoidance and the SSS Thrill and Adventure Seeking scales, loaded almost equally on P and Sociability (E) dimensions, but none of the sensation seeking or impulsivity scales loaded primarily on the E dimension.

In the second study (Zuckerman et al., 1991), the number of subjects was doubled and number of variables reduced by selecting only the best markers for the more narrow factors in the first study. Orthogonal factor rotations were used in this study to sharpen the contrast between factors. As shown in Figure 2, all of the sensation seeking and impulsivity scales loaded primarily on the P-ImpUSS dimension. As in the previous study, H. J. Eysenck and Eysenck's (1976) P scale marks this as what they call the "Psychoticism" dimension.

For a variety of reasons (Zuckerman, 1989, 1991), I would prefer the clinical label "Psychopathy" for P rather than Psychoticism. I believe that the clinical extreme of the dimension is represented by the antisocial personality, rather than being inclusive of all forms of psychosis, such as schizophrenia and psychotic depression. Unlike the first of these two studies, in which the Eysenck Personality Questionnaire (EPQ; H. J. Eysenck & Eysenck, 1975) Extraversion (E) scale was a pure marker for the sociability dimension, in the second study E had a strong secondary loading on the P-ImpUSS dimension.

The results presented thus far are from the factor analyses in which three factors were rotated. Actually, analyses were done with six-, five-, and four-factor rotations, as well as the three-factor rotation. At the six-factor level, impulsivity and sensation seeking scales formed separate factors, but these factors merged at the five-factor level to form the single P-ImpUSS factor, which remained essentially unchanged in the four- and three-factor analyses. The five-factor model was superior to the six- and four-factor models in terms of replicability of the factors comparing male and female samples, and therefore it was used as the basis for a model of personality. Apart from the sociability and impulsive unsocialized sensation seeking dimensions, three other major factors were found: neuroticism-anxiety, aggression-hostility, and activity.

A QUESTIONNAIRE MEASURE OF P-IMPUSS

The data from the Zuckerman et al. (1991) study of personality scale factors were used to develop the initial form of a questionnaire to measure the five factors. Item responses from all of the scales used in the study except the EPQ-R were correlated with factor scores for all of the subjects. On the basis of this preliminary analysis, many of the items were rewritten and a second and third form of the scale evolved. The last form was based on a factor analysis of the items in the second form. The final form, called the Zuckerman–Kuhlman Personality Questionnaire, Form III (ZKPQ III) (Zuckerman, Kuhlman, Joireman, Teta, & Kraft, 1993), contains five scales, each consisting of the items most highly related to their respective factors. The scales were internally reliable, as shown in subsequent studies.

The factor analysis of items, like the previous factor analysis of scales, produced a broad factor (ImpSS) combining items from sensation seeking and impulsivity scales. The items constituting this scale are listed in Table 3. The fact that all of these items loaded on a single common factor suggests the close link between the traits of sensation seeking and impulsivity. Most of the impulsivity items indicate a tendency to get into tasks or situations without much planning or thinking about possible negative outcomes. Some of the sensation seeking items show the same tendency (e.g., No. 34). The sensation seeking items are a mixture of disinhibition (e.g., No. 95), experience seeking (e.g., No. 34), thrill and adventure seeking (e.g., No. 45), and boredom susceptibility (e.g., No.

TABLE 3
Items From the Zuckerman–Kuhlman Personality Questionnaire, Form III, ImpSS Scale

Impulsivity type items

1. I tend to begin a new job without much advance planning on how I will do it. (T)
6. I usually think about what I am going to do before doing it. (F)
9. I very seldom spend much time on the details of planning ahead. (T)
14. I often do things on impulse. (T)
29. Before I begin a complicated job I make careful plans. (F)
39. I enjoy getting into new situations where you can't predict how things will turn out. (T)
84. I often get so carried away by new and exciting things and ideas, I never think of the possible complications. (T)
89. I am an impulsive person. (T)

Sensation Seeking type items

24. I like to have new and exciting experiences and sensations even if they are a little frightening, unconventional, or illegal. (T)
34. I would like to take off on a trip with no preplanned or definite routes. (T)
45. I like doing things just for the thrill of it. (T)
50. I tend to change interests frequently. (T)
55. I sometimes like to do things that are a little frightening. (T)
61. I'll try anything once. (T)
65. I would like the kind of life where one is on the move and traveling a lot. (T)
70. I sometimes do "crazy" things just for fun. (T)
71. I like to explore a strange city or section of town by myself, even if it means getting lost. (T)
81. I prefer friends who are excitingly unpredictable. (T)
95. I like "wild" uninhibited parties. (T)

Note. T = scored if answered "True." F = scored if answered "False."

81). They express a desire or preference for excitement, change or variety, and unpredictable situations or people, even if potentially dangerous.

IMPULSIVITY AND SENSATION SEEKING IN OTHER CLASSIFICATIONS OF PERSONALITY TRAITS

H. J. Eysenck (chapter 4, this volume) believes that the subtraits of broad impulsivity and sensation seeking are differentially related to his three major dimensions of personality, some belonging to one and some to another dimension. However, in his procrustean assignment of the broad traits to Psychoticism (P), Extraversion (E), and Neuroticism (N) (H. J. Eysenck & Eysenck, 1985), he assigned sensation seeking and venturesomeness to E and impulsivity to P. Sensation seeking was also regarded as a subtrait of impulsivity in the form of *risk-taking* in the earlier impulsivity scales and *venturesomeness* in the later form (S. B. G. Eysenck, chapter 8, this volume). Venturesomeness, however, consists mainly of items of the thrill-and-adventure or physical risk-taking types and does not include the type of experience seeking, disinhibition, and boredom susceptibility items that are more relevant to the socialization or P dimension of personality. This may be why Eysenck conceives of sensation seeking as belonging primarily to the E dimension, although acknowledging that it may represent a combination of P and E tendencies.

In Norman's (1963) original version of the five-factor model of personality, a bipolar subtrait called "Adventurous–Cautious" seemed to combine both sensation seeking (adventurous) and impulsivity (cautious) in a single subtrait classified under the major trait of extraversion. Another major trait called "Conscientiousness," however, includes traits like "Persevering–Quitting" and "Fickle" that could reflect the changeability of impulsive sensation seekers. The "Imaginative–Simple" subtrait of the major trait called "Culture" might reflect some of the experience-seeking subtrait of sensation seeking.

Goldberg's (1990) version of what has come to be known as the "big five" stresses the "surgency" aspect of extraversion, including lack of restraint and inhibition, as well as "playfulness" (which includes "adventurousness") and courage. Although this would seem to place sensation seeking in the extraversion dimension, there are other subtraits that might correspond to aspects of sensation seeking placed in other factors. Examples are "Recklessness" under conscientiousness and "Curiosity" under intellect.

Costa, McCrae, and Dye (1991) separate sensation seeking and impulsivity in their five-factor model and include impulsivity as a "facet" of neuroticism and "Excitement-Seeking" as a facet of extraversion. Despite the classification model on which these researchers based their five-factor questionnaire, three of the four subfactors of the SSS are most highly related to their major factor, called "Openness to Experience," whereas the disinhibition subscale of the SSS is more highly related to their "Conscientiousness" factor (Angleitner & Ostendorf, in press; Costa & McCrae, 1992). However, the new ImpSS scale of the ZKPQ is primarily related to the Costa and McCrae conscientiousness scales and not at all related to their Openness to Experience scales (Zuckerman et al., 1993).

IMPULSIVITY AND SENSATION SEEKING IN TEMPERAMENT MODELS

Temperament often refers to those personality traits that are more biologically determined and are expressed in stylistic-expressive aspects of personality rather than being oriented toward specific needs or goals

(Strelau, 1983). Thus, need for achievement might be regarded as a personality trait, whereas impulsivity would be regarded as a trait of temperament. Sensation seeking has commonly been regarded as both a temperament trait, because of its strong heritability (Fulker, Eysenck, & Zuckerman, 1980) and many biological correlates (Zuckerman, 1984; Zuckerman, Buchsbaum, & Murphy, 1980), and a basic trait of personality. However, unlike most other putative traits of temperament, sensation seeking is not readily identifiable in infants and young children, although it may be equivalent to the general tendency to approach novel stimuli or people (Zuckerman, 1990). Thomas and Chess (1977) have called this the *Approach or Withdrawal* dimension of infant and child temperament. In infants and young children, the dimension is defined by ratings of reactions to novel stimuli, such as new foods, toys, or persons. The dimension may also correspond to what Kagan et al. (Kagan, Reznick, Clarke, Snidman, & Garcia-Coll, 1984; Kagan, Reznick, & Snidman, 1988) called the inhibited versus noninhibited dimension of temperament, although they stressed the sociable versus shy aspect of the dimension and equated the trait with adult extraversion versus introversion.

Buss and Plomin (1975) included impulsivity as one of four major dimensions of temperament, with sensation seeking as a subtrait of the broader trait. Later, they excluded impulsivity, reducing their big four to a big three (emotionality, activity, and sociability) because of doubts about its factorial unity and heritability (Buss & Plomin, 1984). Correlations among the subscales of impulsivity (including sensation seeking, persistence, inhibitory control, and decision time) were low, but this may have been due to the abbreviated five-item form of these scales. In view of the subsequent evidence for the heritability and biological basis for the trait (to be discussed) and evidence for its centrality in all of its forms in the P dimension of personality (discussed earlier), their decision was probably misguided, and is being reconsidered by Buss (1991).

Strelau's (1983) theory of temperament is based on neo-Pavlovian hypothetical properties of the central nervous system. The three major properties include strength of excitation, strength of inhibition, and mobility of nervous processes. Impulsivity, regarded as a component of extraversion, is supposedly related to the strength of excitation property, but conceptually one could argue that it is the result of a weakness of the strength of inhibition trait. Sensation seeking is regarded as most closely related to the strength of excitation, characterized by a strong nervous system, low sensitivity to stimulation, and high activity (directed toward the increase of stimulation). It might also be argued that high sensation seekers are characterized by a high degree of mobility of nervous processes, considering their predilection to shift from one stimulus or activity to another and their general changeability (Zuckerman, 1979). Furthermore, the disinhibition form of sensation seeking, in particular, should have a negative relationship with the strength of inhibition in Strelau's system.

The actual relationships between questionnaire scales designed to measure the Strelau (1983) constructs and sensation seeking scales have been recently summarized (Zuckerman, in press). Sensation seeking is positively related to strength of excitation and mobility and negatively related to strength of inhibition. Similar but stronger correlations were found using a revised and psychometrically improved temperament inventory for the Pavlovian constructs. The correlations of two tests like the SSS and the Strelau Temperament Inventory (STI) are instructive, but broader analyses including other measures of temperament and personality are necessary to properly define the place of the subscales within dimensions of personality.

Angleitner and Ostendorf (in press) performed a factor analysis using the scales of the SSS, the Strelau Temperament Inventory—Revised (STI–R) (Ruch, Angleitner, & Strelau, 1991), and the Dimensions of Temperament Survey (DOTS; Windle & Lerner, 1986). The analysis was based on the Thomas and Chess (1977) traits of temperament, Buss and Plomin's (1975) EASI scales of temperament, and the NEO five-factor scales developed by Costa and McCrae (1992). When they included the NEO personality scales, Angleitner and Ostendorf found a dimension identified as "openness to experience" that included the following: openness to experience, three of the four sensation seeking subscales, a strong loading for the mobility scale and a moderate loading for the strength of excitation scale from the STI, two of the impulsivity scales from the EASI (short decision time and sensation seeking), strong loadings for approach/withdrawal and flexibility/rigidity scales, and moderate negative loadings from the rhythmicity (regularity of habits) scales of the DOTS. When the NEO scales were excluded, the results for this factor were similar, except that now it included all of the SSS subscales, and strength of excitation had a higher loading than mobility on the factor. Rhythmicity scales formed a separate factor.

These results show that sensation seeking and at least some forms of impulsivity (particularly of the narrow kind) are most clearly identifiable with strength of excitation, approach, and flexibility in the temperament dimensions. Although narrow impulsivity and mobility demonstrate some overlap with the extraversion factor, and disinhibition sensation seeking tends to load negatively on the agreeableness factor, the results provide evidence for a primary factor of temperament and personality in which impulsivity and sensation seeking are components; this factor is relatively independent of extraversion and neuroticism factors. Eysenck's P scale provides a strong marker for the factor. The factor is also associated with antisocial or asocial tendencies and, at the opposite pole, responsibility, restraint, and socialization.

PSYCHOBIOLOGY OF IMPULSIVITY AND SENSATION SEEKING

Genetics

The association of the traits of sensation seeking and impulsivity could be based on some common biological determinants. H. J. Eysenck (chapter 4 this volume) describes the results on the heritability of impulsivity and sensation seeking on the basis of the questionnaire scales given to large heterogenous adult twin samples. Buss and Plomin (1975) found mixed evidence for the inheritance of impulsivity, using parental ratings of the trait, with strong heritability for boys on some aspects of the broad trait but less for girls. The studies reviewed by H. J. Eysenck (1983; chapter 4, this volume) show somewhat stronger heritability for sensation seeking than for impulsivity, but both traits show significant heritability. Since heritable aspects of impulsivity are amply covered in Eysenck's chapter (chapter 4, this volume), I refer the reader to that chapter for more in-depth discussion. In addition to genetic evidence, there are several physiological and biological areas important to impulsivity that will now be discussed.

Neurology and Neuropsychology

In the 1950s, there was a great deal of interest in psychosurgery for the amelioration of severe and chronic mental illness, particularly schizophrenia. Local ablations were also done for the relief of focal epilepsy. Veterans with brain lesions produced by war wounds were also studied with the hope of learning something about brain–behavior relationships. Many of these studies were poorly controlled, and few used precise assessment methods (Zuckerman, 1991). However, they offered some interesting and initial observations regarding behavioral changes associated with acquired cerebral dysfunction.

Patients in whom the connections between the orbitofrontal cortex and subcortical areas in the limbic system were severed by undercutting the cortex showed reductions in anxiety and depression if these affectual states were elevated prior to the operation. However, these operations also increased impulsivity. Miller (1985) and Miller and Milner (1985), using more precise laboratory techniques, found that patients with right frontal and temporal lobe lesions did more impulsive guessing in a recognition task. Grafman, Vance, Weingarten, Salazar, and Amin (1986) reported that veterans with brain wounds in right or bilateral orbitofrontal areas had higher scores on Eysenck's P scale, a marker for the ImpUSS dimension, as previously discussed. McIntyre, Pritchard, and Lombroso (1976) found that epileptics with right temporal lobe foci tended to respond more quickly than controls on the MFFT, a behavioral measure of impulsivity (see chapter 14 by Fink & McCown, this volume, for a description of the MFFT).

These studies suggest that both frontal and temporal lobes, particularly orbitofrontal areas and the right hemisphere, are involved in the control of behavior and that lesions or impairments in these areas tend to disinhibit behavior, resulting in characteristic impulsivity. These areas of the cortex are immediate sources of cortical input into what Gray (1982) has described as the septohippocampal Behavioral Inhibition System. Septal lesions in rats produce animals that are very responsive to reward, but that have little capacity to restrain behavior in pursuit of the reward. They also show difficulty in inhibiting responses associated with punishment, and demonstrate enhanced stimulation seeking behavior. Gorenstein and Newman (1980) associated this "septal syndrome" with the "disinhibitory psychopathology" characteristic of humans with antisocial personality disorders.

Gray (1987) conceived of the septohippocampal system as most active in persons with strong anxiety and least active in persons characterized by psychopathy (antisocial personality). However, the problem with this idea of a bipolar trait, running from anxiety at one extreme to psychopathy at the other, is that among normal populations anxiety and psychopathic tendencies (P-ImpUSS) constitute two independent dimensions of personality (Zuckerman et al., 1988, 1991).

Gray (1987) also postulated another dimension of personality called *impulsivity*, which he identified with an approach system. Both dimensions represent combinations of high E and P, according to Gray, but impulsivity is also associated with high N (neuroticism), whereas psychopathy is related to low N or the absence of anxiety. It may be more parsimonious to explain the two types of impulsivity in terms of one dimension related to impulsivity, sensation seeking, and psychopathy, with anxiety as an independent dimension with no intrinsic relationship to psychopathy. This definition would be more in line with the current diagnostic standards for the antisocial personality disorder from the third revised edition of the *Diagnostic and Statistical Manual of Mental Disorders* (American Psychiatric Association, 1987).

Psychopharmacology

Both impulsivity and sensation seeking traits tend to be associated with low levels of the enzyme monoamine oxidase (MAO), although the correlations are weak (generally between $-.2$ and $-.3$) and not significant in some studies (Schalling, Edman, & Asberg, 1983; Schalling, Edman, & Oreland, 1988; Zuckerman, 1991). Extraversion has also been related to low levels of MAO in some studies, but these studies used older forms of the Eysencks' questionnaires, in which the E scale measured impulsivity as well as sociability. It may have been the impulsivity component of the E scale that was related to MAO.

MAO is lower in men than in women and increases with age in both blood platelets and the brain, consistent with higher sensation seeking scores in men and a decline in sensation seeking with age. MAO has been inversely related to activity in newborn infants, assertive social activity, aggressiveness, sexuality in monkeys and humans, the use and abuse of alcohol and drugs, and antisocial behavior in humans (Zuckerman et al., 1980). Both normal and hyperactive children with low MAO make quicker decisions and commit more errors on the MFFT test of impulsivity (Shekim et al., 1984, 1986).

MAO regulates the monoamine systems in the brain by metabolizing excess neurotransmitter substance in the neuron or synaptic cleft. The type B MAO, usually measured from blood platelets, is particularly involved in the catabolism of the neurotransmitter dopamine. However, some investigators have suggested that MAO is a positive indicator of serotonergic activity.

The neurotransmitter serotonin seems to serve the function, among others, of behavioral inhibition (Cloninger, 1987; Crow, 1977; Gray, 1982, 1987; Panksepp, 1982; Soubrie, 1986; Stein, 1978; Zuckerman, 1984, 1991). These researchers generally suggest that serotonin serves the arousal function associated with the emotion of fear and anxiety. However, Panksepp (1982) views serotonin as involved in the general inhibition of all emotional systems (including fear), and Soubrie (1986) believes that serotonin serves behavioral inhibition only.

According to my model of the psychobiology of personality (Zuckerman, 1991), a weak serotonergic system is associated with both brain and behavioral disinhibition, the central feature of the P-ImpUSS dimension of personality. Gray (1987) maintained that the neurotransmitter dopamine is the pharmacological basis of the impulsivity or approach dimension (including high E and P), and in my model (Figure 3) this is also the case.

Dopamine is essential for the intrinsic reward effects of stimulation in limbic reward areas and is the source of the euphoria produced by stimulant drugs like amphetamines and cocaine (Zuckerman, 1987). It may also be the source of pleasure that high sensation seekers find in certain kinds of risky activities (Zuckerman, 1979). However, the same drugs that activate the dopamine system also activate the noradrenergic system. This system, originating in the locus coeruleus and ascending to all parts of the neocortex, seems to serve a general activation function in response to "signals of biological significance" or to intense and novel stimuli in general (Grant, Aston-Jones, & Redmond, 1988). Gray (1982, 1987) and Redmond (1987) have suggested that this system is partic-

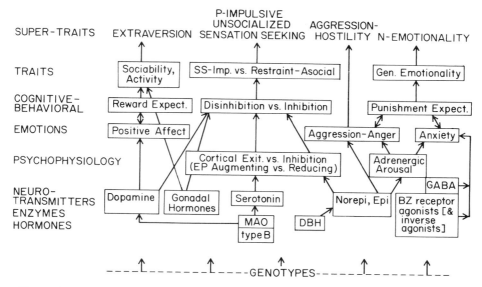

Figure 3. A psychobiological model for personality. (From *Psychobiology of Personality* (p. 407) by M. Zuckerman, 1991, New York: Cambridge University Press. Copyright 1991 by Cambridge University Press. Reprinted by permission.)

ularly connected with anxiety, serving an alarm function set off by stimuli associated with punishment. The specificity of the arousal system to one emotion is questionable in view of the fact that the system can be activated by non-noxious stimuli as well as aversive stimuli (Aston-Jones & Bloom, 1981). However, a sensitive or overarousable noradrenergic system may be a necessary component of anxiety traits, and low levels of activity in the system may characterize the ImpUSS trait, as shown in Figure 4. According to this view, the neurotransmitter systems are not independent biological traits but may have agonistic or antagonistic effects on the same behaviors. Dopamine and serotonin are known to have antagonistic effects, as in male sexual behavior, where dopamine stimulates sexual behavior and serotonin inhibits it. As shown in Figure 4, serotonin and MAO are involved in the inhibition of behavior. The type B MAO studied in primates regulates the dopamine systems, and high levels of MAO in dopamine neurons may produce inhibition by over-regulation and degradation of dopamine. Although type B MAO does not regulate serotonin neurons, it is in highest concentrations in the same brain areas that have high concentrations of serotonin.

The general arousal produced by the noradrenergic system may be a factor in anxiety arousal but may be inhibitory to approach and the P-ImpUSS trait, which results from the general behavioral tendency. The interactions among hormones, enzymes, and neurotransmitters are complex, and their mediating roles in transmitting the plan in the genotype

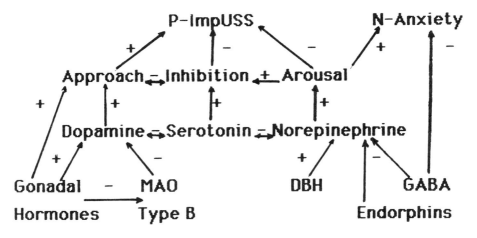

Figure 4. Specific interactions of behavioral and underlying neurotransmitter, enzyme, and hormonal functions in the traits of impulsive Unsocialized Sensation Seeking (ImpUSS) and Neuroticism–Anxiety (N-Anx).

to the behavioral trait levels are largely unknown. Other neurotransmitters, enzymes, and hormones are bound to be involved in the process.

Psychopaths (antisocial personalities) suffer from a lack of inhibition and planning, but it is not obvious that they have a strong approach or pleasure system. In fact, Cleckley (1976) noted that they do not seem to derive much enjoyment from activities such as normal sexual behavior, and it has been speculated that they seek unusual and sometimes antisocial forms of stimulation to arouse themselves to an optimal level (Quay, 1965). This would also explain why so many psychopaths abuse drugs that stimulate the catecholamine systems. These drugs may supply what nature has deprived them of, namely, rich and responsive dopamine and opiate reward systems.

SUMMARY

The trait of impulsivity is primarily a failure of inhibition and anticipation of negative consequences. The source of this deficit may lie more in the lack of inhibitory neurotransmitters than in an excess of transmitters mediating approach behavior. Temperament and heritability studies suggest that impulsivity is in part genetically determined. At the same time, evidence regarding acquired impulsivity secondary to central nervous system dysfunction suggests that cortical mediation of impulsiveness can be modified and that particular cerebral structures are localized for impulsiveness or disinhibition. Thus, on a genotypic and molar structural level, both genetics and particular structures within the central nervous system are implicated in the expression of impulsive behavior.

On a more molecular and neurochemical level, serotonergic and dopaminergic systems are related to impulsivity, and it is likely that a weak serotonergic system is associated with brain and behavior disinhibition. However, certain neurotransmitters are important in behavioral disinhibition, and others in reward systems. It is presently uncertain whether impulsivity is primarily related to a lack of inhibitory neurotransmitters or to an excess of neurotransmitters that mediate approach behavior.

Large-scale, longitudinal studies seeking to integrate genetic, neuropsychological, and neurochemical contributions to behavioral manifestations of impulsivity are presently lacking. However, a broad theoretical and practical perspective is needed to encompass the biological and behavioral complexities of impulsiveness. Impulsivity is the product of complex and interacting factors, many identified and many more unknown. As such, it represents an intriguing area for further research focused on the precise interrelationships among psychobiological variables.

REFERENCES

American Psychiatric Association. (1987). *Diagnostic and statistical manual of mental disorders* (3rd ed., rev.). Washington DC: Author.

Anderson, G., & Brown, R. I. F. (1984). Real and laboratory gambling, sensation seeking, and arousal. *British Journal of Psychology, 75,* 401–410.

Angleitner, A., & Ostendorf, F. (in press). Temperament and the Big Five factors of personality. In. C. F. Halverson, G. A. Kohnstamm, & R. P. Martin (Eds.), *The developing structure of temperament and personality from infancy to adulthood.* Hillsdale, NJ: Erlbaum.

Aston-Jones, G., & Bloom, F. E. (1981). Norepinepherine-containing neurons in behaving rats exhibit pronounced responses to nonnoxious environmental stimuli. *The Journal of Neuroscience, 8,* 887–900.

Barratt, E. S., & Patton, J. H. (1983). Impulsivity: Cognitive, behavioral, and psychophysiological correlates. In M. Zuckerman (Ed.), *Biological bases of sensation seeking, impulsivity, and anxiety* (pp. 77–116). Hillsdale, NJ: Erlbaum.

Buss, A. H. (1991). The EAS theory of temperament. In J. Strelau & A. Angleitner (Eds.), *Explorations in temperament: International perspectives on theory and measurement* (pp. 43–60). New York: Plenum Press.

Buss, A. H., & Plomin, R. (1975). *A temperament theory of personality development.* New York: Wiley.

Buss, A. H., & Plomin, R. (1984). *Temperament: Early developing personality traits.* Hillsdale, NJ: Erlbaum.

Cleckley, H. (1976). *The mask of sanity* (5th ed.). St. Louis, MO: Mosby.

Cloninger, C. R. (1987). A systematic method for clinical description and classification of personality. *Archives of General Psychiatry, 44,* 573–588.

Costa, P. T., Jr., & McCrae, R. R. (1992). Four ways five factors are basic. *Personality and Individual Differences, 13,* 861–865.

Costa, P. T., Jr., McCrae, R. R., & Dye, D. A. (1991). Facet scales for Agreeableness and Conscientiousness: A revision of the NEO personality inventory. *Personality and Individual Differences, 12,* 887–898.

Crow, T. J. (1977). Neurotransmitter-related pathways: The structure and function of central monoamine neurons. In A. N. Davidson (Ed.), *Biochemical correlates of brain structure and function* (pp. 137–174). San Diego, CA: Academic Press.

Eysenck, H. J. (1983). A biometrical–genetical analysis of impulsive and sensation seeking behavior. In M. Zuckerman (Ed.), *Biological bases of sensation seeking, impulsivity and anxiety* (pp. 1–27). Hillsdale, NJ: Erlbaum.

Eysenck, H. J., & Eysenck, M. W. (1985). *Personality and individual differences: A natural science approach.* New York: Plenum Press.

Eysenck, H. J., & Eysenck, S. B. G. (1975). *Manual of the Eysenck Personality Questionnaire (Junior and Adult).* London: Hodder & Stoughton.

Eysenck, H. J., & Eysenck, S. B. G. (1976). *Psychoticism as a dimension of personality.* New York: Crane, Russak.

Eysenck, S. B. G., & Eysenck, H. J. (1977). The place of impulsiveness in a dimensional system of personality description. *British Journal of Social and Clinical Psychology, 16,* 57–68.

Eysenck, S. B. G., & Eysenck, H. J. (1978). Impulsiveness and venturesomeness: Their position in a dimensional system of personality description. *Psychological Reports, 43,* 1247–1255.

Fowler, C. J., von Knorring, L., & Oreland, L. (1980). Platelet monoamine oxidase activity in sensation seekers. *Psychiatry Research, 3,* 273–279.

Fulker, D. W., Eysenck, S. B. G., & Zuckerman, M. (1980). The genetics of sensation seeking. *Journal of Personality Research, 14,* 261–281.

Gerbing, D. W., Ahadi, S. A., & Patton, J. H. (1987). Toward a conceptualization of impulsivity: Components across the behavioral and self-report domains. *Multivariate Behavioral Research, 22,* 357–379.

Goldberg, L. R. (1990). An alternative "description of personality": The big-five factor structure. *Journal of Personality and Social Psychology, 59,* 1216–1229.

Gorenstein, E. E., & Newman, J. P. (1980). Disinhibitory psychopathology: A new perspective and a model for research. *Psychological Review, 87,* 301–315.

Grafman, J., Vance, S. C., Weingarten, H., Salazar, A. M., & Amin, D. (1986). The effects of lateralized frontal lesions on mood regulation. *Brain, 109,* 1127–1148.

Grant, S. J., Aston-Jones, G., & Redmond, E. (1988). Responses of primate locus coeruleus neurons to simple and complex sensory stimuli. *Brain Research Bulletin, 21,* 401–410.

Gray, J. A. (1982). *The neuropsychology of anxiety: An enquiry into the functions of the septohippocampal system.* New York: Oxford University Press.

Gray, J. A. (1987). The neuropsychology of emotion and personality. In S. M. Stahl, S. D. Iverson, & E. C. Goodman (Eds.), *Cognitive neurochemistry* (pp. 171–190). Oxford, England: Oxford University Press.

Jackson, D. N. (1974). *Personality Research Form manual.* Goshen, NY: Research Psychologists Press.

Kagan, J. (1966). Reflection–impulsivity: The generality and dynamics of conceptual tempo. *Journal of Abnormal Psychology, 71,* 17–24.

Kagan, J., Reznick, J. S., Clarke, C., Snidman, N., & Garcia-Coll, C. (1984). Behavioral inhibition to the unfamiliar. *Child Development, 55,* 2212–2225.

Kagan, J., Reznick, J. S., & Snidman, N. (1988). Biological bases of childhood shyness. *Science, 240,* 167–171.

McIntyre, M., Pritchard, P. B., & Lombroso, C. T. (1976). Left and right lobe epileptics: A controlled investigation of some psychological differences. *Epilepsia, 17,* 377–386.

Miller, L. (1985). Cognitive risk-taking after frontal or temporal lobectomy: I. The synthesis of fragmental visual information. *Neuropsychologia, 23,* 359–369.

Miller, L., & Milner, B. (1985). Cognitive risk-taking after frontal or temporal lobectomy: II. The synthesis of phonemic and semantic information. *Neuropsychologia, 23,* 371–379.

Norman, W. T. (1963). Toward an adequate taxonomy of personality attributes: Replicated factor structure. *Journal of Abnormal and Social Psychology, 66,* 574–583.

Panksepp, J. (1982). Toward a general psychobiological theory of emotions. *Behavioral and Brain Sciences, 5,* 407–422.

Quay, H. C. (1965). Psychopathic personality as pathological stimulation seeking. *American Journal of Psychiatry, 122,* 180–183.

Redmond, D. E., Jr. (1987). Studies of locus coeruleus in monkeys and hypotheses for neuropsychopharmacology. In H. Y. Meltzer (Ed.), *Psychopharmacology: The third generation of progress* (pp. 967–975). New York: Raven Press.

Ruch, W., Angleitner, A., & Strelau, J. (1991). The Strelau Temperament Inventory—Revised (STI-R): Validity studies. *European Journal of Personality, 5,* 287–308.

Schalling, D., & Edman, G. (1987). *Personality and vulnerability to psychopathology: The development of the Karolinska Scales of Personality (KSP).* Stockholm, Sweden: Karolinska Institute.

Schalling, D., Edman, G., & Asberg, M. (1983). Impulsive cognitive style and the inability to tolerate boredom. In M. Zuckerman (Ed.), *Biological bases of sensation seeking, impulsivity, and anxiety* (pp. 125–147). Hillsdale, NJ: Erlbaum.

Schalling, D., Edman, G., & Oreland, L. (1988). Platelet MAO activity associated with impulsivity and aggressivity. *Personality and Individual Differences, 9*, 597–605.

Shekim, W. O., Bylund, D. B., Alexson, J., Glaser, R. D., Jones, S. B., Hodges, K., & Perdue, S. (1986). Platelet MAO and measures of attention and impulsivity in boys with attention disorder and hyperactivity. *Psychiatry Research, 18*, 179–188.

Shekim, W. O., Hodges, K., Horowitz, E., Glaser, R. D., Davis, L., & Bylund, D. B. (1984). Psychoeducational and impulsivity correlates of platelet MAO in normal children. *Psychiatry Research, 11*, 99–106.

Soubrie, P. (1986). Reconciling the role of central serotonin neurons in human and animal behavior. *Behavioral and Brain Sciences, 9*, 319–364.

Stein, L. (1978). Reward transmitters: Catecholamines and opioid peptides. In M. A. Lipton, A. DiMascio, & K. F. Killam (Eds.), *Psychopharmacology: A generation of progress* (pp. 569–581). New York: Raven Press.

Strelau, J. (1983). *Temperament, personality, activity*. San Diego, CA: Academic Press.

Thomas, A., & Chess, S. (1977). *Temperament and development*. New York: Bruner/Mazel.

Windle, M., & Lerner, R. M. (1986). Reassessing the dimensions of temperament individuality across the life span: The Revised Dimensions of Temperament Survey (DOTS-R). *Journal of Adolescent Research, 1*, 213–230.

Zuckerman, M. (1974). The sensation seeking motive. In B. A. Maher (Ed.), *Progress in experimental personality research* (Vol. 7, pp. 79–148). San Diego, CA: Academic Press.

Zuckerman, M. (1979). *Sensation seeking: Beyond the optimal level of arousal*. Hillsdale, NJ: Erlbaum.

Zuckerman, M. (1983). Sensation seeking and sports. *Personality and Individual Differences, 4*, 285–292.

Zuckerman, M. (1984). Sensation seeking: A comparative approach to a human trait. *Behavioral and Brain Sciences, 7*, 413–471.

Zuckerman, M. (1987). Biological connection between sensation seeking and drug abuse. In J. Engel, L. Oreland, D. H. Ingvar, B. Pernon, S. Rössner, & L. A. Pellborn (Eds.), *Brain reward systems and abuse* (pp. 165–176). New York: Raven Press.

Zuckerman, M. (1989). Personality in the third dimension: A psychobiological approach. *Personality and Individual Differences, 10*, 391–418.

Zuckerman, M. (1990). The psychophysiology of sensation seeking. *Journal of Personality, 58*, 313–345.

Zuckerman, M. (1991). *Psychobiology of personality*. New York: Cambridge University Press.

Zuckerman, M. (in press). *Behavioral expressions and biological bases of sensation seeking*. New York: Cambridge University Press.

Zuckerman, M., Buchsbaum, M. S., & Murphy, D. L. (1980). Sensation seeking and its biological correlates. *Psychological Bulletin, 88,* 187–214.

Zuckerman, M., Eysenck, H. J., & Eysenck, S. B. G. (1978). Sensation seeking in England and America: Cross-cultural, age, and sex comparisons. *Journal of Consulting and Clinical Psychology, 46,* 139–149.

Zuckerman, M., Kuhlman, D. M., & Camac, C. (1988). What lies beyond E and N? Factor analyses of scales believed to measure basic dimensions of personality. *Journal of Personality and Social Psychology, 54,* 96–107.

Zuckerman, M., Kuhlman, D. M., Joireman, J., Teta, P., & Kraft, M. (1993). A comparison of three structural models for personality: the Big Three, the Big Five, and the Alternative Five. *Journal of Personality and Social Psychology, 65,* 757–768.

Zuckerman, M., Kuhlman, D. M., Thornquist, M., & Kiers, H. (1991). Five (or three) robust questionnaire scale factors of personality without culture. *Personality and Individual Differences, 12,* 929–941.

6

A FAMILY THEORY OF IMPULSIVITY

LUCIANO L'ABATE

Families exert a powerful influence on human development, providing both the genetic composition and the social context for development. (Sorenson & Rutter, 1991, p. 861)

The central thesis of this chapter is that impulsive behavior is learned in the family according to a linear model of abusive/apathetic and reactive/repetitive parental or caretaking styles. These two styles result from families whose responses are typically characterized by immediacy and oppositionality. This abusively reactive style tends to produce impulsive behavior that generalizes from the family to the outside world, thus displacing itself on targets external to the family. Socially acceptable impulsivity includes hyperactivity, drivenness, and Type A (coronary-prone) behavior patterns. Socially unacceptable qualities of impulsivity include criminal behavior; addictions; and physical, sexual, and emotional abuses (L'Abate, Farrar, & Serritella, 1992). Impulsivity suggests a pattern of reacting immediately without thinking about the possible consequences of one's actions. Hence, impulsivity represents a form of discharge in which the individual has inadequate control, resulting in an inability to delay gratification or to withstand internal tensions or hurts.

Several other contributors to this volume (e.g., H. J. Eysenck, S. B. G. Eysenck, Zuckerman, and Barratt) emphasize a view or "explanation" of impulsivity from factors *within* the individual. This bottom-

93

up view attempts to identify antecedents or causes for impulsivity from within the individual on the basis of physiological or other "internal" characteristics. This focus on impulsivity as innate, or even genetically produced, behavior de-emphasizes the importance of an external context and essentially places the individual in a vacuum. The viewpoint argued in this chapter, on the other hand, is the converse of this, an explanation from the top-down position.

Although the bottom-up position maintains that the individual's physiological or characterological traits are responsible for impulsivity, the present view sees impulsivity as acquired or learned behavior. Impulsivity occurs when a family is characterized by two qualities: *immediacy*, as assessed through reactivity, which is by its very nature *oppositional*. All three of these characteristics, immediacy, reactivity, and oppositionality, are produced by a family background of an abusively apathetic, or neglectful, style in which contradictory marital and parental styles tend to produce a much greater percentage of defeats than victories. Describing these families in a systematic manner is a complex task, but it is the responsibility of a theory to accomplish such a goal.

REQUIREMENTS FOR A THEORY OF IMPULSIVITY

Before the theory of impulsivity to be discussed in this chapter is introduced, one needs to consider certain requirements for such a theory and address pretheoretical assumptions present in most theories of personality development and interpersonal competence.

Among the many potential requirements for a theory of impulsivity, at least six factors are salient. They include contextuality, relationality, verifiability, integrative function, interventional relevance, and reducibility. *Contextuality* suggests that personality development is interconnected with, not isolated or independent from, the behavior of other intimates. In interdependent relationships, every significant person influences others and is influenced by them, both positively and negatively. No member of this unit is indifferent to other members. This view of interdependency differs from traditional views that consider personality development in a vacuum. It conceptualizes interdependence as the inevitable consequence of intimate transactions within the family.

Contextuality is closely related to the second requirement for a theory of impulsivity, *relationality*. One function of a theory is to guide the selection of the most relevant units of behavior to observe, understand, predict, and control. In the past, a variety of distinctions were used to distinguish and select units of behavior: molar–molecular, proximal–distal, internal–external (or public–private), observable–inferred, and, more recently, vacuum–inferred and nonrelational–relational.

The present theory is particularly concerned with *molar-relational* behavior that is both proximal and distal, public and private, observable and inferred. This theory rejects the vacuum view of personality development that is held by most traditional theories of personality. Thus, the units of behavior of interest to the present theory are relationships among human beings and settings in which they live and function, primarily the family setting. These relationships are a function of interactions and transactions among people who share a common history and a common household. We are both the products of and the generators of intimate relationships in an infinite spiral.

Although the importance of relationships is emphasized, one cannot have a theory for individuals that is not valid or applicable to dyads and couples. Nor can one have a theory for dyads that is not valid and applicable to families. A comprehensive and valid theory is needed that is applicable to individuals, couples, and families in relation to their relevant settings. For instance, many theorists prefer intrapsychic dynamics, inferred concepts (such as self-esteem), or characterological traits as explanatory genotypes. In contrast, I argue that relational rather than inferred or internal constructs are needed to understand impulsivity. For example, mourning the loss of a loved one does not depend on levels of self-esteem or particular personality characteristics; instead, we mourn the loss of a person because of their importance to us.

Additionally, it is necessary to distinguish intimate relationships from superficial, short-lived, or trivial ones. Traditional monadic psychology and many personality theories (at least in the United States) are largely based on the psychology of the college sophomore in short-lived, contrived, and even trivial experimental situations. In other words, our knowledge base is rooted in a vacuum view of the individual, with all potential "confounds" controlled for through experimental or statistical design. Instead of such studies, we need to focus on marital, parental, and sibling relationships as the most enduring and influential ones among adults. Each of these relationships has differing degrees of impact on personality development; hence, a great deal of information can be derived from them. In order of greater to lesser degrees of influence, marriage is followed by parent–child relationships and relationships with one's family of origin, in-laws, siblings, relatives, friends, neighbors, co-workers, and, finally, strangers.

The third requirement for a theory of impulsivity is *verifiability*. If a theory is not testable, how useful can it be? Many personality and psychotherapy theories are untestable because demonstrativeness and verifiability are not criteria for professional applications. Usefulness, aesthetic appeal, and ease of dialectic applicability are the hallmarks of a profession. On the other hand, demonstrativeness and the context of justification (accountability and verifiability) are the hallmarks of science.

We need both scientific and professional criteria without assuming mutually exclusive, extreme positions in either direction (L'Abate, 1986).

A theory of impulsivity needs to be testable through models derived from the theory. Because models are theory-derived, they offer a way of testing the theory in a modest and somewhat restricted manner. Models may also operate by themselves, separate and isolated from any theory. Eventually, however, even the latter type of model will need to be reconciled with existing theoretical frameworks. Although the present theory may be completely verified, some of its models have been verified as well, and evidence from other sources will be used to support its validity.

The fourth requirement, *integration*, essentially calls for a theory to integrate past formulations. Integration will be discussed in more detail later with regard to the ERAAwC model of impulsivity.

The fifth requirement for a theory of impulsivity, *interventional relevance*, reminds us that a theory needs to be evaluated both empirically and clinically, either in diagnostic evaluations or in interventions. If a theory cannot be applied to a variety of situations, what value does it have? A theory cannot be valid in the laboratory and invalid in real-life situations. It needs to be valid in the home as well as in the hospital, the clinic, and the laboratory. The theory should also apply to primary, secondary, and tertiary prevention (L'Abate, 1990a).

The sixth requirement, *reducibility*, is met by using psychological terms that are both interactional and relational for inferred and observable relationships. Love and negotiation, the two cornerstones of the present theory, for example, imply the presence of a partner whom one can love and with whom one negotiates. Both sets of abilities are vital for individual, dyadic, and familial functioning. In fact, most family dysfunctions and impulsivity are the outcomes of deficits in one or both abilities. The impulsive individual finds it difficult to care for others and is practically unable to negotiate, unless coerced into it.

PRETHEORETICAL ASSUMPTIONS

Impulsivity is the expression of stunted and distorted personality development accompanied by inadequate interpersonal skills. Both developmental and structural aspects are necessary to understand impulsivity (L'Abate, 1976, 1985, 1986, 1990a, 1990b, in press-a).

The theory discussed in this chapter attempts to clarify the manner in which interplay among intimate interactions generates either the development or the fixation of competencies. In this theory, the intimate context, especially the family, is perceived as the basic social framework for the learning of impulsive behavior. Family structure and function are an integrative entity involving multiple dimensions interrelated in a myriad of configurations. Some are healthier than others, but all can be classified on a continuum ranging from health to dysfunction. The idea of a continuum aids in understanding that hidden or not-so-hidden resources can be found in dysfunctional families and relationships as well as in their healthy counterparts. Impulsivity is found somewhere in the middle of this continuum but tends to fall within the dysfunctional range.

Four pretheoretical assumptions usually present in theories of interpersonal competence and personality development must be addressed prior to expanding on the theory. These assumptions are (a) the experiencing–expressing continuum, ranging from the receptive obtainment, processing, and assimilation of events through observations, contact, and eventual expressive confrontation; (b) levels of interpretation, ranging from directly observable and recordable events to the least observable and, therefore, inferred and hypothetical constructs; (c) definitions of personality; and (d) styles in intimate relationships.

The Experiencing–Expressing Continuum

This continuum is described best by the ERAAwC model (L'Abate, 1986). Optimal functioning in intimate contexts relies on at least five different sets of skills. The first set of skills is Emotionality (E), or the phenomenological, receptive experience of events (input). These experiences are the focus of humanistic approaches stressed by the existential and experiential theories and interventions. We need to experience (i.e., get in touch), express, and share (i.e., exchange) feelings and emotions before we can even begin most interpersonal processes in intimate relationships. It is only after such sharing has occurred that we can determine if problem solving is required by the situation. If it is necessary, we can then rely on cognitive brainstorming of reciprocal strategies based on realistic rewards and costs.

The second set of skills is that of Rationality (R), or the logical processing of emotional events (throughput). These skills are typically embraced by psychoanalytic, rational–emotive, reality, and cognitive–behavioral theories and therapies. These skills assume that, as a result of brainstorming about options, one course of action may appear to be superior to others.

Activity (A) is the expression of behavior (output) and constitutes the third set of skills. It is most often emphasized by behavioral schools and therapies that include reinforcements and conditionings. It describes a course of action that may be put into effect and focuses on overt behavior.

The fourth set of skills involves Awareness (Aw), or the process of becoming aware of one's emotions, cognitions, and behaviors through nonverbal experiences such as touch, movement, and body awareness.

Exercises for enhancing awareness are an integral part of Eastern approaches and Gestalt therapies, and increased awareness serves a feedback function in providing information regarding the relative rewards and costs of A. Such feedback may produce eventual revisions in maintaining the same course of action. Corrective, change-oriented feedback is based on an awareness of the internal context (how we feel about the issue) and of external realities defining the issue.

The final set of skills involves Context (C), or the immediate setting (past, present, and future) that surrounds one's experience and expression, as found internally in intimate relationships where external, or environmental, manipulation is present. This skill is essential to the family therapy movement, as well as to community psychology.

The ERAAwC model is flexible and useful in various ways. First, it fulfills the requirement of being integrative, in the sense that most theoretical and therapeutic schools can be placed along a continuum of experiencing–expressing, giving a rationale to the repeated finding that no single school of therapy is "better" than another. This outcome occurs because all therapies impinge, in unique but mostly positive ways, upon different ranges of the same experiencing–expressing continuum. The classification of theoretical schools and therapeutic movements depends on the theory's location on the continuum. For example, humanism emphasizes E, psychodynamic theory stresses R, behaviorism stresses A, and Gestalt stresses Aw. In contrast, the family therapy and community psychology movements emphasize C.

The ERAAwC model is also diagnostic because it facilitates the identification and classification of relationships on the basis of which component is most emphasized. For example, in some intimate relationships one partner prefers E, whereas the other prefers R, thus producing basic and conflictual polarizations. An individual who short-circuits feelings may rely either on overly rational obsessions (R) or on impulsive behavior (A). In the latter situation, not only E but also R is automatically avoided, short-circuited, and bypassed. A couple may be polarized when the husband is overly reliant on R while his wife overrelies on E.

The ERAAwC model is also interventive, allowing one to see that successful negotiations among intimates require a process of expressing how each participant feels about emotional issues or events (E) and of processing this cognitively, as in brainstorming (R). Next, a course of action must be chosen that seems most beneficial to those involved in the decision-making process (A), and monitoring, through awareness, the long-term effects of that course of action (Aw) must occur. Finally, observation of internal and external contexts must take place in order to evaluate the outcome of the course of action, with concomitant decisions regarding whether to continue or to change the behavior (C).

Levels of Interpretation

Most theories possess explicit or implicit assumptions regarding the interpretation of behavior. For example, humanistic and behavioral theories assume only one level of behavior and make few or no assumptions about underlying causes or antecedents. The psychoanalytic school, on the other hand, makes assumptions about unconscious, preconscious, and conscious levels of functioning.

Developmental competence theory differentiates between descriptive and explanatory levels. Descriptive levels are composed of the self-presentation and phenotypical sublevels. *Self-presentation* refers to the initial, superficial impression one wishes to present in brief social situations. In the case of the impulsive individual, for example, this sublevel is often characterized by a knowledge of how to conform to short-lived, superficial situations and therefore by the ability to occasionally make a good impression. The second sublevel, labeled the *phenotypical* sublevel, represents how we interact in intimate relationships. The proof of control is found at this second sublevel, in prolonged and committed relationships, which the impulsive individual has experienced in an incomplete and distorted way. A president of a company may appear "very nice" at work but be a rigid martinet at home. One can observe and record both sublevels if one has access to the president's home and office, where comparisons between the two behaviors could be made.

The explanatory level can also be divided into two sublevels, a genotypical sublevel and the individual's developmental history. The former represents how one views oneself as defined by the continuum of likeness (Table 1), described below, which determines different styles in intimate relationships. The latter represents the transgenerational transmission of behavior from one's family of origin and its repetition in one's family of procreation. Both sublevels are inferred or hypothetical because neither is directly observable, in contrast with the two descriptive sublevels. Consistencies and inconsistencies among all four of these sublevels can be explained by a thorough knowledge of one's generational and developmental history, as well as by knowledge of one's present situation (L'Abate, 1976, 1990a, 1990b, 1992, in press-a).

Levels of Interpretation and the Experiencing–Expressing Continuum

Levels of interpretation and the continuum of experiencing–expressing are related by the fact that each component of the ERAAwC model needs to be considered at different levels of interpretation, as shown schematically in Table 1. From this perspective, impulsivity can be viewed as an extreme emphasis on immediate A in which dealing with

TABLE 1
Levels of Interpretation and the Experiencing–Expressing Continuum

Levels of interpretation	Experiencing		(Processing)		Expressing
	Emotionality	Rationality	Activity	Awareness	Context
Description					
Presentational	Verbal/nonverbal	Logical/illogical	Impulsive, balanced, compulsive, obsessional	Restricted, wide, constricted	Used/denied
Phenotypical	Approach/ Avoidance	Discharge/ delay	Abusive/apathetic, Reactive/repetitive, Conductive/creative		Used/denied
Explanation					
Genotypical	Hurt/happiness (intimacy)	Self-concept	Likeness continuum	Dialogue/ monologue	Used/denied

Historical antecedents
  Generational: family of origin, grandparental influences
  Developmental: parental practices, sibling influences, traumas
  Situational: financial, occupational, societal reverses

E, R, Aw, and C is avoided as much as possible. The emotional repertoire of the impulsive individual is limited. Painful or hurtful affect is avoided, whereas the major feelings allowed and expressed are frustration and anger when the individual does not get his or her way (L'Abate, in press-a). Rationality is limited because the temporal perspective of past, present, and future is defective. The immediate present is what counts, and the past and future are of no consequence. Awareness is limited to immediate pleasure and gratification, whereas context is denied. An in-depth explanation of this table is beyond the scope of this chapter. However, some of its contents will be explained here, and more detailed information can be found elsewhere (L'Abate, in press-a).

Definition of Personality

The definition of personality that is basic to this theory is derived from Foa' and Foa's (1974) resource exchange theory. These authors maintained that there are six classes of resources exchanged between individuals: love, status, information, services, money, and goods. These classes can be reduced to three modalities through which personality is expressed, observed, and measured. In the first modality, love and status are combined to obtain being or presence, defined as being emotionally available to oneself and to loved ones. By combining information and services, one obtains doing or performance, the second modality. Finally, by combining money and goods, one obtains having or production. Thus, personality is what a person is, what a person does, and what a person has. Doing and having can be further combined to obtain power. Although presence is nonnegotiable in that we do not negotiate how much we love someone and how important that individual is to us, power is negotiable in functional relationships. However, it is usually either not negotiated or negotiated unsuccessfully in dysfunctional relationships. In this theory, status has been changed to the attribution of importance, and love has been changed to intimacy. Although the impulsive individual uses performance for the purpose of obtaining production, she or he is unable to *be* emotionally present, available, and intimate with her- or himself and with loved ones.

However, this definition is one-sided unless these three modalities are seen in interaction with the five major settings of home, work, leisure, transit, and transitory settings. For example, presence is more relevant to the home than to work. Performance and production are more important to work and only secondary to home and other settings. Although leisure activities can occur in a variety of settings (like the company manager putting a golf ball in the office), transit settings represent all of the ways to travel from one setting to another (cars, buses, planes, airports, bus depots), whereas transitory settings represent shopping malls, stores, bars, barber shops, and beauty salons. Each setting has its peculiar task demands that are specific to it and that share little overlap with the others, even though certain characteristics of an individual may generalize from one setting to another. For example, an impulsive individual may be emotionally volatile in most settings because impulsivity has generalized from the home to other settings.

Styles in Intimate Relationships

Development through the life span can be seen from at least two perspectives. One perspective views development through emphasizing the individual, marital, and familial life cycles (L'Abate, in press-a). The other perspective views development as differentiation—that is, the level of structural complexity, specificity, and subdivision of the component parts of one's genotype. The latter perspective constitutes the explanatory level underlying the descriptive level. One's genotype, of course, is the outcome of generational, developmental, and situational events in one's life. Thus, differentiation refers to the internal self-definition, one's genotype, which in terms of the relational requirement of the theory needs to be couched in relational terms, as in the continuum of likeness derived from social comparison theory (L'Abate, 1976, 1986, in press-a).

Personality differentiation takes place early in life along a continuum of comparison, either explicit or implicit, based on the similarity/dissimilarity of close and available models. Instead of an either/or dimension, it is useful to conceptualize differentiation, including processes of imitation and identification, as the outcome of a lifelong process of comparison, conceptualized originally by social comparison theory (Suls & Miller, 1977; Suls & Wills, 1991). The process of comparison with others takes place along a dialectical dichotomy of similarity/dissimilarity differentiating along sameness and oppositeness, with the two extremes of symbiosis and autism. These six ranges produce a psychophysical, bell-shaped continuum of likeness (Table 1) ranging from symbiosis to sameness and similarity, on the one hand, and continuing from dissimilarity or differentness to oppositeness and ultimately autism or alienation, on the other (L'Abate, 1989). According to this continuum, we constantly compare ourselves with our parents, spouses, children, friends, and even our enemies. The impulsive individual is characterized by operating in the sameness–oppositeness ranges and reverting to the symbiotic and autistic ranges under stressful conditions.

Three different and distinct styles in intimate relationships can be derived from these six ranges (L'Abate, 1986, in press-a). Referred to as the ARC model, they consist of two dysfunctional and one functional style of relating. By combining symbiosis ("I am you") with alienation ("I am not, I do not exist"), one obtains the *abusive*/apathetic (AA;

neglectful) style. By combining sameness (demands for blind conformity) with oppositeness (rebellion), one obtains the reactive/repetitive (RR) style. By combining similarity with difference, one obtains the conductive/creative (CC)style (L'Abate, 1976, 1986).

In the AA style, physical, substance, sexual, and verbal abuses derive from a context of helplessness and hopelessness and may be manifest in a downward spiral of poverty, hopelessness, extreme psychopathology, and violence. The AA style leads to deteriorations in relations and break-downs in communication, resulting in experiences consisting largely of defeats. There is neither intimacy nor problem solving, except sporadically in rare, accidental, and short-lived circumstances.

The RR style is based on revengeful rebuttals and manipulatively coercive patterns of response. These response patterns can be found in 40% to 60% of most family interactions (yes–no, right–wrong, black–white, true–false). For instance, with few exceptions, most parent–child and husband–wife interactions are of a reactive nature. This style fur-nishes the normative context for repetitive sameness, with spirals of pos-itive differentiation upward (toward creativity) or negative differentiation downward (toward abuse or apathy). In this style, the ability to withstand stress and use intimacy as a buffer and defense against it is more pro-nounced than in AA. However, the outcome is still unsatisfactory because the RR style increases rather than reduces stress. It may do so to a somewhat lesser degree than in the AA style, but it is still enough to decrease the level of functioning and of coping resources in relationships. Impulsivity is the outcome of intimate relationships that are characterized by the AA style, the RR style, or both.

The CC style indicates commitment to change and improvement through various positive actions. Unlike reactors, who "explode" in in-teractive situations, conductors "keep cool" in the decision-making pro-cess. Relationships characterized by this style usually follow a plan or a "score" (hence a second meaning for "conductors") that allows those involved to think ahead, requesting and obtaining relevant information before reaching a decision or pursuing action. This style is found in relationships where coping strategies are characterized by democratic de-cision making and gainful interactions, as well as change-oriented plans. These conductive relationships are characterized by an equality of im-portance but a difference in functions. They tend to produce reciprocity and intimacy, with positively prolonged outcomes for the family, in spite of stresses and traumas. These CC relationships are more resistant and resilient to stress than are those characterized by RR or AA styles. The impulsive individual has never seen or experienced CC relationships. Consequently, she or he does not know how to behave in any style other than AA, RR, or both.

Immediacy, dichotomous thinking, impulsivity, discharge, and con-tradictory variability from one extreme of abuse to the other extreme of apathy or neglect are characteristic of the first two styles (AA and RR). In contrast, spontaneity, control, and the ability to delay are the major characteristics of the CC style. In the CC style, the individual possesses self-control and respects loved ones enough to allow them to give relevant information before responding to them. Conductors process information without getting emotionally embroiled and are therefore able to follow a score. Reactors, on the other hand, tend to explode and are unable to follow a score or plan.

Intimacy, defined as the sharing of hurts and of fears of being hurt, is most likely to be found in the CC style. Intimacy occurs sporadically in the RR style and rarely in the AA style (L'Abate, in press-a). Thus, stress is managed more successfully in the CC style because it is shared equally among various family members. The impulsive individual has never experienced intimacy and is fearful of it, especially if intimacy is defined as the sharing of hurts and of fears of being hurt. The qualities of vulnerability in being hurt by others, fallibility in hurting others, and neediness in having to rely on those very ones who have hurt or who have been hurt are completely denied. Being vulnerable, fallible, and needy would place the impulsive individual in a position of dependency that is completely unacceptable and that is therefore denied. Indeed, it is this denial of dependency that attracts the codependent individual to the impulsive one (L'Abate & Harrison, 1992). One could define the impulsive individual as a "hurt collector" in the sense that, even though hurts are denied and avoided, they are inevitably stored inside but acted out immediately (L'Abate, 1986, 1993).

According to the continuum of likeness, the AA, RR, and CC styles are transmitted developmentally from and through families of origin to individuals, as well as intergenerationally from individuals back to their families of procreation. This transmission process proceeds through transition points along stages of the family life cycle that may produce either (a) conflicts, fixations, and failures to advance or (b) the ability to advance normatively, progressively, and sequentially in development and differentiation. The impulsive individual has learned both AA and RR styles well and is unable to contemplate or learn any other style unless taught step-by-step, as discussed later in the therapeutic implications.

TOWARD A THEORY OF IMPULSIVITY

The two cornerstone assumptions of the theory discussed in this chapter involve space and time. Space is defined by distance within the dimensions of approach–avoidance. Examples of the approach dimension include both attachment to and dependency on others. Examples of the avoidance dimension include abandonments, adoptions, and rejections that lead to withdrawal and isolation. Hence, the assumption of space eventually develops into the ability to love. This ability is composed of (a) the attribution of importance to self and others (loved ones) and (b) intimacy, the ability to share hurts and fears of being hurt (L'Abate, 1986, 1993, in press-a).

Time is defined by control within the extremes of discharge–delay. Externalizations like hyperactivity, Type A behavior, drivenness, anger, and aggression are examples of discharge. Internalization like sadness, worry, obsessions, and phobias are examples of delay. Control is eventually necessary to learn negotiation skills. Impulsive behavior may vary in its approach–avoidance tendencies except in the drive to avoid painful experiences and approach pleasurable ones. However, in the realm of time the tendency is to the extreme of discharge because, by definition, impulsivity is one form of discharge and represents an inability to delay.

The Attribution of Importance: A Model of Selfhood

In the impulsive individual's family, the importance of family members is continuously discounted to enhance the one who does the discounting (L'Abate, 1985). Importance is attributed negatively to the other, to the self, or to everybody in the family through a no-self position. The first case is characterized by the belief "I am important; you are not important" and thus leads to selfishness. The second condition, in which the importance of self is discounted, is expressed as "I am not important; you are more important than I am." This thinking leads to a position of selflessness. In the no-self position, on the other hand, the primary view is "I am not important and neither are you." Hence, these attributions lead directly to learning to inflict interpersonal defeats in an either/or dichotomous fashion. Winning is accomplished at someone else's expense ("I win, you lose") or not at all ("I lose, you lose"). When importance of the self is asserted at the expense of someone else, no one is aware that a victory achieved through a parallel defeat perpetrated against other family members results in ultimate defeat.

In most families of impulsive individuals, the mother often assumes the selfless position, in which no lines are ever drawn to protect and assert one's importance. At the same time, the father or other masculine authority figure assumes the selfish position, most of the time winning at the woman's expense. The stereotypical model for the selfless mother is found in the so-called codependent individual, who will allow any abuse, addiction, criminality, and neglect as a result of the denial of self-importance. The affirmation of the husband's importance, in combination with a denial of self-importance, leads to an inability to set limits on men in general and on the husband and male children in particular (L'Abate & Harrison, 1992). The child born and raised in this kind of familial context very soon learns to follow the rules of this game. If the child is a boy, in four out of five cases he will stereotypically identify with and follow in the male authority figure's selfish path. If the child is a girl, she will tend to follow the woman's selfless path in four out of five cases. If other paths are chosen, they represent the no-self position. The selfful position ("I am important, you are important, we both must win because we are both interdependent") is unknown and unavailable for modeling in these families (L'Abate, in press-a).

This classification parallels and is isomorphic with the continuum of likeness and styles in intimate relationships reviewed earlier. The selfful position shows itself through the CC style. The selfish–selfless polarization, found mostly in conflictful couples, exhibits itself through the RR style. The AA style parallels the no-self position.

Intimacy: The Sharing of Hurts and Joys

The second component in the ability to love is intimacy, already referred to in various parts of this chapter. Intimacy is rarely experienced in families producing impulsive behavior. Although most definitions of intimacy refer to a variety of affective, cognitive, and action-based behaviors, the specific definition emphasized in the present theory involves the sharing of hurts and fears of being hurt. One needs to be "strong" enough to be weak, and weak people are typically not strong enough to acknowledge personal fallibilities. This conclusion implies that only functional or selfful individuals, couples, and parents characterized by the CC style can be intimate. They admit to having needs and allow themselves to be vulnerable or fallible. Individuals, couples, and parents characterized by RR and AA styles cannot make such acknowledgments and do not allow themselves to be intimate because they do not have the experiential repertoire to experience and express hurt feelings. Although the selfless individual, usually a woman, may admit to qualities of vulnerability, fallibility, and neediness, these admissions will be seen as signs of "weakness" and "dependency" by the selfish, impulsive male partner, who will deny any personal weakness. Therefore, he will not know how to deal with what are perceived as "weaknesses" in intimate others. Tears and crying, the most congruent expressions of hurt, when not used manipulatively, will be the trigger for impulsive anger and withdrawal. Simply put, hurts are used to produce further hurts.

The child raised in this kind of atmosphere will soon learn to model parental behavior. The boy will model after the selfish father, whereas the girl will model after the selfless mother. Gender reversals in these roles do occur, of course. However, the ratio of four to one typically prevails (L'Abate, in press-a). In single-parent families headed by women, this model is still applicable. In this circumstance, mothers are not able to set limits on rebellious and oppositional sons and may not need to set limits on daughters, who conform to them as submissively as their mothers historically submitted to masculine figures.

The Ability to Negotiate Performance and Production

In the families of impulsive individuals, being present (important and intimate) is confused and fused with the ability to negotiate. Doing (performance) and having (money and goods) are substitutes for *being*. However, neither of these resources is really negotiated because the structure and process of negotiation are either nonexistent or severely defective. The structure of negotiation that includes bargaining, decision making, and problem solving consists of two roles, authority and responsibility. Authority refers to who makes the decision, whereas responsibility indicates who implements the decision. In functional families, both roles are shared equally and flexibly by both adult caregivers. In dysfunctional families, the man either may want the authority but is not willing to carry out responsibilities, or may abdicate authority altogether. Decisions may be orchestrational, referring to once-in-a-lifetime choices such as changing jobs and moving to another town, or instrumental, referring to routine everyday choices. In functional families, these decisions are made with a minimal amount of upheaval because everybody participates in the decision-making process. In dysfunctional families, most choices or decisions are usually the source of anger, conflict, and uproars. The content of these decisions is related to the information and services allowed and shared in the house (doing, or performance) and to money and goods (having, or production).

The process of negotiation depends on the various abilities necessary to negotiate. This ability is a multiplicative function of (a) the level of functioning of the system (individual, couples, family) as defined by the ARC model (*Ill*, illness or dysfunctionality); (b) the presence and use of the various components of negotiation, which ideally should follow an ERAAwC sequence if the outcome is to be successful (*Skill*); and (c) the motivation and willingness to negotiate, which is a function of the personal priorities of the individuals involved in the negotiation (*Will*). Thus, the ability to negotiate, known as the negotiation potential, is equal to Ill $\times$ Skill $\times$ Will. Priorities may be vertical, referring to the developmental sequence of self, partner, parent, families of origin, and extended family. This includes siblings, work, friends, hobbies, and leisure. Horizontal priorities, on the other hand, refer to the relative importance of the settings previously mentioned: home, work, leisure, transit, and transitory settings.

Thus, there are four possible positions vis-à-vis presence, importance and intimacy, and power, as shown in Figure 1. On the horizontal axis

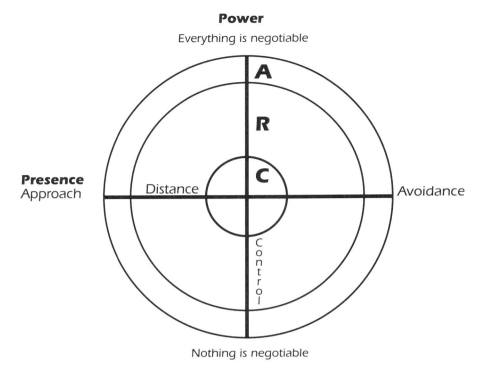

Figure 1. A model for distance and negotiation of three levels of functionality.

of intimacy, one can use the continuum of likeness and the three different ARC styles. On the vertical axis of power, one extreme exists where everything is negotiable. This extreme is illustrated by the codependent, permissive position discussed earlier, where no limits are set on self or others. This, of course, is the extreme of selflessness, and leads to depressions, dependencies, obsessions, anxieties, and phobias. The opposite extreme, selfishness, is illustrated by the authoritarian dictator or martinet who makes all decisions and expects others in the family to carry them out. This leads to character disorders, acting out, aggression, and criminal behavior. Impulsive behavior grows out of conflictual and contradictory demands and practices made by the presence of either one or two caregivers who not only contradict each other, but are contradictory within themselves.

Because of the two-valued (either/or) logic of undifferentiated systems, two possibilities are available to children raised under this kind of regime: submit quietly and develop internalizations of all of the experienced hurts, or rebel loudly and develop oppositional behavior that will promote an immediate externalization of any hurtful experience. Hyperactivity, impulsivity, and Type A personality all result from the externalization of hurts. The first possibility is usually displayed by females, whereas the second is usually found in males.

EVIDENCE TO SUPPORT THIS THEORY OF IMPULSIVITY

A theory can be supported in three different ways. First, support can be empirically independent of the theory but epistemologically and inferentially supportive of it, as in the case of research into the areas of love and negotiation. Second, support may be empirically related to the theory, as in the research generated by Foa' and Foa's (1974) resource exchange theory, from which the present theory was partially derived. Finally, evidence to support a theory can be derived directly from the theory itself (L'Abate, in press-a; L'Abate & Wagner, 1985, 1988). This section will review inferential, rather than related or derived, evidence to support a model of impulsivity as learned behavior emanating from a specific familial context. This research deals with character-disordered children and their families, where impulsivity seems to be one of the major characteristics.

There are many lines of research that directly or indirectly support the foregoing framework regarding the development of impulsivity in a family context. The first is the work of Wahler and Dumas (1989), who reviewed the literature on dysfunctional mother–child interactions that are apparently largely influenced by "stress-induced deficiency in maternal attention." Mothers who live amidst numerous stressors (i.e., poor socioeconomic status[SES], little or no emotional support from the husband, many children) do not seem to be "in synchrony" with cues offered by the children's behaviors. This limited response repertoire is probably responsible for inadequate and contradictory child-raising practices that result in extremes of either internalizations or externalizations. In line with this conclusion, Dumas and Gibson (1990) observed 47 families with conduct-disordered children and found a significant interaction between marital depression and the conduct-disordered child's behavior toward all members of the family. Children whose mothers were distressed, but not depressed, tended to be more compliant (i.e., less rebellious) toward and less aversive to their mothers than their fathers. The opposite pattern applied to children whose mothers were not distressed.

Another line of research is the work of Robins (1991), who found that conduct disorders are more common in boys than in girls. Additionally, hyperactivity was a behavioral predictor of the disorder and either the precursor or the correlate of impulsivity. Children of criminal and alcoholic parents are much more likely to develop such disorders. Parental discord, large family size, and neglectful, erratic, or severe disciplinary practices were also associated with a higher incidence of these disorders.

These children were described as irritable and uncooperative from an early age, characterized first by frequent arguments, stubbornness, and temper tantrums, with subsequent oppositional behaviors. In more aggressive children, who were characterized by fire-setting and stealing, these behaviors ultimately resulted in truancy, vandalism, and substance abuse. Tendencies toward lying were present at all ages.

Other research supporting the present model relates to the manner in which adolescents perceive and react to parental conflict. Enos and Handal (1986) found that adolescents' psychological adjustment and satisfaction with social life were significantly related to the level of perceived conflict in the family. Borrine, Handal, Brown, and Searight (1991) corroborated this conclusion with additional data, and it was further supported by Jouriles and his associates (Jouriles et al., 1991). These authors found that the relationship between parental reports of marital adjustment and child conduct problems was significantly stronger in clinic-referred families compared with parents of nonclinic children, and stronger in families of lower SES than in families of higher SES.

The significance of the marital relationship to a child's adjustment was supported by additional analyses of 1,200 mothers of 3-year-old boys (Jouriles et al., 1991). Child-rearing disagreements between marital partners predicted a greater variety of behavior problems than did global marital adjustment. Smetana, Yau, Restrepo, and Braeges (1991) added to this conclusion by specifying that parent–adolescent conflicts "generally occur over the everyday details of family life, such as doing chores, getting along with others, regulating activities, and doing homework" (p. 1007). These conflicts, however, were greater in married than in divorced families. Smetana et al. proposed that in divorced families adolescents are granted greater autonomy than in marriage-intact families. Furthermore, the adolescents are no longer subject to observing parents argue, unlike children of married couples. Taken as a whole, this line of research appears to support the view that marital and parent–child relationships are characterized by reactivity, where the greater the marital conflict, the greater the chances of maladjustment in the child. This maladjustment tends to increase either internalizing or externalizing tendencies in children of reactive marital relationships.

Yet another line of research related to maternal characteristics that affect the way the child will be raised can be found in the work of Hammen and her co-workers (Hammen, 1991; Hammen, Burge, & Adrian, 1991). She first found that

> children of unipolar depressed mothers seem to be especially subjected to negative and avoidant interaction styles shown by their mothers. The mothers displayed critical and disconfirmatory communications with their children and engaged in behaviors during the Conflict Discussion task that reflect low involvement and resistance

to task-focused resolution of the conflict. In turn, such negative and withdrawn patterns are associated with diagnoses and dysfunction in the children, both in the short term and at 6-month follow-ups, suggesting that they may be fairly typical and general styles of interaction that have enduring consequences. (Hammen, 1991, p. 171)

In spite of these conclusions, Hammen could not conclude that "depressive illness is itself the cause of dysfunctional parenting" (p. 171). However, she failed to study the manner in which the marital relationship could affect the mother's self-concept and parenting style. If the marital relationship is characterized by abusively apathetic and neglectful behavior on the husband's part, resulting in the mother's assuming most of the domestic responsibilities, including child-rearing ones, there is a greater likelihood of ineffective (AA and RR) parenting styles in both her and the husband.

Support for a conclusion of systemic rather than internal effects is found in the work of the Family Research Consortium (Patterson, 1990). A summary of this work by Radke-Yarrow (1990) concludes, "From the data we have reported here, the environments of children of depressed mothers are distinguishable from the environments of children of well mothers on each of the rearing dimensions: impairments with respect to affective relationships and control practices" (p. 182). The two factors that seem relevant here, closeness and control, are the same ones expounded by the present theory. Hops, Sherman, and Biglan (1990) found that

> among the depressed families, age was positively correlated with rates of dysphoric affect for both boys and girls . . . older girls would be more negatively influenced by having a depressed mother. . . . These results may be indicative of a restrictive range of affective behavior in these girls, *similar to that displayed by their mothers*. . . . As girls enter the adolescent phase of their lives, during which they become more independent and display more adult-like behaviors, they may be more greatly influenced, adversely or positively, by their mother's behavior and level of distress. (p. 201; emphasis added)

From this conclusion and other results, Hops et al. (1990) conceptualized "aversive" behavior as dividing into depressive and aggressive types (p. 202). Aversive behavior, as behaviorally defined, would have both abusive–apathetic and reactive–repetitive characteristics. The descriptive language may be different, but the behavior and its outcomes are similar. It would not be an unreasonable assumption to conclude that boys would tend toward externalization and aggression, two characteristics present in impulsive behavior, as the result of a selfish self-definition. Girls would tend toward internalization and depression, which would be the outcome of a selfless self-definition, as predicted from the present theory. Of course, these conclusions would hold true even after divorce, when the single mother is unable to set limits for boys as well as she can do for girls (Patterson & Forgatch, 1990). The fact that adolescents are granted greater autonomy, as dicussed above, would support the loosening of boundaries and the pushing of limits, especially in boys.

As Sigel and Blechman (1990) reflected on the work of the Consortium, they were able to include a "system paradigm" (p. 294) that articulated the possible effects of the marital relationship and parental child-rearing practices on the child's ultimate functionality. They noted that not all daughters "developed depressive symptoms" and that not all boys developed acting-out behaviors. However, they seemed to lack a sense of direction in pursuing an answer to the whole issue of variability in individual differences. Perhaps a selfhood model derived from the present theory may clarify the role of individual differences in determining why some girls act out while some boys become depressed.

An even greater systemic emphasis, which included husbands and fathers of depressed mothers of "noncompliant" boys, is found in the work of Forehand and his associates (Forehand, Lautenschlager, Faust, & Graziano, 1986; Forehand & McCombs, 1988; Forehand, Thomas, Wierson, Brody, & Fauber, 1990; Tesser, Forehand, Brody, & Long, 1989; Thomas & Forehand, 1991; Wierson, Armistead, Forehand, Thomas, & Fauber, 1990). *Noncompliance*, like *aversive*, is another term used by behavioral researchers and clinicians that is similar to, if not synonymous with, impulsive oppositeness and rebelliousness. In the present theory, this behavior would result from a likeness continuum defined by inept, contradictory, and uncritical demands for conformity (sameness) in the child by parents who are simultaneously modeling reactive behaviors with each other and demanding nonreactive behavior on the part of their child. As a result, repetitiveness is the ultimate outcome of such reactivity (L'Abate, 1986, in press-a). It would be impossible to summarize the detailed results of Forehand's impressive and extensive research. However, a significant relationship was repeatedly found among maternal depression, marital maladjustment, and noncompliant behavior in clinic-referred boys, in a manner that would be predicted from the present theory.

THERAPEUTIC AND PARATHERAPEUTIC IMPLICATIONS

The model discussed in this chapter serves to describe and prescribe necessary loving and negotiation skills informally in the therapist's office, more formally through systematically written homework assignments, or through structured enrichment programs for couples and families to be administered by paraprofessionals (L'Abate, 1990a; L'Abate & Weinstein, 1987; L'Abate & Young, 1987). The impulsive individual, of course, has never been subjected to situations where he or she could learn loving and negotiation skills. For this kind of individual, talk is cheap. It is used as a smoke screen to dissimulate, manipulate, and con the therapist, to fill the need to win at someone else's expense. Consequently, face-to-face therapy with the impulsive individual is either a power struggle culminating in defeating and frustrating outcomes or a complete waste of time. One does not learn to think before acting on the basis of face-to-face interaction with an interested professional. The outcome is going to be defeat because the impulsive individual does not know any other outcome. Consequently, I have abandoned the use of traditional "verbal" therapy with character disorders and impulsive individuals. If they want change for themselves, they have to work for it by completing in writing a Social Training program of 18 to 22 homework lessons especially designed for people with high Pd scores on the MMPI-2. This program takes place without face-to-face contact with the therapist, who gives reactions to each completed lesson through the mail. These individuals have to pay the full fee prior to being accepted for this form of treatment. If and when they finish the program, defined by completion of all of the lessons, plus three MMPI-2 administrations (before the program, after completion, and at a 3-month follow-up), they are refunded a third of the original fee.

Two cases of acting-out individuals with high Pds treated in this fashion were reported in L'Abate (1992). Other cases treated in my private practice remain unpublished. However, one case involved a middle-aged woman of considerable financial means who had been in therapy intermittently for 10 years with me, my wife, and another well-known family therapist. When it was evident that immediacy was the outstanding characteristic that could not be lowered verbally, an MMPI profile revealed a scaled score of 95 on the Pd scale. After this result, the patient agreed to undertake the Social Training program, administered through the mail, without relying on face-to-face contact. At the end of the program, her score on the same scale was within the normal scaled range.

This self-administered written approach has many advantages and relatively few disadvantages in treating impulsive individuals. In the first place, the interpersonal context is taken away. Individuals have to interact through the written medium, which forces them to think before they answer. Additionally, if they do not complete the program, the whole fee is lost, a concrete consequence to which they have agreed through the written pretreatment contract. The cost-effective nature of this approach allows the therapist to tailor costs to the financial conditions of the client and the family. Third, this approach allows a therapist to help through face-to-face contact the remainder of the family system that is amenable to such contact. In cases of dual diagnosis (Evans & Sullivan, 1990), this approach allows the therapist to treat one condition face-to-face while the other condition is treated through written homework assignments (L'Abate, 1986, 1992).

Fourth, the energy and time consumed by these individuals when treated through face-to-face contact is saved. They are usually frustrating individuals to help, producing many failures and disappointments. The therapist's energy and time can be used to assist those who can benefit by that kind of contact, usually internalizers. Fifth, this approach can take place even when the therapist is ill or otherwise unavailable, provided that written feedback is either available from a co-worker or delayed until the therapist returns. In cases where the client does not read or write, a tape recorder can be used as the medium of exchange, or a volunteer can be found who will do the reading and writing. Finally, there is never going to be a sufficient number of professionals to treat impulsivity in its multifarious manifestations at various stages of the life cycle, especially puberty and adolescence. From a preventive, paratherapeutic viewpoint, we need to rely on mass-produced approaches based on the written medium that can be supervised by middle-level professionals or paraprofessionals (L'Abate, 1986, 1990a, 1991, 1992, in press-b; L'Abate, Boyce, Fraizer, & Russ, 1992; L'Abate & Platzman, 1991).

The major disadvantage of this approach lies in the possibility of the impulsive individual trying to deceive the therapist, as shown in one of the two published case studies mentioned earlier, where the subject was too submissively and obsequiously conscientious in his writing. In spite of external controls, completing the program was not sufficient for at least this individual, who was rearrested for allegedly robbing a bank 4 years after completing the Social Training program (L'Abate, 1992). Thus, excessive submissiveness and obsequiousness may be as counterproductive as too much assertiveness. If an impulsive individual wants to con a therapist, it can be done in writing as well as verbally.

CONCLUSION

Impulsive behavior is learned in a home characterized by abusively apathetic, contradictorily neglectful, and repetitively reactive marital, parental, and sibling-to-sibling relationships. In this atmosphere, the child learns to react immediately to obtain what is desired for gratification. Any tension, displeasure, and disappointment that may decrease the sense of self-importance is denied, blocked, and avoided by jumping into thoughtless action. Self-importance is asserted at the expense of intimate others, and people are usually found where limits can be pushed. For example, the impulsive (selfish) man will usually find a submissively dependent (selfless) woman who allows him to push limits to meet selfish needs for gratification. Gratification is achieved through doing (performance) and having (production) to get what cannot be received through being, a competence that is not present in the impulsive individual's repertoire. The impulsive individual is unable to experience deep emotions and in fact avoids emotional experience by jumping immediately into action, responding to pressures and stresses through an avoidance of feelings as well as of cognitions that may in any way delay immediate discharge. Discharge is the major avenue of expression, without thought to the possible consequences and costs of an immediate response.

REFERENCES

Borrine, M. L., Handal, P. J., Brown, N. J., & Searight, H. R. (1991). Family conflict and adolescent adjustment in intact, divorced, and blended families. *Journal of Consulting and Clinical Psychology, 59*, 753–755.

Dumas, J. E., & Gibson, J. A. (1990). Behavioral correlates of maternal depressive symptomatology in conduct-disorder children: Systemic effects involving fathers and siblings. *Journal of Consulting and Clinical Psychology, 58*, 877–881.

Enos, D. M., & Handal, P. J. (1986). The relation of parental marital status and perceived family conflict to adjustment in White adolescents. *Journal of Consulting and Clinical Psychology, 54*, 820–824.

Evans, K., & Sullivan, J. M. (1990). *Dual diagnosis: Counseling the mentally ill substance abuser.* New York: Guilford Press.

Foa', U., & Foa', E. (1974). *Societal structures of the mind.* Springfield, IL: Charles C Thomas.

Forehand, R., Lautenschlager, G. J., Faust, J., & Graziano, W. G. (1986). Parent perceptions and parent–child interactions in clinic-referred children: A preliminary investigation of the effects of maternal depressive moods. *Behaviour Research and Therapy, 24*, 73–75.

Forehand, R., & McCombs, A. (1988). Unraveling the antecedent–consequence conditions in maternal depression and adolescent functioning. *Behavior Research and Therapy, 26*. 399–405.

Forehand, R., Thomas, A. M., Wierson, M., Brody, G., & Fauber, R. (1990). Role of maternal functioning and parenting skits in adolescent functioning following parental divorce. *Journal of Abnormal Psychology, 99*, 278–283.

Hammen, C. (1991). *Depression runs in families: The social context of risk and resilience in children of depressed mothers.* New York: Springer-Verlag.

Hammen, C., Burge, D., & Adrian, C. (1991). Timing of mother and child depression in a longitudinal study of children at risk. *Journal of Consulting and Clinical Psychology, 49*, 341–345.

Hops, H., Sherman, L., & Biglan, A. (1990). Maternal depression, marital discord, and children's behavior: A developmental perspective. In G. R. Patterson (Ed.), *Depression and aggression in family interaction* (pp. 185–208). Hillsdale, NJ: Erlbaum.

Jouriles, E. N., Murphy, C. M., Farris, A. M., Smith, D. A., Richters, J. E., & Waters, E. (1991). Marital adjustment, parental disagreements about child rearing, and behavior problems in boys: Increasing the specificity of the marital assessment. *Child Development, 62,* 1424–1433.

L'Abate, L. (1976). *Understanding and helping the individual in the family.* New York: Grune & Stratton.

L'Abate, L. (1985). Descriptive and explanatory levels in family therapy: Distance, defeats, and dependence. In L. L'Abate (Ed.), *Handbook of family psychology and therapy* (pp. 1218–1245). Monterey, CA: Brooks/Cole.

L'Abate, L. (1986). *Systematic family therapy.* New York: Brunner/Mazel.

L'Abate, L. (1989). *The likeness continuum in intimate relationships: Theory and research.* Unpublished manuscript.

L'Abate, L. (1990a) *Building family competence: Primary and secondary prevention strategies.* Newbury Park, CA.: Sage.

L'Abate, L. (1990b). A theory of competencies $\times$ setting interactions. *Marriage and Family Review, 11,* 253–269.

L'Abate, L. (1991). The use of writing in psychotherapy. *American Journal of Psychotherapy, 45,* 87–98.

L'Abate, L. (1992). *Programmed writing: A self-administered approach for interventions with individuals, couples and families.* Pacific Grove, CA: Brooks/Cole.

L'Abate, L. (1993). *Hurt: The fundamental, but neglected, human emotion.* Manuscript submitted for publication.

L'Abate, L. (in press-a). *A theory of personalitiy development.* New York: Wiley.

L'Abate, L. (in press-b). Writing and computer-assisted therapy. In S. R. Sauber (Ed.), *Managed mental health care: Clinical applications for practitioners.* New York: Brunner/Mazel.

L 'Abate, L., Boyce, J., Fraizer, L., & Russ, D. (1992). Programmed writing: Research in progress. *Comprehensive Mental Health Care, 2,* 45–62.

L'Abate, L., Farrar, J. E., & Serritella, D. A. (Eds.). (1992). *Handbook of differential treatments for additions.* Needham Heights, MA: Allyn & Bacon.

L'Abate, L., & Harrison, M. G. (1992). Treating codependency. In L. L'Abate, J. E. Farrar, & D. A. Serritella (Eds.), *Handbook of differential treatments for addictions* (pp. 286–307). Boston, MA: Allyn & Bacon.

L'Abate, L., & Platzman, K. (1991). The practice of programmed writing (PW) in therapy and prevention with families. *American Journal of Family Therapy, 19,* 1–10.

L'Abate, L., & Wagner, V. (1985). Theory-derived, family-oriented test batteries. In L. L'Abate (Ed.), *Handbook of family psychology and therapy* (pp. 1006–1032). Monterey, CA: Brooks/Cole.

L'Abate, L., & Wagner, V. (1988). Testing a theory of developmental competence in the family. *American Journal of Family Therapy, 16,* 23–35.

L'Abate, L., & Weinstein, S. (1987). *Structured enrichment programs for couples and families.* New York: Brunner/Mazel.

L'Abate, L., & Young, L. (1987). *Casebook of structured enrichment programs for couples and families.* New York: Brunner/Mazel.

Patterson, G. R. (Ed.). (1990). *Depression and aggression in family interaction.* Hillsdale, NJ: Erlbaum.

Patterson, G. R., & Forgatch, M. S. (1990). Initiation and maintenance of process disrupting single-mother families. In G. R. Patterson, (Ed.), *Depression and aggression in family interaction* (pp. 209–245). Hillsdale, NJ: Erlbaum.

Radke-Yarrow, M. (1990). Family environments of depressed and well parents and their children: Issues of research methods. In G. R. Patterson (Ed.), *Depression and aggression in family interaction* (pp. 169–184). Hillsdale, NJ: Erlbaum.

Robins, L. N. (1991). Conduct disorder. *Journal of Child Psychology and Psychiatry, 32,* 193–213.

Sigel, I. E., & Blechman, E. (1990). Reflections: A conceptual analysis and synthesis. In G. R. Patterson (Ed.), *Depression and aggression in family interaction* (pp. 281–313). Hillsdale, NJ: Erlbaum.

Smetana, J. G., Yau, J., Restrepo, A., & Braeges, J. L. (1991). Adolescent–parent conflict in married and divorced families. *Developmental Psychology, 27,* 1000–1010.

Sorenson, S. B., & Rutter, C. M. (1991). Transgenerational patterns of suicide attempt. *Journal of Consulting and Clinical Psychology, 59,* 861–866.

Suls, J. M., & Miller, R. L. (Eds.). (1977). *Social comparison processes: Theoretical and empirical perspectives.* Washington, DC: Hemisphere.

Suls, J. M., & Wills, T. A. (Eds.). (1991). *Social comparison: Contemporary theory and research.* Hillsdale, NJ: Erlbaum.

Tesser, A., Forehand, R., Brody, G., & Long, M. (1989). Conflict: The role of calm and angry parent–child discussion in adolescents. *Journal of Clinical and Social Psychology, 8,* 317–320.

Thomas, A. M., & Forehand, R. (1991). The relationship between paternal depressive mood and early adolescent functioning. *Journal of Family Psychology, 4,* 260–271.

Wahler, R. G., & Dumas, J. E. (1989). Attentional problems in dysfunctional mother–child interactions: An interbehavioral model. *Psychological Bulletin, 105,* 116–130.

Wierson, M., Armistead, L., Forehand, R., Thomas, A. M., & Fauber, R. (1990). Parent–adolescent conflict and stress as a parent: Are there differences between being a mother or a father? *Journal of Family Violence, 5,* 187–197.

7

A NEW LOOK AT IMPULSIVITY: HIDDEN ORDER BENEATH APPARENT CHAOS?

TERRY MARKS-TARLOW

Most behavioral scientists of the past 20 years have tended to view personality in terms of relatively stable traits that reflect temporal and situational consistency (Brody, 1988). Statistics, often from the family of general linear methods, are chosen by such researchers to measure the degree of linear prediction from one point in time or one variable to another. Unexplained variance in the data is believed to reflect an error of some sort at the level of theory, experimental design, or measurement. There is usually an underlying assumption that although complex, "behavior exhibits considerable temporal stability when measured reliably, that is by summing repeated observations and thereby reducing the error of measurement" (Anastasi, 1988, p. 566). Presumably, the more that is known about relevant variables, and the more precisely those variables are translated into measurable empirical constructs, the more accurate will be our predictions about how people think, feel, and behave.

Yet, it is apparent from other chapters in this volume that there is little agreement among researchers and clinicians about the definition, nature, and measurement of "impulsivity." As a symptom of psychological disorder, it appears rather ubiquitous, common to many *Diagnostic and Statistical Manual* (American Psychiatric Association, 1987) diagnoses. These include attention deficit disorder (Zametkin & Borcherding, 1989), borderline personality disorder (Dahl, 1990), substance abuse and dependency disorders, and other disorders of impulse control, such as kleptomania, pyromania, and intermittent explosive disorder. Sometimes (Lowe & Eldredge, chapter 10, this volume), the term *impulsivity* refers to behavioral symptoms that may include excessive gambling, eating, or expression of sexual or aggressive impulses. In the cognitive literature, impulsivity often refers to a style of distractibility, lack of attention, or self-reflection (Kagan, Rosman, Day, Albert, & Phillips, 1964). Many researchers conceptualize impulsivity as problematic or causally linked to psychopathology (Zuckerman, 1987). However, others suggest that some forms of impulsivity may actually be a sign of healthy psychological functioning (Dickman, this volume). Speculation, theory, and research on the topic have focused on certain facets of impulsivity, but little consensus exists, and a comprehensive theoretical and explanatory framework has not yet been developed.

There may be good reasons for this lack of agreement. Perhaps the wrong lens has been used for viewing impulsivity, one that causes errors in basic assumptions. Both researchers and clinicians have attempted to rein in the heterogeneity and variability, in search of stable traits, consistent correlates, and situational predictability. Yet, the very essence of impulsivity may be quite the opposite—involving instability, variability, and fundamental unpredictability. Perhaps impulsivity is a sign of a system in chaos. The purpose of this chapter is to explore this possibility on a theoretical level. A new lens is provided for examining impulsivity, one that uses an interdisciplinary paradigm that has emerged over the past 20 years—that of chaos theory, fractal geometry, and nonlinear dynamics.

Much use can come from translating the language and concepts of these new sciences to psychology in general and the domain of impulsivity in particular. Most importantly, the staggering degree of human complexity and infinite degree of human variability may be understood and quantified in a way not previously possible. In the latter part of this chapter, some applications of this new paradigm to the study of psychology and impulsivity will be elucidated. Implications for methodology will be discussed, and a broad model will be presented that includes both evolutionary and developmental components. However, some basic terms, concepts, and current areas of application of chaos theory, fractal geometry, and nonlinear dynamics will first be briefly reviewed.

WHAT IS CHAOS THEORY?

Chaos theory is a name applied to a loose collection of theories regarding the complexities of nonlinear systems (Casti, 1990). Its emergence represents the onset of a genuine paradigm revolution in science, as understood and popularized by Thomas Kuhn (1970). Although the mathematics of chaos theory are often formidable, the major concepts are not as difficult. Perhaps the two most important tenants of chaos theory are simply the following: (a) There is often order operating in data generated from complex systems, even systems that appear to behave randomly; and (b) even knowing this order, it is often impossible to make accurate predictions regarding the system's behavior.

The radical implications of these statements are clear when contrasted with those held during the past 300 years of scientific thinking. Dating back at least since the time of Newton's laws of force and motion, and articulated explicitly by Laplace, the fundamental assumption of scientific inquiry has been of an orderly, predictable world, operating mechanistically, in clockwork fashion (Davies, 1992). The underlying assumption was one of linearity: If we just know each relevant variable, the formulas that guide their interactions, and the starting position for each, we can analyze, calculate, and predict the position of every variable in the universe indefinitely into the future. This line of thinking dismissed our inability to predict highly complex phenomena like the weather, by claiming either that we do not possess enough information about critical variables or that there are so many variables and such complexity that it is impractical, although still theoretically possible, to take them all into account. What could not be predicted was relegated to "error" and viewed primarily as a technical problem of little theoretical interest.

Kuhn (1970) noted that preparadigmatic emergence is characterized by anomalous data that predates a paradigm's rise to prominence. This is certainly true regarding the development of chaos theory, although interestingly, anomalies have existed more in the domain of theoretical progress than in the data. Theoretical advances predicting chaos theory are found in the works of Hadamard, Duhem, and Poincaré, all mathematicians who made substantial contributions before the first decade of the 20th century (Ruelle, 1991). By the 1940s, difficulties with mathematical descriptions of fluid turbulence suggested insurmountable problems with explanations of dynamical systems provided by general linear models. Somehow, however, the linear model limped along grudgingly, surviving challenges from cybernetics, information theory, and other attempts to banish it to lesser prominence.

Chaos theory is generally considered to have been born in 1961 with research on long-range weather prediction by meteorologist Edward Lorenz (Gleick, 1987). While using a computer to simulate patterns of weather, Lorenz inadvertently discovered that the slightest degree of difference in the initial position of his variables eventually threw off his calculations hopelessly. This property is known as "sensitive dependence on initial conditions." It means that the slightest trace of a deviation in the starting position of any variable will eventually send the entire system careening off in a completely different direction. Lorenz proposed that our difficulty in predicting complex behavior in many systems is not just measurement error or lack of knowledge about critical variables. Instead, he concluded something radically different—that for certain kinds of chaotic phenomena, such prediction is fundamentally impossible.

To date, chaos has been studied in diverse phenomena (Briggs & Peat, 1989; Crutchfield, Farmer, Packard, & Shaw, 1986; Cvitanović, 1984; Gleick, 1987; Peterson, 1988; Ruelle, 1991; Schroeder, 1991; Stewart, 1989). Chaotic processes may be found in the irregularity of a dripping faucet, the flow of eddies in a stream of water, turbulence in the air, the pattern of smoke rising, and the spread of epidemics and forest fires. Chaos theory has also been applied to economic systems, including the stock market (Arrow & Pines, 1988), and, to a lesser extent, to the life sciences. Problems in population ecology, in particular, have been fruitfully elucidated with an extremely simple model from chaos theory, beginning with a classic paper by May (1976). Regardless of the domain involved, the same kinds of processes occur repeatedly, across different media, time frames, and size scales. Chaos has been discovered on every scale, from the macro-level movement of bodies in the solar system (Sussman & Wisdom, 1992) to the micro-level movement of wave particles in the quantum world (Gutzwiller, 1992).

One of the most important insights to emerge is that complex behavior is not necessarily the result of many interacting variables. Even a simple system with as few as three variables can, under certain conditions, demonstrate extremely complex, unpredictable behavior (Ruelle, 1991). Despite this, it is important to remember that chaos theory remains deterministic. Although fundamentally unpredictable, chaotic phenomena still display a degree of orderliness capable of complete description by mathematical formulas (Stewart, 1989). It is this last point that is the most counterintuitive and that, not surprisingly, requires a background in mathematics to actually prove, although a simple illustration is available in Gleick (1987).

Strange Attractors, Fractal Structure, and Dimensionality

To examine the orderliness underlying chaotic phenomena, I will begin with the definitions of a few key terms. Readers interested in a more detailed and comprehensive introduction to chaos theory are referred to the excellent books by Briggs and Peat (1989), Ruelle (1991), and Stewart (1989), as well as to the popular account by Gleick (1987).

The behavior of any system can be mapped as a trajectory through phase space according to critical input variables called *control parameters*. *Phase space* consists of all possible states of a system, given one's control parameters. The *trajectory* represents the movement of the system over

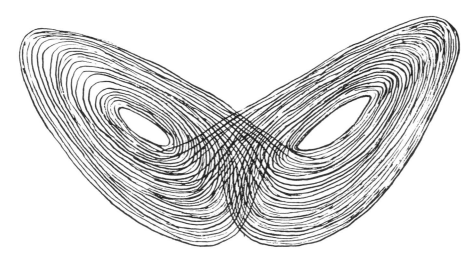

Figure 1. Lorenz attractor.

time. In the language of orthodox dynamics, every system is attracted to an end state called an *attractor*. In classical dynamics, an attractor may be a point if the system reaches a static equilibrium point. An example of this would be a swinging pendulum that gradually comes to rest. Or an attractor may consist of a circle, called a *limit cycle*, if there is periodic behavior that never reaches an end point, such as a pendulum that swings eternally in a frictionless environment. Finally, traditional attractors can take more complicated forms, like a donut shape called a *torus*, if two systems are coupled, meaning brought together, such as a pendulum driven by a motor.

Each of these attractors describes regular, stable, and predictable behavior. In contrast, a chaotic system displays a different kind of trajectory through phase space, one that moves erratically on an ever-changing path. This is called a *strange attractor*. Strange attractors come in many different shapes, depending on the nature of the system. The first attractor was named after Lorenz and was initially used to describe patterns of turbulence in weather (see Figure 1). It looks a bit like the wings of a butterfly. Strange attractors capture the deterministic orderliness underlying a chaotic system. Although the specific trajectory may not be predictable, the parameters of the overall shape are bounded and delimited.

Essentially, a strange attractor reveals the process dynamic of a chaotic system as it unfolds over time. But there is another way to visualize the irregular structure of a chaotic system: That is to analyze its fractal composition as it is frozen in an instant in time. In other words, a cross-section sliced through a chaotic attractor will reveal fractal architecture. Fractal geometry is a branch of mathematics named and described by

Figure 2. Koch snowflake (fractal).

Benoit Mandelbrot (1977). *Fractals* are dynamic shapes easily created on the computer by taking the same mathematical formula and feeding its output back in as input again and again. By assigning certain colors to ranges of numbers, this simple, recursive process is easily visualized, producing extremely complex shapes (fractals), such as the Koch snowflake pictured in Figure 2.

Every fractal may be described mathematically by a number known as its *fractal dimension*. This number captures the degree of irregularity in the shape. Fractal dimensions are fascinating because they describe the degree to which an irregular shape falls between ordinary Euclidean dimensions. For example, a straight line is represented by the Euclidean dimension of 1. A broken line, which occupies more space than a point but less than an unbroken line, may be represented by a fractional dimension of .6309. A squiggly line wending its way through a plane, as in the case of a coastline, may be represented by a fractal dimension of 1.5287. Similarly, a piece of Swiss cheese, existing somewhere between a two-dimensional plane and a three-dimensional solid, might have a fractal dimension of 2.7268. The higher the fractional number, the more complex the shape, which in nonlinear fractals such as the Mandelbrot set (see Figure 3) indicates a greater degree of randomness or disorderliness. Despite being bounded shapes, fractals can be infinitely complex.

Fractals exhibit a remarkable property called *self-similarity*. This means that a small section of a fractal that is magnified will eventually

Figure 3. Mandelbrot set (fractal).

reproduce the whole shape, either identical to the original or approximately so, depending on whether any degree of randomness is introduced into the formula. Fractals are very useful for modeling the kinds of complex, irregular shapes we see around us in everyday life, both inorganic and organic (Jurgens, Peitgen, & Saupe, 1990; Sander, 1987; Schroeder, 1991). The shapes of clouds and mountains, the irregularity of coastlines or electrical discharges, and the growth of some crystals are all fractal. Each of these shapes displays the property of self-similarity across different size or time scales, also known as *scale invariance*. Fractals are particularly likely to occur at boundary points, the nonlinear remnant of chaotic forces shaping an environment. Fractal algorithms are especially useful for modeling the growth and evolution of dynamic processes.

In human physiology, fractal structure is evident in the branching patterns of ducts, the lungs, the small intestines, blood vessels, and the dendritic structure of some neurons (R. Abraham, 1983; Garfinkel, 1983; Glass & Mackey, 1988; Goldberger, Rigney, & West, 1990; Hao, 1984;

Figure 4. Fractual branching pattern.

Skarda & Freeman, 1987). Fractal branches or folds in organs greatly amplify the surface area available. This facilitates absorption in the intestine; distribution and collection by blood vessels, bile ducts, and bronchial tubes; and information processing by the nerves (Goldberger et al., 1990). Partly because of their redundancy and irregularity, fractal structures are robust and resistant to injury (see Figure 4).

Systems Far From Equilibrium

To understand the kinds of systems that demonstrate fractal architecture and chaotic behavior, we turn to the pioneering work of nobel laureate Ilya Prigogine and his colleagues (e.g., Nicolis & Prigogine, 1977; Prigogine & Stengers, 1984; Scott, 1991). Prigogine is a chemist interested in a certain class of chemical reactions called Belvzov-Zhabotinsky (BZ) reactions. These have a peculiar property that appears to contradict the second law of thermodynamics. This law states that the irreversible arrow of time steadily drives the universe toward a greater and greater state of disorganization, called *entropy*. In contrast, BZ reactions display spontaneous organization with increasing, not decreasing, levels of order and complexity over time. These systems tend to remain stable until they reach a critical juncture, known as a *bifurcation point*, where they change unpredictably. Prigogine (Nicolis & Prigogine, 1977; Prigogine, 1980) provided important clues to this mystery by differentiating between closed systems that exist at equilibrium conditions and open systems that exist at conditions far from equilibrium.

To understand this distinction, imagine a sealed container with two chambers, one containing gas at high pressure, the other gas at low pressure. If the partition separating the two is removed, an irreversible process occurs by which the gas molecules intermingle until they eventually reach a stable pressure that is consistent throughout the container. The system is closed, exhibiting entropy as its pressure reaches equilibrium. By contrast, imagine a system that is not sealed off from its surroundings. Because there is a constant exchange of resources with the outside environment, such a system does not tend toward a static equilibrium point. Instead, this system is open, able to dispel entropy into the environment through the exchange of matter, energy, and information. In the process, the system stabilizes itself and spontaneously builds structure through a process called *self-organization*. This open system appears to defy the law of entropy only by existing in conditions far from equilibrium. Thus, Prigogine and his colleagues concluded that this type of complex behavior is demonstrated by open systems that exist in far from equilibrium conditions.

LIVING SYSTEMS POISED ON THE EDGE OF CHAOS

It appears that these same principles apply not just to chemical processes, but to biological and social ones as well. Bak and Chen (1991) suggested that large interactive systems naturally evolve toward a critical state in which a minor event can eventually trigger catastrophic change. They speculated that such self-organized criticality could explain large fluctuations in economic markets or catastrophic changes in ecosystems, such as the disappearance of dinosaurs. Kauffman (1991, 1993) proposed that the force of natural selection, which Darwin identified as acting on random genetic mutations, may not be the sole source of order in the process of evolution. Self-organization, spontaneously arising in certain complex systems, is a kind of "antichaos" through which a higher degree of order suddenly crystallizes in a relatively less ordered system. If self-organization does play a role in genetic evolution, this hitherto unrecognized force helps demystify the appearance of nature stumbling along blindly, creating order on an ad hoc basis, against staggering odds.

Kauffman (1991, 1993) suggested that the concept of self-organization also helps illuminate how the genetic blueprint directs ontogeny in different species. *Ontogeny* is the mysterious process by which a single fertilized cell divides into two identical cells, then four, eight, and so on. Eventually, this leads to cell specialization and the development of a fully differentiated organism with many distinct cell types arranged in a highly precise and complex organization. Kauffman proposed that the genetic systems controlling the ontogeny of all species lie in a narrow, phase transition between zones of chaos and order. Freeman (Freeman, 1991; Skarda & Freeman, 1987) found aperiodic, self-organized chaos in the background brain activity during olfaction. He postulated that chaos enables the olfactory system to exist in a critical state of readiness where nerve assemblies are acutely sensitive to input, able to respond dramatically to a weak stimulus, and highly responsive to trial and error.

Although one might assume intuitively that orderliness in physiological functioning is a sign of health and chaos a sign of sickness, evidence is rapidly accumulating that this is far from the case (Garfinkel, 1983; Glass & Mackey, 1988; Goldberger et al., 1990; Scott, 1991). Many systems in our bodies depend on aperiodicity for proper functioning. For example, epilepsy represents a regularity and entrainment of brain waves that is highly dangerous. Some kinds of cardiac arrest or difficulty are preceded by a similar kind of pathological periodicity. Normal functioning in both the circulatory and nervous systems depends on a certain degree of variability and irregularity. These findings contradict the previous assumption of homeostasis in internal functioning, which, for example, would describe variations of heart rate as transient responses to a fluctuating environment. Recent research (Goldberger et al., 1990) has indicated that even in the absence of environmental fluctuations, variability in heart rate indicates neither a sick nor an aging heart, but is a hallmark of healthy functioning.

As a general principle, living systems appear poised on the edge of chaos (Goodwin & Saunders, 1989). This enables the most efficient functioning, by allowing rapid mobilization into action and rapid recovery from it. Weakly chaotic systems are best able to cope with the exigencies of an unpredictable and changing environment and to coordinate highly complex behavior. In physiological as well as perceptual functioning, living organisms exhibit the greatest plasticity, adaptability, and flexibility when they are poised at the edge of chaos. The same can be said for psychological functioning as well (F. Abraham, Abraham, & Shaw, 1991). Ego psychologists, such as Blanck and Blanck (1974), describe the neurotic individual's ego as overstructured, containing rigid defenses in need of dismantling to achieve greater flexibility and adaptability to environmental input. Rigid patterns of highly predictable behavior, thoughts, or emotions are often considered unhealthy by clinicians, a sign of pathology. Examples include highly ritualized behavior or obsessive thinking in an obsessive–compulsive individual; the act of stereotyping individuals according to preconceived ideas; and fixed cycles of emotion evident in a manic–depressive. In each case, the individual's thoughts, actions, and emotions exist in a relatively closed system, essentially unresponsive to open exchange according to particulars of the environment.

Impulsivity—A Healthy Response to a Chaotic World?

Perhaps impulsivity in the human species is essentially chaotic in nature, arising as a natural response to fundamental nonlinearity evident at every level of existence. Major life conditions and events, both external and internal, from the weather to the occurrence of puberty or death, follow an unpredictable, frequently nonlinear course. The capacity for an individual to exhibit aperiodicity behaviorally and to respond to environmental aperiodicity in kind might therefore be adaptive. Technically, chaos often arises at points of phase transition, when a system, poised at a bifurcation point, must "choose" between different process structures (e.g., Scott, 1991). As is true for various physiological and perceptual systems, a weakly chaotic behavioral system would confer maximal plasticity to the infant for responding to environmental variability. It could also serve as a "neutral" starting point for later environmental entrainment. This characterization is consistent with Schwalbe's (1991) dynamical description of living organisms as engaged in a continuous process of energy dissipation. He proposed that infants exist in an "impulse phase," with bodies that spontaneously and "blindly" dissipate energy through random movement and vocalization. This serves as a biomaterial basis for the later development of an ordered, complex, and conscious sense of self.

Whether there is measurable chaos in the early expression of impulses is an empirical question that remains to be tested. One possibility for doing so would be a multiple-point time series that plots the frequency of occurrence of some operational definition of impulsivity (e.g., the crying of an infant). The data could be analyzed for the presence of a strange attractor, plus its fractal dimension. Preliminary empirical evidence already exists (Hannah, 1990) that normal mood variation in adults is chaotic in nature, revealing the presence of both strange attractors and fractal dimensionality. For example, Hannah (1990) measured the intensity of mood variation in 3 undergraduates, using a self-rated, adjective checklist (Memorial University Mood Scale) completed every waking 15 minutes over a 2–week time period. All 3 subjects displayed similar patterns, with the data yielding low dimensional chaos and strange attractors of a Lorenzian variety.

In many ways, one intuitively considers impulsivity healthy in the infant and even the toddler. There is a clear age factor implicit in the concept of impulsivity. Whereas uncontrollable expression of the sucking impulse is applauded in the newborn babe, it is frowned on in the adult alcoholic. Only as children mature toward adulthood is impulse control expected and impulsivity considered a problem. Indeed, as much research has shown (e.g., Gordon, 1979; Klinteberg, Magnusson, & Schalling, 1989), individuals of different ages score differently on paper-and-pencil measures of impulsivity, with a general decrease in impulsive scores associated with maturity.

The Development of Impulse Control

Perhaps one could consider the development of impulse control a process of self-organization occurring in an open, selective system that exists in conditions far from equilibrium. A similar point—albeit in more biological and cognitive language—has been raised by Daruna and Barnes (chapter 2, this volume). Certainly, the infant is an open system in constant exchange with the outside world (Stern, 1985). There is intake of matter, energy, and information through the ingestion of food and physical, mental, and social stimulation. There is expulsion of matter, energy, and information through the deposit of waste products and the expression of emotions, thoughts, and actions.

Infants are selective systems, in that special feedback mechanisms, consisting of internal channels of information flow, enable them to react to the effects of their own action. Through iterative, self-organizing processes that selectively receive, amplify, and reduce information about the environment and its effects on the organism, the integrity and stability of the system is established and preserved. Far from equilibrium conditions exist because self-renewal depends critically on this dynamic flux. Appearances of steady states and stable organization are not static events, but emerge instead as dynamic processes resulting from continual interaction at physical, chemical, biological, psychological, and social levels. There is no static end point of development, with death truly the only certainty in life. However, even death is not a point of static equilibrium, in that its timing remains unpredictable, and open exchange with the environment extends long after its occurrence, in the form of decomposition.

Recent infant research (Greenspan, 1991) has suggested that the newborn's world is not quite the "booming, buzzing confusion" described by William James (1890/1950). Nevertheless, compared with adult sophistication, the infant is much less organized in every way. Neonates live at the whim of disorganized impulses and rely on a caregiver to help organize external and internal worlds in a reliable, stable, and predictable fashion (Stern, 1985). Over time, internal structure builds as this capacity for organizing both the internal and external world becomes internalized in the young child. Along these lines, theorists as diverse as Sullivan (1953) and Kernberg (1976) have suggested a spiraling process of development in which biological processes become organized into impulses, which later are organized into internal, visual images. These, in turn, are further organized by language and linked through predominant affective dispositions. This spiraling process of development eventually enables the emergence of a conscious sense of self from these early seeds.

Schwalbe (1991) has used the language and concepts of dynamical systems theory to illuminate the mechanism through which this may occur. He has described the process of autogenesis in which a mature self develops from biomaterial rudiments. Through recursive processes of feedback, the end products of experience are literally fed back into the system, serving as the starting point for the next cycle. Each level builds from that which preceded it, yet is an emergent property of the whole. Self-organization of neural networks leads to the selective capture of information by the body. This stimulates the self-organization of impulses by imagery, and later the organization of imagery by language. In this way, processes of self-organization that occur in an open, selective system (here, consisting primarily of the infant and caregiver) may eventually lead to the development of self-reflection and self-control as by-products of an increasingly conscious and ordered sense of self.

IMPULSIVITY AND PATHOLOGY

Impulsivity may be considered pathological when, for genetic, traumatic, psychological, or social reasons, self-organizational capacities of the individual do not proceed in a normal fashion. In that case, one could hypothesize that cognitive and behavioral responses would remain relatively more chaotic and unpredictable. Sensitive dependence on initial conditions might explain how the tiniest disturbances can completely disrupt attention and concentration in hyperactive children. This is consistent with the obvious difficulty with learning in some impulsive children, as discussed by Fink and McCown (chapter 14, this volume). Because of the absence of self-organized internal structure, such children may be less capable of environmental entrainment or of utilizing iterative feedback mechanisms, from which stable, ordered, and complex processes of self-reflection and self-control emerge. Although controversial, some empirical evidence does exist that impulsive individuals are less responsive to environmental feedback than are nonimpulsive individuals (see chapter 4 by H. J. Eysenck, this volume, for a discussion of this issue). Finally, this formulation provides a possible explanation for the effectiveness of cognitive behavioral treatment strategies such as self-evaluation, self-control, and self-instructional training (see Fink & McCown, chapter 14, this volume).

In adults, pathological impulsivity might represent an inability to contain the chaos of disruptive emotions, again because of the absence of the self-organized complexity of internal structure. Hannah (1990) has already established that mood itself is chaotic in nature. Some people may not be able to tolerate this inner chaos, especially if it is inconsistent with cognitive structures that define the self. In this case, the chaos of intense mood could itself be disorganizing, threatening the fragmentation of self. This might explain why strong positive, and not just negative, emotions can trigger an impulsive episode in some individuals.

Chaos and Crisis

The challenge to contain and work with internal chaos may be why crisis is such an important time for change and growth. Crisis is a time of great instability and disorganization, characterized by high emotionality, fluid defenses, and the breakdown of ordinary modes of coping (McCown & Johnson, 1993). Clinically, this is a time when individuals with addictive proclivities are most likely to express or escalate impulsive behavior. Certain periods of development, such as adolescence, are characterized by tremendous hormonal and emotional upheaval. Perhaps the instability and internal chaos of this period is why adolescents are generally more prone toward impulsive, acting-out behavior.

In the language of dynamical systems theory, people in crisis may be exhibiting a chaotic phase transition characterized by sensitive dependence on initial conditions. In crisis, a very small input can have tremendous impact on an individual, whether toward growth or deterioration. As discussed previously, chaos in dynamical systems is often a product of the same forces that create process structures and give rise to self-organization. Crises represent unpredictable bifurcation points that lead in the best of conditions to a higher degree of self-organization through the internalization of new process structures. Recent data from families in crisis support this often observed phenomena (McCown & Johnson, 1993). However, if an individual has insufficient internal structure to contain chaotic emotions, self-organizational processes may be disrupted instead, leading to regression and further breakdown of internal structure. Acute psychosis may represent an extreme case of this.

Chaos and the Addictions

In the case of the various addictions, impulsivity may indicate an attempt to avoid the disorganizing effect of chaotic emotions. Alcohol and substance abuse may represent attempts to self-medicate by blunting or eradicating disruptive emotions. Behavioral addictions to such activities as sex, gambling, or fire setting could represent attempts to dissipate energy directly into the environment through taking action. The aim of all of these behaviors would be to bypass the threat of disintegration of self-structures, should the individual be forced to hold chaotic emotions internally. This formulation is certainly consistent with common descriptions of overeaters using food to "stuff" emotion, alcoholics "drowning sorrows" with booze, or impulsively aggressive individuals "blowing off steam."

Impulsivity and Borderline Pathology

The formulation of impulsivity discussed in this chapter may provide some explanation for personality disorders characterized by high levels of impulsivity, such as borderline personality disorder. Such individuals could be said to possess a relative absence of internal structure, also evident in more "primitive" defenses, such as acting out, splitting, and somaticizing. In contrast, individuals with a greater degree of internal structure are believed to possess more sophisticated defenses, such as repression, rationalization, and intellectualization.

At the same time, individuals with borderline personalities are characterized by a high degree of cognitive, emotional, and behavioral complexity, which can challenge even the brightest and most experienced clinician (Kernberg, 1976). The relative absence of internal self-structures might appear to contradict this kind of complexity, but not if we look to chaos theory for guidance. Remember that even the simplest equation, with very few variables, can generate an extraordinary degree of complexity under conditions of chaos. Thus, one might hypothesize that the instability and impulsivity of the borderline personality are due to the continual presence of chaos in the intrapsychic organization of the individual. This in turn represents the failure of natural, self-organizational processes in the development of a self-system.

The borderline personality appears to be a system in a perpetual state of chaos. Sensitive dependence on initial conditions would explain why these individuals appear so entirely unpredictable and why they go careening off from one state to another with such abrupt changes in between. It would also explain the heightened hypersensitivity of these individuals to the most minute or seemingly insignificant therapeutic input. In the future, it might be possible to devise a test that quantifies the degree of cohesiveness of self-structures versus their degree of chaotic fragmentation. Such a measure might yield relevant strange attractors and fractional dimensionality to visually portray and quantify the self-system of any individual.

PROCESS VERSUS STRUCTURE MODELS

This perspective is consistent with the neoanalytic theory of self-psychology (e.g., Kernberg, 1976; Kohut, 1971; Stolorow, Brandchaft, & Atwood, 1987). Invariant organizing principles, plus self-object functions believed to be served by others in the environment, could each be seen as self-organizing attempts of an open, selective system in an effort to build and maintain cohesive internal self-structures. The value of a perspective informed by nonlinear dynamics and related sciences lies in the relative absence of an artificial dichotomy between a process versus a structure model of the psyche. Whereas self-psychology theory talks about the organization of experience according to certain internal structures, a more dynamical approach need not draw any distinction between structure and process, because structure emerges in self-organized fashion in the course of dynamic processes such as ongoing experience. One can discuss process or structure depending on the object of focus, much as one can talk about dynamic strange attractors or the more static fractal architecture and still be addressing the same phenomenon.

CHAOS AND CREATIVITY

Finally, a dynamical systems formulation appears to shed some light on the difference between pathological impulsivity and other forms considered both normal and healthy. Perhaps certain kinds of creativity involve a kind of impulsivity, requiring that one live on the edge of each moment with complete abandon and spontaneity. Examples include the improvisation of a comedian or musician (see Dickman, chapter 9, this volume for further discussion of this point). Such talents require immediate or impulsive response in a moment, with little forethought or planning. The behaviors may at times appear identical to pathologically impulsive behaviors, such as the comedian's "aggressive" insult. However, there is a major difference in that the self-organized, internal structures are intact in the comedian. Although the individual is choosing to act with abandon, he or she is nonetheless capable of choosing otherwise and exerting self-control should the need arise. In contrast, pathologically impulsive individuals, lacking the requisite self-organization, may be unable to choose to stop their behavior, and subsequently experience a sense of loss of control.

A second difference is that unlike pathologically impulsive people, creative individuals are not oriented toward avoiding the chaos of disruptive emotions through dissipating energy into the environment. In fact, quite the opposite is true. Extremely creative individuals often tolerate a higher degree of internal chaos and ambiguity than their noncreative counterparts. They remain in an unstable, often excruciating, transition zone, poised on the edge of chaos, awaiting the emergence of a bifurcation point into an unpredictable, new order. Perhaps such tolerance for internal instability accounts for the frequent, though controversial, association of creativity with forms of psychopathology, evident in artists such as Vincent van Gogh and Paul Gauguin (Gedo, 1983).

This is not to imply, however, that both forms of impulsivity—healthy and unhealthy—cannot exist simultaneously in the same individual, albeit in different domains. Many artists, such as Ernest Hemingway, Jackson Pollock, and Jack London have been addicted to alcohol. Others, such as Fyodor Dostoyevski and Henry Miller, have been addicted to gambling or sex. Jungian analyst Linda Leonard (1990) has suggested that a flight or fall into the hellfire of addiction intimately relates to the fire and passion of the creative process. In both, there is descent into chaos and the unknown underworld of the unconscious. In both, there is psychological death, pain, and suffering. According to Leonard, the difference is that the addict is "held hostage" by the addiction, whereas the artist chooses to go down into the unknown realm, even if the choice may feel destined. Leonard noted that in the case of addicted artists, there is a kind of "double descent" that renders their situation complicated, with a different resolution for each individual. Some, such as Tennessee Williams and Carson McCullers, can continue to create despite their addiction; others, like Eugene O'Neill and John Cheever, must give up their addiction in order to create; and still others, such as Jackson Pollock and Jack London, have died young because of their addictions. Perhaps in some creative individuals, the self-organizing structures afforded by the creative process are sufficiently resilient to withstand destabilizing forces of the addiction, whereas in others they are not.

UTILITY OF THE PARADIGM

In summary, the application of the new sciences of nonlinear dynamics, fractal geometry, and chaos theory appears to be fruitful for psychology in general and the phenomenon of impulsivity in particular. A new methodology is available that is capable of quantifying the variance, variability, and aperiodicity of human behavior. An advantage of a dynamical systems perspective is that it is nonreductive, allowing for the simultaneous consideration of multiple levels of focus, including the biological, psychological, and social. This means that perturbations in any level of organization affect what occurs on all others and that causality can operate in either direction (e.g., upward from the biological or downward from the social). The broad formulation of impulsivity in terms of chaotic and unstable dynamical systems allows for incredible heterogeneity in its manifestation. A possible mechanism of explanation is thus provided that helps to unify diverse clinical observations and empirical findings. The resulting model includes both evolutionary and developmental features, and sheds light on process versus structure models of psychological development and psychopathology. A distinction can be made between healthy and unhealthy forms of impulsivity, which helps illuminate the nature of creativity more generally. Whether or not impulsivity actually displays quantifiable, chaotic features is of course an empirical issue that remains to be tested. Hopefully, this chapter will stimulate further theorizing and research to either confirm or disconfirm the array of ideas presented here.

REFERENCES

Abraham, F., Abraham, R., & Shaw, C. (1991). *A visual introduction to dynamical systems theory for psychology.* Santa Cruz, CA: Aerial Press.

Abraham, R. (1983). Dynamical models for physiology. *American Journal of Physiology, 245,* 467–472.

American Psychiatric Association. (1987). *Diagnostic and statistical manual of mental disorders* (3rd ed., rev.). Washington, DC: Author.

Anastasi, A. (1988). *Psychological testing* (6th ed.). New York: MacMillan.

Arrow, K., & Pines, D. (Eds.). (1988). *The economy as an evolving complex system.* Redwood City, CA: Addison-Wesley.

Bak, P., & Chen, K. (1991, January). Self-organized criticality. *Scientific American,* pp. 46–53.

Blanck, G., & Blanck, R. (1974). *Ego psychology: Theory and practice.* New York: Columbia University Press.

Briggs, J., & Peat, F. (1989). *Turbulent mirror: An illustrated guide to chaos theory and the science of wholeness.* New York: Harper & Row.

Brody, N. (1988). *Personality: In search of individuality.* San Diego, CA: Academic Press.

Casti, J. (1990). *Searching for certainty: What scientists can know about the future.* New York: Morrow.

Crutchfield, J., Farmer, J., Packard, N., & Shaw, R. (1986, December). Chaos. *Scientific American,* pp. 46–57.

Cvitanović, P. (1984). *Universality in chaos.* Bristol, England: Adam Hilger.

Dahl, A. (1990). Empirical evidence for a core borderline syndrome. *Journal of Personality Disorders, 4,* 192–202.

Davies, P. (1992). *The mind of God: The scientific basis for a rational world.* New York: Simon & Schuster.

Freeman, W. (1991, February). The physiology of perception. *Scientific American,* pp. 78–85.

Garfinkel, A. (1983). A mathematics for physiology. *American Journal of Physiology, 245,* 455–466.

Gedo, J. (1983). *Portraits of the artist.* New York: Guilford Press.

Glass, L., & Mackey, M. (1988). *From clocks to chaos: The rhythms of life.* Princeton, NJ: Princeton University Press.

Gleick, J. (1987). *Chaos: Making a new science.* New York: Penguin Books.

Goldberger, A. L., Rigney, D. R., & West, B. J. (1990, February). Chaos and fractals in human physiology. *Scientific American,* pp. 43–49.

Goodwin, B., & Saunders, P. (Eds.). (1989). *Theoretical biology: Epigenetic and evolutionary order from complex systems.* Edinburgh, Scotland: Edinburgh University Press.

Gordon, M. (1979). The assessment of impulsivity and mediating behaviors in hyperactive and non-hyperactive boys. *Journal of Abnormal Child Psychology, 7*(3), 317–326.

Greenspan, S. (1991). Clinical assessment in infancy and early childhood. In J. Wiener (Ed.), *Textbook of child & adolescent psychiatry.* Washington, DC: American Psychiatric Press.

Gutzwiller, M. (1992, January). Quantum chaos. *Scientific American,* pp. 78–84.

Hannah, T. (1990). Does chaos theory have application to psychology: The example of daily mood fluctuations? *Network, 8,* 3.

Hao, B. (1984). *Chaos.* Singapore: World Scientific.

James, W. (1950). *The principles of psychology.* New York: Dover. (Original work published 1890)

Jurgens, H., Peitgen, H., & Saupe, D. (1990, August). The language of fractals. *Scientific American,* pp. 60–67.

Kagan, J., Rosman, B., Day, D., Albert, J., & Phillips, W. (1964). Information processing in the child: Significance of analytic and reflective attitudes. *Psychological Monographs, 78*(1, Whole No. 578).

Kauffman, S. (1991, August). Antichaos and adaptation. *Scientific American,* pp. 78–84.

Kauffman, S. (1993). *Origins of order: Self organization and selection in evolution.* Oxford, England: Oxford University Press.

Kernberg, O. (1976). *Borderline conditions and pathological narcissism.* New York: International Universities Press.

Klinteberg, B., Magnusson, D., & Schalling, D. (1989). Hyperactive behavior in childhood and adult impulsivity: A longitudinal study of male subjects. *Personality and Individual Differences, 10*(1), 43–50.

Kohut, H. (1971). *The analysis of self.* New York: International Universities Press.

Kuhn, T. (1970). *The structure of scientific revolutions* (2nd ed.). Chicago: University of Chicago Press.

Leonard, L. (1990). *Witness to the fire: Creativity and the veil of addiction.* Boston: Shambhala.

Mandelbrot, B. (1977). *Fractal geometry of nature.* New York: Freeman.

May, R. (1976). Simple mathematical model with very complicated dynamics. *Nature, 261,* 459–467.

McCown, W., & Johnson, J. (1993). *The treatment resistant family: A consultation/crisis intervention model.* Binghamton, NY: Haworth Press.

Nicolis, G., & Prigogine, I. (1977). *Self-organization in nonequilibrium systems.* New York: Wiley.

Peterson, I. (1988). *The mathematical tourist: Snapshots of modern mathematics.* San Francisco: Freeman.

Prigogine, I. (1980). *From being to becoming.* San Francisco: Freeman.

Prigogine, I., & Stengers, I. (1984). *Order out of chaos: Man's new dialogue with nature.* New York: Bantum Books.

Ruelle, D. (1991). *Chance and chaos.* Princeton, NJ: Princeton University Press.

Sander, L. M. (1987, January). Fractal growth. *Scientific American,* pp. 94–100.

Schroeder, M. (1991). *Fractals, chaos, power laws: Minutes from an infinite paradise.* San Francisco: Freeman.

Schwalbe, M. (1991). The autogenesis of self. *Journal for the Theory of Social Behaviour, 21,* 269–295.

Scott, G. (Ed.). (1991). *Time, rhythms, and chaos in the new dialogue with nature.* Ames, IA: Iowa State University Press.

Skarda, C. A., & Freeman, W. J. (1987). How brains make chaos in order to make sense of the world. *Behavioral and Brain Sciences, 10,* 161–195.

Stern, D. J. (1985). *The interpersonal world of the infant: A view from psychoanalysis and developmental psychology.* New York: Basic Books.

Stewart, I. (1989). *Does God play dice? The mathematics of chaos.* Cambridge, MA: Basil Blackwell.

Stolorow, R., Brandchaft, B., & Atwood, G. (1987). *Psychoanalytic treatment: An intersubjective approach.* Hillsdale, NJ: Analytic Press.

Sullivan, H. (1953). *The psychiatric interview.* New York: Norton.

Sussman, G., & Wisdom, J. (1992). Chaotic evolution of the solar system. *Science, 257,* 56–62.

Zametkin, A., & Borcherding, B. (1989). The neuropharmacology of attention-deficit hyperactivity disorder. *Annual Review of Medicine, 40,* 447–451.

Zuckerman, M. (1987). Biological connection between sensation seeking and drug abuse. In J. Engel & L. Oreland (Eds.), *Brain reward systems and substance abuse* (pp. 165–176). New York: Raven Press.

II

CURRENT RESEARCH AND SPECIAL POPULATIONS

8

The I_7: DEVELOPMENT OF A MEASURE OF IMPULSIVITY AND ITS RELATIONSHIP TO THE SUPERFACTORS OF PERSONALITY

SYBIL B. G. EYSENCK

This chapter is concerned with tracing the development of work on the construct of impulsivity and its relation to the three "superfactors" of extraversion, neuroticism, and psychoticism. Beginning with modifications of the Eysenck Personality Inventory (EPI; H. J. Eysenck & Eysenck, 1964), the continued refinement of the construct and measurement of impulsivity will be described, with particular emphasis on the psychometric validation and development of a theory-based measure of impulsivity. Research into related constructs such as delinquency, criminality, antisocial behavior, venturesomeness, and empathy will be reviewed, and the relationship between these constructs and impulsiveness will be discussed. Finally, present and future areas of research will be highlighted.

THE MEASUREMENT OF IMPULSIVITY:
THE DIMENSIONAL APPROACH

The chapter in this volume by H. J. Eysenck (chapter 4) indicated two major challenges in the measurement of impulsivity: (a) the factorial composition and complexity of the trait and (b) its relation to the major dimensions of personality, namely E (Extraversion–Introversion), N (Neuroticism–Stability), and P (Psychoticism–Conformity). These dimensions are measured by the Eysenck Personality Questionnaire (EPQ; H.J. Eysenck & Eysenck, 1975), which grew out of earlier instruments such as the EPI.

Several changes occurred in our approach to the assessment and categorization of dimensions of personality when the EPI was upgraded to the EPQ. First and foremost, the Psychoticism (P) factor was introduced (H. J. Eysenck & Eysenck, 1976; see Zuckerman, chapter 5, this volume, for a discussion of the nature of this factor). Second, the number of Lie (L) factor items was increased, thus improving the validity of this measure considerably. Third, small improvements were made to the Neuroticism (N) factor. Finally, item changes were made to the Extraversion (E) factor. These changes have substantial importance to the construct of impulsivity.

Not surprisingly, the introduction of the P factor and the subsequent changes to the E factor affected the measurement of impulsiveness. What we found when we factor analyzed our data was that some of those items on the EPI-E scale that were labeled "impulsiveness" actually loaded on the P scale, rather than on extraversion. This was not altogether unexpected, because it had always been noticeable that whereas sociability, liveliness, and activity items maintained high loadings on the E scale, the impulsiveness ones on EPI forms both A and B loaded less highly on the E scale. However, because there was no P scale for these items to identify with, the "next best thing" was the E factor, to which these items aligned on the EPI, albeit with lower loadings throughout. When P became an alternative factor for impulsiveness to identify with, some, although not all, items readily switched to P.

CONSTRUCTION OF AN IMPULSIVENESS QUESTIONNAIRE

It was at this point that we decided to look at the concept of impulsiveness more closely, with a view to establishing an Impulsiveness questionnaire to complement the EPQ. Our first attempt at this was an analysis of the EPQ items, adding three varying sets of impulsiveness items given to three different populations (S. B. G. Eysenck & Eysenck, 1977). The Impulsiveness items were changed from one study to the next according to the psychometric results from the earlier studies. Factor analyses of each study and for each gender were calculated (i.e., six factor analyses) using varimax and promax rotations (Hendrickson & White, 1964). The nature of the items led us to label them ImpulsivenessN (ImpN, where N refers to narrow), Risk-taking, Non-planning, and Liveliness. When we examined the correlations of P, E, N, and L with the four impulsiveness (ImpN, Risk-taking, Non-planning, and Liveliness) factors we had identified, the results were suggestive but not satisfactorily clear. In brief, ImpN was related (in all three studies) to P and N (positively) and to L (negatively). Risk-taking showed a positive relation to E and P. Non-planning was positively related to P and negatively to N, suggesting that high N scorers plan and high P scorers do not. Liveliness clearly correlated positively with E and negatively with N.

Relation of Impulsivity to Sensation Seeking

At this point, Marvin Zuckerman decided to spend a few months working with us at the Institute of Psychiatry, and our next question naturally concerned the position of his Sensation-Seeking Questionnaire items with respect to impulsiveness (Zuckerman, 1979). We therefore embarked on a joint study (S. Eysenck & Zuckerman, 1978) in which we combined Zuckerman's scales with our impulsiveness ones to clarify the relationships among these factors.

It became obvious that our Liveliness factor hardly belonged to impulsiveness and all of these items were dropped, being much closer factorially and contentwise to extraversion (i.e., the sociability side of extraversion). Our hierarchical factor analysis now gave us a startlingly simple result, which was a two-factor solution in which the first factor contained impulsiveness items such as *Are you an impulsive person?* or *Do you often do things suddenly?*, whereas the other factor contained risk-taking and sensation-seeking items. We therefore christened these two factors Impulsiveness and Venturesomeness and embarked on our next study to pursue the relationship of these factors to our superfactors of P, E, N, and L (H. J. Eysenck & Eysenck, 1978).

Empathy, Impulsivity, and Venturesomeness: Components of the I₅

When constructing the Impulsiveness (Imp) and Venturesomeness (Vent) scales, we were concerned that these items might be somewhat similar in content, and decided to include buffer items to relieve the monotony. However, it seemed a good idea to use a meaningful scale while we were at it, and empathy as a concept appealed to us as having possible relations to some of our factors. Hence, we introduced a scale of Empathy (Mehrabian & Epstein, 1972), giving us a 63-item questionnaire of Imp, Vent, and Empathy (Emp) items. (This was called the I_5 because we had tried several previous inventories, the I_5 being considered the best one to use in this study.)

Our concept of Imp and Vent can best be described by analogy to a driver who steers his car around a blind bend on the wrong side of the road. The driver who scores high on Imp never considers the danger he might be exposing himself to and is genuinely surprised when an accident occurs. The driver who scores high on Vent, on the other hand, considers the position carefully and decides consciously to take the risk, hoping no doubt for the "thrill" of the sensation-seeking arousal caused by what he hopes will be merely a "near miss." Thus, our P and E dimensions seemed involved, and the next series of studies occupied us with mapping out the relations between Imp, Vent, and Emp and P, E, N, and L. We also worked at improving the actual Imp, Vent, and Emp scales psychometrically.

Results of the 63-item (I_5) questionnaire study suggested that both Vent and Imp correlated positively with P and E, whereas N correlated positively with Imp and Emp and negatively with Vent. Men scored higher on Vent and lower on Emp than did women, with no difference on Imp. Reliabilities (alpha coefficients) of Imp (.85 for men and .82 for women), Vent (.79 and .78, respectively) and Emp (.65 and .64) were reasonable, although Emp was somewhat inferior to the other two. At this early stage, Imp and Vent were correlated .41 for men and .32 for women, which concerned us somewhat and was one weakness we tried to overcome in later studies.

Meanwhile, it seemed desirable to look at impulsiveness in children as well, and to that end we undertook a Junior I_5 study, this questionnaire also containing 63 items (S. B. G. Eysenck & Eysenck, 1980). Although the same three factors of Imp, Vent, and Emp appeared in the factor analysis of our items, reliabilities were somewhat lower: .78 for boys and .75 for girls for Imp, .71 and .73, respectively, for Vent, and only .54 and .68 for Emp. Interestingly, the intercorrelations with P, E, N, and L for the children began to point to a clearer solution than that obtained in the adult study: Namely, Imp correlated .37 and .34 with P in boys and girls, respectively, but only .18 and .16, respectively, for E, the strongest correlations here being with N at .44 and .40 for boys and girls, respectively. Although Vent correlated with P (.31 and .28, respectively, for boys and girls), a rather higher correlation with Vent was observed with E (.57 and .62, respectively), the correlation with N being negligible (−.07 and .02, respectively). We thus observed both similarities and differences between adult and junior I_5 questionnaire responses, and began new studies to try to clarify the situation.

Cross-Validational Studies

Our research group has always been active in cross-cultural validation of its instruments and constructs. Our work with the measurement of impulsivity is no exception. The Imp and Vent studies were taken further by a Canadian replication (Saklofske & Eysenck, 1983) in which both the I_6 (or interim inventory) and the Junior EPQ were applied to Canadian children. Reliabilities were .78 and .83 for boys and girls, respectively, on Imp, .77 and .82 for Vent, and .79 and .68 for Emp. Imp and Vent were modestly correlated at .17 and .23. Once again, the S. B. G. Eysenck (1981) results were replicated in that P and Imp were correlated .50 for boys and .53 for girls, E and Vent were correlated .59 and .64, respectively, N and Emp were correlated .32 and .20, and P and Emp were correlated −.43 and −.20). N and Imp were correlated .47 for boys and .56 for girls.

Finally, one last study on the Junior I_6 and the J.EPQ was run, and minor item scoring changes were made (S. B. G. Eysenck, Easting, & Pearson, 1984). Reliabilities and means for different ages were computed, as were intercorrelations of Imp, Vent, and Emp. Incidentally, the Imp–Vent correlations that had concerned us earlier were pleasingly low overall, at .17 for boys and .26 for girls, which does not seem overly high.

IMPULSIVITY AND CRIMINALITY: I_5 STUDIES

Meanwhile, the possibility that criminality and antisocial behavior might be involved in impulsive or venturesome behavior led us to conduct a study with delinquents using the I_5 questionnaire as well as the EPQ (S. B. G. Eysenck & McGurk, 1980). In addition to a confirmation of the structure of the Imp, Vent, and Emp factors, the following results were obtained. Reliabilities (alpha coefficients) for delinquents were .73, .67, and .63 for Imp, Vent, and Emp, respectively, while remaining at .85, .79, and, .65 for controls. Interestingly, delinquents scored significantly higher than the controls on P, E, N, L, and Imp, significantly lower on Emp, and were not differentiated on Vent. Intercorrelations still saw a .41 correlation between Imp and Vent for controls and .30 for delinquents. Moreover, although Imp was correlated with P, E, and N (and incidentally L), the highest correlation was with P (.52 for controls and .47 for delinquents), the correlation with E was .39 and .12, respectively, that with N was .38 and .24, and that with L was −.43 and −.45. Vent, however, correlated .33 and .13, respectively, with P, but more highly, .46 and .38, respectively, with E.

The picture thus emerging was one in which Imp was aligning mainly with P, somewhat with E and N, and negatively with L. Vent was largely connected to E.

As far as Emp was concerned, there was a positive (.33) relationship with N for the control group (.40 for delinquents). There was a negative (−.40) correlation with P for offenders but none (−.05) for the control group.

With promising results emerging from the study with delinquents, our interest became focused on antisocial behavior, and a study was conducted that looked at impulsiveness and antisocial behavior in normal schoolchildren (S. B. G. Eysenck, 1981). Three questionnaires were used: the Junior I_6 (77 items), the Junior EPQ (H. J. Eysenck & Eysenck, 1975), and the Anti-Social Behaviour Scale, devised originally by Gibson (1967), modified by Allsopp (1975) and Allsopp and Feldman (1974, 1976), and further adapted and used by Powell (1977).

Once again, upon factor analyzing the 77-item I_6, the same dimensions of Imp, Vent, and Emp were obtained, the scales comprising 23, 23, and 20 items, respectively. Alpha coefficient reliabilities were as before (.80 and .82 on Imp for boys and girls, respectively, .79 and .79, respectively on Vent, and .68 and .67 on Emp).

When P, E, N, L, and Anti-social Behaviour (ASB) were scored, these variables were correlated with Imp, Vent, and Emp. S. B. G. Eysenck (1981) gave a table of these intercorrelations for boys and girls. These confirm previous findings that Imp relates mainly to P (.31 and .47 for boys and girls, respectively) and only .10 and .22 for E, although .39 and .48 for N. Vent, however, correlated .60 and .67 with E and only .02 and .26 for P. Thus, a much clearer alignment of P with Imp, and E with Vent, was now apparent. Emp correlated positively with N for girls (.23) but not for boys (.12), although it correlated negatively with P for boys (−.37) as well as girls (−.29). But the largest correlations by far were with ASB for both sexes, .55 and .51 for P, .28 and .31 for E, −.65 and −.63 for L, .58 and .60 for Imp, .27 and .27 for Vent, and −.29 and −.18 for Emp, for boys and girls respectively. ASB and N were related only .14 and .27, which was an unexpected result.

THE I_7 QUESTIONNAIRE

Further work on the I_6 for adults resulted in the I_7, comprising 54 items (S. B. G. Eysenck, Pearson, Easting, & Allsopp, 1985). The scales now comprised 19 items for Imp, 16 for Vent, and 19 for Emp, and reliabilities were somewhat higher overall than those in the previous study (S. B. G. Eysenck & Eysenck, 1978), being .84 and .83 for men and women, respectively, on Imp, .85 and .84 on Vent, and .69 and .69 on Emp. The alignments of P with Imp and of E with Vent were again, as were the relationships of Emp with N (positively) and P (negatively).

The question arises as to the extent to which the revised P scale (Eysenck, Eysenck, & Barrett, 1985) and the other personality scales of the revised EPQ (EPQ-R) intercorrelate with Imp, Vent, and Emp of the I_7. In a study using the EPQ-R and the I_7, Corulla (1987) supplied the answers. Ninety-two male and 215 female students completed the EPQ-R and the I_7. Our expectation was that a similar relationship between the EPQ-R and the I_7 would obtain as between the EPQ and the I_7. This is essentially what Corulla found in that Imp correlated .51 with P, .46 with E, and .27 with N for male subjects, and .26 with P, .30 with E, and .18 with N for female subjects. Clearly, P, E, and N were all implicated in Imp, but Vent correlated .53 with E but −.14 with P for men and .35 with E and .04 with P for women, suggesting that only E is involved in Vent. Furthermore, Emp correlated positively with N (.31 for men and .36 for women) and negatively with P (−.28 for men and −.33 for women). Imp and Vent correlated .16 for men and .21 for women, which is modest and acceptable. Reliabilities (alpha coefficients) for Imp, Vent, and Emp were .82, .78, and .72, respectively, for men and .84, .78, and .71, respectively, for women. These values agree well with those of other studies using the I_7.

DIRECTION FOR FUTURE RESEARCH

Our findings have left us satisfied that Imp, Vent, and Emp are dimensions that are measurable in both adults and children. Furthermore, our psychometric work has added to the nomological net surrounding the construct of impulsivity and has helped clarify its parameters. With the level of data that we now have, it seems that Imp is a pathological factor, with mainly P and N involved. On the other hand, Vent may be regarded as relating more to E and as comprising Risk-taking and Sensation-seeking items. Both Imp and Vent factors are routinely thought of by laypersons as "impulsivity." However, they are relatively independent and represent largely different behaviors.

Furthermore, our findings demonstrate that there is a definite relation, in children and delinquents, between ASB and P, N, E (somewhat), and Imp. Future research could examine these relationships in greater detail, investigating potential gender differences and the stability of these correlations through time. These findings also suggest that a potential way to modify ASB might be to reduce levels of P and N in subjects with antisocial tendencies. This is an important observation inasmuch as ASB is notoriously difficult to modify directly by psychotherapy or other procedures.

A number of researchers are currently using the I_7 for clinical investigatory purposes, and such efforts will probably continue in the future.

For example, the I_7 is now being used in many "applied" projects involving areas as diverse as drug treatment, psychotherapy attrition, health service compliance, parenting, and academic achievement. One reason the instrument may grow in popularity is the careful attention given to its psychometric qualities. The factor structure of the I_7 appears to be quite stable and can be measured reliably. The construct appears to be valid, perhaps suggesting its appropriateness for clinical applications as well. The psychometric strengths of this instrument, as well as its theoretical underpinnings, suggest that the I_7 may find increasing use in future literature, particularly for researchers interested in modifying dysfunctional impulsive behavior.

REFERENCES

Allsopp, J. F. (1975). *Investigations into the applicability of Eysenck's theory of criminality to the anti-social behaviour of schoolchildren.* Unpublished doctoral thesis, University of London.

Allsopp, J. F., & Feldman, M. P. (1974). Extraversion, neuroticism, psychoticism and anti-social behaviour in schoolgirls. *Social Behaviour and Personality, 2,* 184–190.

Allsopp, J. F., & Feldman, M. P. (1976). Item analysis of questionnaire measures of personality and anti-social behaviour in schoolboys. *British Journal of Criminology, 16,* 337–351.

Corulla, W. J. (1987). A psychometric investigation of the Eysenck Personality Questionnaire (revised) and its relationship to the I_7 Impulsiveness Questionnaire. *Personality and Individual Differences, 8,* 651–658.

Eysenck, H. J., & Eysenck, S. B. G. (1964). Manual of the Eysenck Personality Inventory. London: University of London.

Eysenck, H. J., & Eysenck, S. B. G. (1975). Manual of the Eysenck Personality Questionnaire. London: Hodder & Stoughton.

Eysenck, H. J., & Eysenck, S. B. G. (1976). *Psychoticism as a dimension of personality.* London: Hodder & Stoughton.

Eysenck, S. B. G. (1981). Impulsiveness and antisocial behaviour in children. *Current Psychological Research, 1,* 31–37.

Eysenck, S. B. G., Easting, G., & Pearson, P. R. (1984). Age norms for impulsiveness, venturesomeness and empathy in children. *Personality and Individual Differences, 5,* 315–321.

Eysenck, S. B. G., & Eysenck, H. J. (1977). The place of impulsiveness in a dimensional system of personality description. *British Journal of Social and Clinical Psychology, 16,* 57–68.

Eysenck, S. B. G., & Eysenck, H. J. (1978). Impulsiveness and venturesomeness: Their position in a dimensional system of personality description. *Psychological Reports, 43,* 1247–1255.

Eysenck, S. B. G., & Eysenck, H. J. (1980). Impulsiveness and venturesomeness in children. *Personality and Individual Differences, 1,* 73–78.

Eysenck, S. B. G., Eysenck, H. J., & Barrett, P. (1985). A revised version of the Psychoticism scale. *Personality and Individual Differences, 6,* 21–29.

Eysenck, S. B. G., & McGurk, B. J. (1980). Impulsiveness and venturesomeness in a detention center population. *Psychological Reports, 47,* 1299–1306.

Eysenck, S. B. G., Pearson, P. R., Easting, G., & Allsopp, J. F. (1985). Age norms for impulsiveness, venturesomeness and empathy in adults. *Personality and Individual Differences, 6,* 613–619.

Eysenck, S. B. G., & Zuckerman, M. (1978). The relationship between sensation-seeking and Eysenck's dimensions of personality. *British Journal of Psychology, 69,* 483–487.

Gibson, H. B. (1967). Self-reported delinquency among schoolboys and their attitudes to the police. *British Journal of Social and Clinical Psychology, 6,* 168–173.

Hendrickson, A. E., & White, P. O. (1964). Promax: A quick method for rotation to oblique simple structure. *British Journal of Statistical Psychology, 17,* 65–70.

Mehrabian, A., & Epstein, N. (1972). A measure of emotional empathy. *Journal of Personality, 40,* 525–543.

Powell, G. E. (1977). Psychoticism and social deviancy in children. *Advances in Behavioral Research Therapy, 1,* 27–56.

Saklofske, D. H., & Eysenck, S. B. G. (1983). Impulsiveness and venturesomeness in Canadian children. *Psychological Reports, 52,* 147–152.

Zuckerman, M. (1979). *Sensation seeking: Beyond the optimal level of arousal.* Hillsdale, NJ: Erlbaum.

9

IMPULSIVITY AND INFORMATION PROCESSING

SCOTT J. DICKMAN

Impulsivity is the tendency to act with less forethought than do most individuals of equal ability and knowledge. There exists a body of research on the way impulsive individuals carry out such basic cognitive processes as stimulus encoding, visual search, retrieval from short- and long-term memory, problem solving, and motor control. This chapter will review that body of research, assess the degree to which its findings lend support to each of the major theories of impulsivity, and outline an alternate theory of impulsivity that appears to account better for this body of data than do previous theories.

Impulsivity has primarily been measured by means of self-report inventories. These inventories include the Barratt Impulsivity Scale (Barratt, 1965), the "narrow impulsivity" scale (S. B. G. Eysenck & Eysenck, 1977), and the Impulsivity subscale of the Eysenck Personality Inventory (EPI) Extraversion scale (H. J. Eysenck & Eysenck, 1965).

Impulsivity is one of two closely associated traits that make up the broad personality trait of extraversion, one of the "Big Five" personality traits (McCrae & Costa, 1987). The other component of extraversion is sociability (S. B. G. Eysenck & Eysenck, 1978; Plomin, 1976). Of these two traits, it is impulsivity that appears to account for most of the overall relationships found between extraversion and cognitive functioning (e.g., Dickman, 1985; Dickman & Meyer, 1988; Revelle, Humphreys, Simon, & Gilliland, 1980).

Recently, Dickman (1990) found evidence for the existence of two distinct types of impulsivity: functional and dysfunctional. Both types of impulsivity are characterized by a tendency to act with relatively little forethought. Functional impulsives appear to act with little forethought because they have been rewarded for such behavior. These individuals are lively, adventurous, and willing to take risks, and such characteristics appear to render them productive enough that the sheer quantity of their output compensates for its error proneness.

Dysfunctional impulsives appear to act with relatively little fore-thought in spite of the fact that such behavior generally has negative consequences for them. Although much of the research cited here does not distinguish between functional and dysfunctional impulsivity, the data suggest that the cognitive differences found in these studies are primarily a function of dysfunctional impulsivity. In the studies reviewed here, high impulsives tended to be less productive overall than low impulsives (in terms of such indices as the number of items completed correctly). Such inferior performance is more characteristic of dysfunctional impulsives than of functional impulsives. There are some studies in which the effects of impulsivity appear to be primarily due to functional impulsivity; this can be inferred from the nature of the impulsivity measure used or the overall productivity of high and low impulsives. When this is the case, it will be noted.

There are two dimensions of individual differences that have sometimes been viewed as being identical to impulsivity, which will not be reviewed here. The evidence suggests that although both of these are important dimensions of individual differences, they are different dimensions than the one tapped by the self-report impulsivity measures that have just been discussed.

Kagan and his colleagues have investigated a dimension of individual differences that he has called reflection–impulsivity (Kagan, 1966). In measuring reflection–impulsivity, Kagan and his colleagues have relied almost exclusively on the Matching Familiar Figures Test (MFFT; Kagan, Rosman, Day, Albert, & Phillips, 1964). Although MFFT performance has a number of psychological correlates (see Messer, 1976, for a review), the MFFT does not correlate with either self-report or behavioral measures of impulsivity, whereas these two types of measures do correlate with each other (e.g., Bentler & McClain, 1976). Thus, the MFFT seems to be measuring a different dimension of functioning than do self-report measures of impulsivity.

Another dimension of individual differences related to impulsivity that will not be discussed here is the work of Newman and his colleagues on syndromes of disinhibition (Bachorowski & Newman, 1985; Newman, 1987; Newman, Widom, & Nathan, 1985; Nichols & Newman, 1986; Wallace & Newman, 1990). Although Newman (1987) has suggested that syndromes of disinhibition represent a form of impulsivity, the main scale he has used to measure disinhibition is the Extraversion scale of the Eysenck Personality Questionnaire (EPQ; H. J. Eysenck & Eysenck, 1975). This scale is a revision of the EPI Extraversion scale. Because the main difference between the EPQ and EPI Extraversion scales is that the impulsivity items have largely been removed from the EPQ scale (Rocklin & Revelle, 1981), it does not seem likely that syndromes of disinhibition involve the same psychological characteristics as the ones that are tapped by self-report measures of impulsivity.

ORGANIZATION OF THIS REVIEW

The discussion here of the cognitive characteristics of impulsive individuals is organized from an information-processing perspective. Rather than simply compare the performance of high and low impulsives on particular tasks, an attempt is made to infer the specific cognitive processes in which these individuals differ. As will be seen, this approach eliminates many of the apparent inconsistencies in the literature on impulsivity; such inconsistencies frequently result from the fact that tasks given the same label (e.g., "reaction time" tasks, "vigilance" tasks) tap quite different cognitive processes due to differences in the specific parameters of the tasks.

Arousal Effects

Because two of the major theories of impulsivity, those of Eysenck and Revelle, view differences in arousal as being the source of differences in impulsivity, a number of the studies reviewed here have examined the effects of changes in arousal on the performance of high and low impulsives. These studies have typically taken one of three approaches to studying the effects of arousal: In some studies, arousal has been manipulated by means of the administration of caffeine. In other studies, subjects' current arousal levels have been assessed by means of a self-report measure of arousal devised by Thayer (1967): the Activation–Deactivation Checklist.

In a third group of studies, experimental sessions have been held at different times of day because arousal levels have been shown to vary with the time of day. The interpretation of this latter group of studies is made complicated by the fact that there appear to be at least two qualitatively different types of arousal that have different circadian rhythms.

One type of arousal appears to peak in the late morning or early afternoon and to decline throughout the rest of the day. Self-report measures of arousal follow this circadian rhythm (Thayer, 1967, 1978), as does epinephrine secretion (Akerstedt, 1977; Klein, Hermann, Kuklinski, & Wegmann, 1977).

This type of arousal appears to have an effect on performance on a variety of cognitive tasks. Performance on these tasks is optimal in the late morning or early afternoon. These tasks include mental arithmetic (Laird, 1925), motor control (Blake, 1967), and short-term memory (Blake, 1967; Folkard & Monk, 1980). With regard to short-term memory, it appears to be the storage of information in memory, rather than the retrieval of that information, that peaks in the morning (Blake, 1971; Folkard & Monk, 1980).

One characteristic common to many of the tasks that are sensitive to this type of arousal is that they are demanding in terms of attention. This view is supported by the results of a study by Blake (1967). Blake had subjects carry out the same sort of mental arithmetic task that Laird (1925) had used but allowed them considerable practice, so that the task was no longer as attention demanding; under those circumstances, performance on the task peaked in the evening rather than in the morning.

There is evidence that caffeine increases the ability to carry out attention-demanding tasks, particularly those that require maintaining attention to the task over extended periods of time (e.g., Baker & Theologus, 1972; Hauty & Payne, 1955). Thus, it appears that one of the effects of caffeine is to increase this type of arousal.

The other type of arousal is relatively low in the morning and increases throughout the course of the day, peaking at about 8:00 P.M. Body temperature follows this circadian rhythm (Minors & Waterhouse, 1981). This type of arousal also appears to affect performance on a variety of cognitive tasks; performance on these tasks is optimal in the evening. These tasks include ones that require making simple responses to simple stimuli (e.g., Kleitman, 1963) and tasks that require the retrieval of information from semantic memory (Millar, Styles, & Wastell, 1980). Such tasks appear to be ones that require the immediate processing of information but make few demands on memory or sustained attention (see Hockey & Colquhoun, 1972, for a review of the evidence for circadian rhythms of performance on tasks of this sort).

Most of the studies that have used time of day as an index of arousal have compared high- and low-impulsive subjects' performance in the morning and evening. However, one study (Smith, Rusted, Savory, Eaton-Williams, & Hall, 1991) compared subjects' performance in the morning and afternoon. The time-of-day data from this study are difficult to interpret for two reasons: (a) First, the type of arousal that facilitates sustained attention peaks in the late morning or early afternoon, and therefore performance in morning and afternoon sessions might not have differed because the two sessions were held an equal amount of time before and after the time of peak arousal. (b) Second, the results for the afternoon session may have been confounded by the "postlunch dip," the tendency for performance to go down briefly just after lunch (Colquhoun, 1971). Because of these problems in interpreting performance changes from morning to afternoon, when the Smith et al. (1991) findings are discussed here, the discussion will focus on the data from their morning sessions.

STUDIES OF IMPULSIVITY AND INFORMATION PROCESSING

Simple Encoding

A number of studies have compared high and low impulsives' ability to carry out stimulus encoding of a highly automatized nature. Tasks that tap this type of encoding are ones in which the stimuli are either very simple (e.g., lights of different colors) or very familiar (e.g., letters of the alphabet). High and low impulsives do not appear to differ in their ability to carry out this type of encoding. Dickman (1985) used speeded classification tasks to compare the speed with which high and low impulsives could encode a variety of types of stimuli, including letters of the alphabet. Dickman found no differences between high and low impulsives in the speed of encoding of such stimuli.

Matthews, Jones, and Chamberlain (1989) had high- and low-impulsive subjects attempt to identify digits that had been visually degraded. They found no differences in performance between the two groups. Smith et al. (1991) found no differences between high and low impulsives when subjects were asked to indicate as rapidly as possible which of two lights had been turned on.

In contrast to these findings, Edman, Schalling, and Levander (1985) and Robinson and Zahn (1988) both found that high impulsives were faster on a simple encoding task similar to the one Smith et al. (1991) used. The distinction between functional and dysfunctional impulsivity may help to explain why the findings of Edman et al. (1985) differed from those of the other studies.

It was noted earlier that in most of the studies reviewed here, the association between impulsivity and information processing appears to be due to dysfunctional impulsivity. In the Edman et al. (1985) study, however, the high-impulsive subjects, although less accurate, were actually more productive than the low-impulsive subjects (in terms of the number of correct items completed per second). This suggests that the differences in performance that Edman et al. found were due to functional impulsivity. It may be that individual differences in functional impulsivity were more prominent in the Edman et al. study than in the other studies because Edman et al. used a somewhat different population. Their subjects were 15–16-year-old boys from Swedish public schools, whereas the subjects in all of the other studies reviewed here were adults. Unfortunately, the data that Robinson and Zahn (1988) provided do not permit the calculation of overall performance measures, so it is not clear whether their findings can also be attributed to functional impulsivity.

Arousal. Two of the studies on simple encoding and impulsivity have examined the effects of arousal on the performance of high and low impulsives. Matthews et al. (1989) found that self-reported arousal was associated with superior processing of visually degraded stimuli. They also found an effect of time of day; performance was better in the evening. However, neither arousal nor time of day affected high- and low-impulsive subjects differently. Smith et al. (1991) found no effect of caffeine on the speed of responding to the onset of lights for either group.

Simple Encoding Extended Over Time

Although high and low impulsives do not differ in their ability to carry out simple encoding, when they must continue to carry out such encoding over a significant period of time (e.g., for more than 30 minutes without a break), differences in performance do appear. For example, Thackray, Jones, and Touchstone (1974) had subjects press one of four keys depending on which of the numbers 1 through 4 appeared on a screen. The task lasted for 40 minutes. High impulsives were less accurate. Roessler (1973) had subjects continuously monitor three dials for a period of 50 minutes and respond to changes in those dials. High impulsives were again less accurate.

Arousal. There do not appear to be any studies that have examined the effects of increases in arousal on the performance of high and low impulsives on tasks of this sort.

Continuous Simple Encoding

Typically, in simple encoding tasks, there is an interval between trials of 1–10 seconds. Matthews et al. (1989) had subjects carry out a simple encoding task in which there were no intertrial intervals. Subjects were required to press one of five buttons depending on which of five lights went on, and, as soon as the subject responded, a new stimulus appeared. Thus, the subject was required to maintain attention to the task without a break for the entire 15 minutes that it lasted. Under these conditions, high impulsives made more errors than low impulsives.

Arousal. On the task just described, Matthews et al. (1989) found that self-reported arousal was associated with greater accuracy for low impulsives and lesser accuracy for high impulsives. For speed, there was an interaction between impulsivity, arousal, and time of day. In the morning, self-reported arousal was associated with slower performance for low impulsives, whereas in the evening, self-reported arousal was associated with slower performance for high impulsives.

Complex Encoding

Several studies have compared the performance of high and low impulsives on tasks that required the encoding of complex stimuli. The encoding of such stimuli is more demanding in terms of attention than is the encoding of simple or familiar stimuli. For example, lapses of attention during the process by which the different features of complex stimuli are integrated into a perceptual whole can result in errors in perception, so-called "illusory correlations" (Treisman & Gelade, 1980; Treisman & Schmidt, 1982). It appears that high and low impulsives differ in their encoding of complex stimuli.

Dickman (1985) required subjects to classify stimuli consisting of a number of small capital letters, arranged in such a way that the pattern they formed looked like a large capital letter. The small letters represented the "local" level of each stimulus, whereas the large letter of the alphabet represented its "global" level (see Figure 1). The subjects had to base their responses on both the letter of the alphabet that formed the local level and the letter of the alphabet that formed the global level. Although high- and low-impulsive subjects did not differ in the speed with which they could classify the stimuli according to either the local or the global level alone (a simple encoding task), when they had to classify the stimuli by integrating the information provided by the two levels, high impulsives responded more slowly.

Loo (1979) found that when subjects were asked to identify traffic signs that were embedded in a complex drawing of a street scene, high-impulsive subjects were slower. Similarly, Loo and Townsend (1977) found that high-impulsive subjects performed less well on the Embedded Figures Test (EFT; Witkin, Dyk, Faterson, Goodenough, & Karp, 1962), a task that requires subjects to identify simple geometric figures embedded in more complex ones. Loo and Townsend also found, using self-report measures, that the tendency to make decisions rapidly was associated with poor EFT performance but that adventurousness was not. Because adventurousness is a characteristic of functional impulsives but not dysfunctional impulsives, this suggests that it was dysfunctional impulsivity

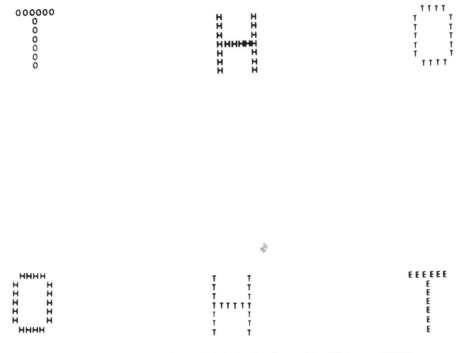

Figure 1. Examples of local/global stimuli used by Dickman (1985).

that was responsible for the overall relationship between impulsivity and the encoding of complex stimuli.

Only one study failed to find a difference between high and low impulsives on a complex encoding task. Davidson (1982), using the EFT, found no differences between high and low impulsives. It is not clear why the findings of this study differed from those of the other studies.

Arousal. There do not appear to be any studies that have compared the effects of arousal on the way in which high and low impulsives carry out complex encoding.

Visual Comparison

Several studies have compared high and low impulsives on tasks that require that they compare two relatively complex stimuli and decide whether the two stimuli were identical. Dysfunctional impulsivity appears to be unrelated to performance on such tasks, although functional impulsivity does show an association with performance.

Dickman (1990) compared the performance of high and low impulsives on a task in which pairs of complex geometric figures made up of Xs were presented and subjects had to decide whether the figures were the same or different (see Figure 2). Functional impulsivity and dys-

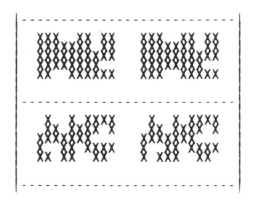

Figure 2. Examples of figure pairs used by Dickman (1990).

functional impulsivity were examined separately. It was found that dysfunctional impulsivity was unrelated to performance on this task. However, individuals high in functional impulsivity tended to be faster and less accurate than did individuals low in functional impulsivity.

Two studies have compared the performance of high and low impulsives on tasks in which they were required to compare stimuli that were presented sequentially rather than simultaneously. Anderson, Revelle, and Lynch (1989) presented subjects with a memory set of one to four words and immediately afterward presented them with a single word. Subjects had to indicate whether the stimulus word was physically identical to any of the members of the memory set. They found that high and low impulsives did not differ in the slope of the function relating reaction time to memory set size. This slope is usually taken to index the speed of comparing the stimulus word with each member of the memory set.

Nichols and Newman (1986) had high and low impulsives indicate whether two complex visual arrays, presented one after the other, were the same or different. They found no relation between impulsivity and either speed or accuracy on this task.

Several investigators have examined the relation between self-report measures of impulsivity and the MFFT, the task used to measure the cognitive style of reflection–impulsivity (Kagan, 1966). The MFFT is a visual comparison task in which subjects must indicate which of a set of simple line drawings is identical to an exemplar (Kagan et al., 1964). Five studies have shown no relation between impulsivity and performance on this task (Barratt, 1981; Barratt, Patton, Olsson, & Zuker, 1981; Davidson, 1982; Gilpin & Larsen, 1981; Malle & Neubauer, 1991).

One study comparing high and low impulsives on the MFFT did show a difference in their performance. Glow, Lange, Glow, and Barnett (1983) found that high impulsives were faster and less accurate than low impulsives on this task. Glow et al. also found that liveliness and risk taking showed a stronger relationship with MFFT performance than did any other impulsivity-related traits. Because traits such as liveliness and risk taking are associated with functional but not dysfunctional impulsivity, the findings of Glow et al. (1983) appear to confirm the finding of Dickman (1990) that it is functional impulsivity that is associated with differences in the process of visual comparison.

Arousal. There appears to be only one study that has examined the manner in which arousal affects the performance of high and low impulsives on visual comparison tasks. Anderson et al. (1989) found that caffeine increased the speed of the process of comparing stimulus words with members of the memory set but that it did not affect high and low impulsives differently.

Spatial Ability

High and low impulsives do not appear to differ in spatial ability. This finding is consistent with the lack of impulsivity-related differences in other perceptual processes that are largely automatic in nature, such as the encoding of simple stimuli. High and low impulsives do not differ on paper-and-pencil measures of spatial ability (Barratt, 1959; Roessler, 1973; Smith et al., 1991) or on the mirror-tracing task (Barratt, 1959), Porteus Maze Test (Barratt, 1967), or pegboard task (Smith et al., 1991).

High and low impulsives did differ in one condition of one of these studies. Smith et al. (1991) showed subjects letters of the alphabet that had been rotated away from the vertical. Some of the letters had been transformed into their mirror images before being rotated away from the vertical, and subjects had to indicate, for each rotated letter, whether it had been mirror transformed first. The data indicated that in order to decide whether the letter had initially been mirror transformed, subjects had to first rotate their mental image of the letter back to the vertical. The performance of high and low impulsives differed in the condition in which the letters had been rotated the farthest away from the vertical; here low impulsives were superior.

Arousal. Smith et al. (1991) found that caffeine had no effect on performance on the pegboard task for either high or low impulsives. Smith et al. also found that caffeine had no effect on performance on the mental rotation task, except for the condition in which the letter had been rotated the farthest from the vertical; in that condition, caffeine helped high impulsives, raising their performance closer to the level of low impulsives.

Visual Search

Studies investigating the visual search process typically require the subject to scan a string of letters (the "search" set) to determine whether certain letters (the "target" set) are present in the string. An attempt is often made to manipulate memory load by varying the number of target letters for which subjects are searching.

Anderson and Revelle (1983), Matthews et al. (1989), and Smith et al. (1991) failed to find differences between high and low impulsives in performance on visual search tasks. In addition, none of these studies found that increasing the memory load affected high and low impulsives' performance differently.

It should be noted that in none of these studies were subjects actually required to maintain the target set in memory. The target letters were always present on the same page or even on the same line as the search strings. As a consequence, subjects could perform the task by rapidly shifting their attention back and forth from the target set to the search set instead of having to maintain the target set in memory. (Subjects in these studies did have to remember which members of the target set they had already detected in the search string that they were currently scanning; however, this information had to be maintained in memory only very briefly, because the average time required to scan a single search string was no more than 10 seconds in any of these studies).

In a variation on the typical visual search task, Anderson and Revelle (1982) had subjects scan prose passages for either "intraword" or "interword" errors. Intraword errors were ones whose presence could be determined on the basis of a single word (e.g., spelling errors). Interword errors were ones whose presence could be determined only by combining information from several words (e.g., grammatical errors) and whose detection therefore was thought to impose more of a memory load. On this task, unlike the other visual-search-type tasks, high and low impulsives did differ. Low impulsives were slower than high impulsives in searching for intraword errors, and the opposite was true for interword errors.

Arousal. The data on the effects of arousal on the visual search performance of high and low impulsives are conflicting. Anderson and Revelle (1983) found that for a visual search task with a small target set, caffeine increased the accuracy of both high and low impulsives, but had no effect on their speed. Smith et al. (1991), also using a small target set, found that caffeine reduced the speed and accuracy of low impulsives, and increased the accuracy and reduced the speed of high impulsives. In Anderson and Revelle's (1982) proofreading study, in the low-memory-load condition (i.e., searching for intraword errors), caffeine increased the speed of high impulsives and reduced the speed of low impulsives.

With a large target set, Anderson and Revelle (1983) found that caffeine reduced the accuracy of both high and low impulsives. Smith et al. (1991) found that caffeine reduced the accuracy of low impulsives, increased the accuracy of high impulsives, and reduced the speed of both groups. Matthews et al. (1989) found no relation between self-reported arousal and the performance of high or low impulsives. In Anderson and Revelle's (1982) proofreading study, in the high-memory-load condition (i.e., searching for interword errors), caffeine increased the speed of high impulsives and reduced both the speed and accuracy of low impulsives. When Anderson and Revelle (1982) asked subjects to search simultaneously for both intraword and interword errors, caffeine increased the speed of high impulsives and reduced the speed of low impulsives.

It is difficult to draw conclusions about the effects of arousal on the visual search performance of high and low impulsives, given the variability of these findings. The most consistent finding was a tendency for arousal to be associated with poorer performance for low impulsives, but even this finding was not completely consistent across studies. The existence of such conflicting findings lends support to the view that visual search tasks can be carried out using different strategies. From this perspective, differences in the findings of different studies would be due to the fact that the wording of the instructions or some other aspect of the experimental procedure affected the proportion of high and low impulsives who adopted each of the available strategies (e.g., maintaining the target set in memory, shifting attention back and forth between the memory and search sets).

Short-Term Memory

A number of studies have compared short-term memory (STM) in high and low impulsives. These studies have generally found no differences between the two groups.

Both Matthews et al. (1989) and Barratt et al. (1981) compared high and low impulsives in digit span performance and found no differences. Craig, Humphreys, Rocklin, and Revelle (1979) found no differences between high and low impulsives in cued recall. And two studies, one by Bowyer, Humphreys, and Revelle (1983) and the other by Anderson et al. (1989), found no differences between high and low impulsives in recognition test performance.

Several studies of the relationship between impulsivity and STM have used the recency effect as an index of STM. When subjects are required to memorize a list of items, the items in the list that are presented first and last are better recalled than the items in the middle of the list (Glanzer & Cunitz, 1966). The items presented last appear to be recalled better because they are still in STM; this is the recency effect. (The items presented first appear to be recalled better because the additional attention they receive increases their likelihood of being stored in long-term memory; this is called the primacy effect and is discussed in the section on long-term memory.) Both Arnold, Petros, Beckwith, Coon, and Gorman (1987) and Erikson et al. (1985) found that the magnitude of the recency effect was identical in high and low impulsives.

Smith et al. (1991) found that low impulsives had a stronger recency effect. This appears to be the only study on impulsivity and STM that has found differences in performance between high and low impulsives. It is not clear why the Smith et al. (1991) study produced findings different from the other studies.

Arousal. Two studies have examined the effects of arousal on the STM of high and low impulsives. Neither of the studies found an effect of arousal on STM for either high or low impulsives. Matthews et al. (1989) found no effect of either self-reported arousal or time of day on the digit span performance of high or low impulsives. Smith et al. (1991) found that caffeine did not alter the magnitude of the recency effect for either group.

Maintaining Information in STM

In the STM studies discussed in the previous section, the total amount of time between the presentation of the first to-be-remembered item and the onset of the recall period was a minute or less. Two studies have examined STM in high and low impulsives when the task required subjects to maintain information in memory over longer periods of time. The greater the length of time for which information must be maintained in STM, the greater are the attentional demands of the task because attention must be allocated to STM for rehearsal, elaborative encoding, or other processes that prevent the information in STM from being lost over time (Glanzer, 1977). In both of the studies discussed in this section, low impulsives performed better than high impulsives.

Smith et al. (1991) had subjects monitor a series of visually presented digits and respond whenever a sequence of digits following the pattern of odd–even–odd appeared. One requirement of this type of task is that subjects continually maintain in STM the last two digits presented. Low impulsives were more accurate on this task.

Barratt (1967) had subjects respond to the onset of one of nine lights by pressing one of nine buttons. There was no simple correspondence between which light went on and which button had to be pressed; subjects were informed of the complex relationship between buttons and lights before the task began and had to maintain that information in memory for the entire duration of the task. High impulsives performed more poorly than low impulsives did.

Arousal. Smith et al. (1991) examined the effects of caffeine on the performance of high and low impulsives on their monitoring task. Caffeine helped performance overall and helped high impulsives more.

Long-Term Memory

Several of the studies on impulsivity and STM that have just been discussed also examined long-term memory (LTM). Their findings for

LTM were similar to their findings for STM. High and low impulsives did not differ. As was noted in the discussion of STM, one index of LTM is the primacy effect, the superior recall of items presented early in a list. Arnold et al. (1987), Erikson et al. (1985), and Smith et al. (1991) found that the magnitude of this effect was identical for high and low impulsives.

Similarly, the failure of Matthews et al. (1989) and Barratt (1981) to find overall differences in performance between high and low impulsives on the digit span test has implications for LTM as well as STM. Although neither of these studies looked at STM and LTM separately, the lack of overall differences in performance indicates that high and low impulsives differed in neither aspect of memory.

Arousal. There appears to be only one study that has compared the effects of arousal on the LTM of high and low impulsives. Smith et al. (1991) found that caffeine did not affect the magnitude of the primacy effect in either group.

Semantic Memory

Several studies have compared high and low impulsives in terms of the speed with which they are able to retrieve information from semantic memory. These studies have generally failed to find differences between the two groups.

Matthews et al. (1989) compared the performance of high and low impulsives on Posner's letter-matching task. In one condition of this task, subjects must decide whether two letters are physically identical (e.g., "A" vs. "A"). In the other condition, subjects must decide whether two letters are identical in their names (e.g., "A" vs. "a"). The difference in speed when the comparison is made on a purely physical basis and when the comparison requires access to the names of the letters is used as a measure of the speed with which the subject can access information stored in semantic memory. Matthews et al. found no differences between high and low impulsives in this regard.

Erikson et al. (1985) compared high and low impulsives on the Vocabulary subtest of the Wechsler Adult Intelligence Scale, a test that requires subjects to retrieve information about word meanings from semantic memory. Erikson et al. found no impulsivity-related differences in performance on this test.

Smith et al. (1991) compared high and low impulsives on a sentence verification task. This task required subjects to retrieve the information from semantic memory that was necessary to determine the truth or falsity of simple sentences such as "A robin is a bird." There were no differences between high and low impulsives on this task.

Anderson et al. (1989) first presented high and low impulsives with one or more category names (e.g., "fruit"). Subjects were then shown a word and had to indicate whether or not that word was a member of one of the previously presented categories (e.g., "apple"). Anderson et al. found no differences between high and low impulsives on this task.

M. Eysenck (1974) presented subjects with the names of five categories and gave them 12 minutes to name as many examples of each of the five categories as they could. High impulsives performed better than low impulsives on this task. However, sociability (the trait that, along with impulsivity, makes up the broad personality trait of extraversion) showed a much stronger relationship with performance than impulsivity did. As noted earlier, it is relatively rare to find that sociability bears a stronger relationship with a particular cognitive process than impulsivity does. Because impulsivity and sociability are highly correlated, it is possible that in this study, the findings for impulsivity were actually due to its association with sociability.

Arousal. Several studies have examined the effects of arousal on the speed with which high and low impulsives are able to retrieve information from semantic memory. Erikson et al. (1985) found that caffeine had no effect on vocabulary test performance for either group. Smith et al. (1991) found that caffeine had no effect on performance on their sentence verification task for either of the two impulsivity groups. And Matthews et al. (1989) found that neither self-reported arousal nor time of day affected speed of access to letter names for high or low impulsives.

Anderson et al. (1989) did find that caffeine had an effect on the speed of access to semantic memory. In their study, caffeine reduced access speed. However, caffeine affected high and low impulsives similarly.

Complex Problem Solving

A number of studies have compared the ability of high and low impulsives to solve complex problems. Such problems typically require subjects to carry out a variety of different cognitive processes, including retrieval of information from semantic memory, manipulation of that information, and storage of the intermediate results of those manipulations in STM.

The findings of these studies have been quite consistent. No impulsivity-related differences in performance have been found on IQ tests (Barratt, 1959; Barratt et al., 1981; Matthews, 1987), on the Scholastic Aptitude Test (Davidson, 1982), on the Graduate Record Examination (GRE; Revelle et al., 1980), on the Differential Aptitude Test (Barratt, 1959; Revelle et al., 1980), on the Lorge-Thorndyke Intelligence Tests (Barratt et al., 1981), on analogies tests (Barratt et al., 1981; Revelle et al., 1980), or on tests of logical reasoning (Smith et al., 1991).

Arousal. Several studies have examined the effects of arousal on the performance of high and low impulsives on complex problem-solving tasks. Smith et al. (1991) found that caffeine helped performance on their test of logical reasoning. However, caffeine did not have different effects on the performance of the two groups.

Both Revelle et al. (1980) and Matthews (1987) examined the effects of arousal and time of day on high and low impulsives' performance on verbal ability measures. Revelle et al. examined the effects of caffeine and time of day on the Verbal Ability subtest of the GRE, whereas Matthews looked at the effects of self-reported arousal and time of day on a verbal IQ test.

For low impulsives, both Revelle et al. (1980) and Matthews (1987) found that arousal hurt performance in the morning and helped performance in the evening. The findings for high-impulsive subjects were less clear-cut. Both Revelle et al. and Matthews found that arousal tended to hurt the performance of high impulsives in the evening; however, this finding was not entirely consistent across the different studies that Revelle et al. reported. Revelle et al. and Matthews differed in their findings on the effects of arousal on high impulsives in the morning; Revelle et al. found that arousal tended to improve the performance of high impulsives at this time of day, whereas Matthews found that arousal tended to hurt the performance of high impulsives at this time.

Motor Control

Three studies have compared the motor performance of high and low impulsives on tasks involving a relatively brief motor response. The two groups do not appear to differ in performance on such tasks.

Dickman and Meyer (1988) examined the effects of increasing the complexity of the motor response on the reaction times of high and low impulsives on a visual comparison task. In one condition, they had subjects respond by pressing a key at the end of the top row of a computer keyboard. In another condition, they had subjects respond by pressing that key and simultaneously pressing the shift key on the opposite side of the keyboard. They found that increasing the complexity of the motor response lengthened reaction times equally for high and low impulsives, indicating that this stage of processing was equally difficult for the two groups.

Cohen and Horn (1974) compared the performance of high and low impulsives on the Stroop Test. One widely accepted explanation for the Stroop effect is that the response of reading the color name is more readily available than is the response of naming the color, and the faster color-reading response interferes with the color-naming response. Cohen and Horn found that high and low impulsives were equally susceptible to response interference on this task.

Matthews et al. (1989) had high and low impulsives tap as quickly as possible for 10 seconds. They found no differences between the two groups in tapping speed.

Arousal. Matthews et al. (1989) also examined the effects of self-reported arousal and time of day on high and low impulsives' ability to tap as rapidly as possible. Neither arousal nor time of day affected tapping speed for either group.

Repetitive Motor Responses

In contrast to the lack of impulsivity-related differences in performance with relatively brief motor responses, high and low impulsives do appear to differ in performance when they must carry out a motor response repeatedly over a significant period of time. For example, Amelang and Breit (1983) asked high and low impulsives to tap as quickly as possible, just as Matthews et al. (1989) had. However, Amelang and Breit asked the subjects to tap for 2 minutes instead of 10 seconds as Matthews et al. did. Under these circumstances, high impulsives tapped more slowly.

When Barratt (1981) asked high and low impulsives to tap at a specified rate for a period several times as long as the one Matthews et al. (1989) used, high impulsives were less accurate, although they tapped more rapidly than did low impulsives. Bachorowski and Newman (1985) asked high and low impulsives to trace a circle as slowly as possible. High impulsives traced the circle more quickly.

Two studies have compared the performance of high and low impulsives on the pursuit rotor task. This task requires that the subject keep a stylus on a mark on a rotating metal disk. The data for this particular task are inconsistent. Barratt (1967) found high impulsives to be less accurate on this task. However, Smith et al. (1991) found no differences between high and low impulsives on this task.

Arousal. Smith et al. (1991) examined the effects of caffeine on the pursuit rotor performance of high and low impulsives. Caffeine did not affect the performance of either group.

THEORIES OF IMPULSIVITY

Several theories have been offered to account for the relationship between self-reported impulsivity and cognitive functioning. Each of these theories makes fairly explicit predictions about the impulsivity-related differences in cognitive functioning that would be expected in some of the areas of research reviewed here, while leaving unclear the nature of the differences that would be expected in others.

Eysenck

Perhaps the most influential of these theories of impulsivity is the one proposed by H. J. Eysenck (H. J. Eysenck & Eysenck, 1985) on the basis of his work on extraversion. As was noted earlier, extraverts are both impulsive and sociable. Because extraverts' impulsivity appears to account for most of the relationships found between extraversion and cognitive functioning, Eysenck's original theory of extraversion has been extended to account for impulsivity-related differences in performance as well.

According to Eysenck, the cognitive differences between introverts and extraverts are due to the fact that introverts (who correspond to low impulsives) are chronically more aroused than extraverts (who correspond to high impulsives). Eysenck does not distinguish between different types of arousal.

According to Eysenck's theory, high and low impulsives will differ in those particular cognitive processes that are known to be affected by arousal. High impulsives will be superior in those cognitive processes that are known to be facilitated by arousal, whereas low impulsives will be superior in those cognitive processes known to be impaired by arousal.

In making predictions about the specific cognitive processes in which high and low impulsives will differ, Eysenck has drawn upon several different bodies of research that have identified cognitive processes influenced by arousal. For example, there is a body of research that suggests that arousal at the time of storage facilitates STM and impairs LTM. According to the "action decrement" theory that Walker (1958) proposed to account for these findings, arousal enhances the process by which memories are consolidated; however, while memories are being consolidated, they are not accessible. Thus, arousal increases long-term memory for information, at the cost of a reduction in immediate access to that information. Eysenck's theory predicts that high impulsives, because of their lower arousal, will be superior in STM.

The data on STM reviewed here do not confirm the prediction made by Eysenck's theory. High impulsives have performed no better than low impulsives in most studies of STM (Anderson et al., 1989; Arnold et al., 1987; Barratt et al., 1981; Bowyer et al., 1983; Craig et al., 1979; Erikson et al., 1985; Matthews et al., 1989). Furthermore, when the STM task has required subjects to maintain the information in STM over an extended period of time, low impulsives have been found to be superior, the opposite of what Eysenck's theory would predict (Smith et al., 1991).

There are several aspects of cognitive functioning in which Eysenck's theory predicts that low impulsives will be superior as a consequence of their higher levels of arousal. Because higher levels of arousal at the time of storage appear to help LTM, Eysenck's theory predicts that low impulsives will show superior LTM performance. However, in the studies reviewed here, high and low impulsives failed to differ in their LTM performance (Arnold et al., 1987; Barratt, 1981; Erikson et al., 1985; Matthews et al., 1989; Smith et al., 1991).

Bakan and others (e.g., Bakan, 1959) have found that higher levels of arousal are associated with superior performance on tasks that require the subject to maintain attention to the task over an extended period of time. Again, Eysenck's theory predicts that the more highly aroused low impulsives will show superior performance on these tasks. The studies reviewed here that have examined performance on tasks requiring sustained attention lend support to Eysenck's theory: Low impulsives have shown better performance on these tasks (Roessler, 1973; Thackray et al., 1974).

A third area of cognitive functioning in which Eysenck's theory predicts that low impulsives will be superior is in the performance of repetitive motor responses. In making his predictions here, Eysenck has drawn upon a body of research that has demonstrated that when making such repetitive motor responses, subjects begin to show momentary pauses in responding sooner than would be expected on the basis of muscle fatigue; these momentary pauses are referred to as *involuntary rest pauses* (IRPs; H. J. Eysenck, 1957). One explanation for the existence of IRPs is that each time a repetitive motor response is made, a little inhibition (i.e., reduction of arousal) occurs. When this inhibition has reduced arousal to a low enough level, the individual must pause in responding until arousal has returned to the minimal level necessary for task performance. Because high impulsives are initially less aroused than low impulsives, Eysenck predicts that they will be more vulnerable to such IRPs, slowing them down on repetitive motor tasks.

The data on repetitive motor tasks do not support Eysenck's predictions. Barratt (1981) found that high impulsives were actually faster than low impulsives on tasks of this sort—the opposite of what Eysenck would predict. Although Amelang and Breit (1983) did find that high impulsives were slower overall than low impulsives, when these investigators looked specifically at the frequency of IRPs (brief pauses in responding), they found no differences between the two groups.

Eysenck's theory also implicitly makes predictions about the performance of high and low impulsives on tasks for which there is no a priori theory about the effects of arousal on performance. Because Eysenck's theory holds that high and low impulsives differ in arousal, whenever it can be demonstrated empirically that performance on a particular task is affected by arousal, Eysenck must predict that high and low impulsives will differ in their performance on that task.

The data here are inconsistent with Eysenck's theory. Arousal appears to affect performance on visual comparison tasks, but high and low impulsives do not differ in their performance on such tasks (e.g., Anderson et al., 1989). Arousal also affects performance on tasks requiring access to semantic memory and tasks requiring logical reasoning, but high and low impulsives do not differ in their performance on such tasks (e.g., Anderson et al., 1989).

Eysenck's theory also makes predictions about differences in the way the performance of high and low impulsives will be affected by increases in arousal. In making these predictions, Eysenck has drawn upon the Yerkes–Dodson law (Yerkes & Dodson, 1908). According to this law, both very low levels of arousal and very high levels of arousal are associated with impaired performance. Because high impulsives are chronically underaroused, increases in arousal should improve their performance. Because low impulsives are chronically highly aroused, increases in arousal should impair their performance (or, at least, not improve it as much as it does the performance of high impulsives).

The data for several types of cognitive processes are consistent with Eysenck's theory. Smith et al. (1991) found that for the most difficult condition of their mental rotation task, as well as for the simple encoding task in which subjects had to identify sequences of odd–even–odd digits, arousal either hurt low impulsives or helped them less than it did high impulsives.

The data on visual search also provide some support for Eysenck's theory. Arousal tends to hurt low impulsives on visual search tasks, although the data are not entirely consistent (Anderson & Revelle, 1983; Smith et al., 1991). As noted before, the data on the effects of arousal on visual search performance in high impulsives are too conflicting to yield a clear interpretation.

There are some data on the effects of arousal on performance that are clearly not consistent with Eysenck's theory. On Matthews et al.'s (1989) simple encoding task, in which subjects had to respond continuously for 10 minutes, arousal was associated with greater accuracy for low impulsives and lower accuracy for high impulsives. In addition, when this task was carried out in the evening, arousal reduced the speed of high impulsives more than it did the speed of low impulsives. Similarly, Revelle et al. (1980) reported several studies using complex problem-solving tasks in which, during the evening at least, arousal hurt the performance of high impulsives and helped the performance of low impulsives.

Overall, the studies reviewed here provide only limited support for Eysenck's theory of impulsivity. Although the data on the effects of arousal on performance are somewhat consistent with Eysenck's theory, that theory does not appear to account very well for differences in the performance of high and low impulsives in the absence of arousal manipulations.

Revelle

Revelle (Humphreys & Revelle, 1984; Revelle et al., 1980) has proposed a theory of impulsivity that is similar to Eysenck's in that it attributes impulsivity-related differences in cognitive functioning to differences in arousal and does not distinguish between different types of arousal. Revelle's theory differs from Eysenck's in two main ways.

First, according to Revelle, high and low impulsives do not differ in their chronic levels of arousal, but rather in the nature of their circadian rhythms of arousal. According to Revelle, low impulsives reach their peak level of arousal earlier in the day than do high impulsives, so that in the morning their arousal level is higher than that of high impulsives, and in the evening it is lower. Because most psychological research is conducted during the day, the performance of low impulsives in these studies would be expected to reflect a higher level of arousal than would the performance of high impulsives.

The other difference between Revelle's theory of impulsivity and that of Eysenck lies in their respective assumptions about the effects of arousal on information processing. According to Revelle, arousal has two specific effects on information processing. Arousal facilitates short-term information transfer (SIT) and impairs STM.

According to Revelle, SIT is the process of encoding a stimulus, associating an arbitrary response with that stimulus, and then executing that response. No appreciable retention of information is required. An example of a task that primarily involves SIT would be a simple reaction time task in which subjects have to press a button as soon as possible after the onset of a light.

According to Revelle, changes in arousal will only affect performance on a particular task if performance on that task is "resource limited" (Norman & Bobrow, 1975). The task must make significant demands in terms of SIT or STM in order for arousal-produced changes in the resources required for those types of cognitive processes to affect performance.

Revelle has also proposed that on tasks that require both SIT and STM, increases in a subject's level of arousal will initially improve performance by increasing SIT resources. However, as the subject's level of arousal rises further, the increasing impairment of STM will eventually cause performance to decline.

Revelle's theory makes several specific predictions about the cognitive processes investigated in the studies reviewed here. These predictions assume that most of the studies that did not examine time-of-day effects specifically were held during the day rather than in the evening.

According to Revelle, low impulsives will perform better on tasks that primarily require SIT because of their higher level of arousal. The data on the performance of high and low impulsives on tasks that primarily require SIT are not consistently in line with this prediction. On complex encoding tasks, high and low impulsives did differ in the way Revelle would predict (Dickman, 1985; Loo, 1979). However, the data for simple encoding tasks are mixed, with low impulsives showing superior performance when the task was lengthy (Roessler, 1973; Thackray et al., 1974) but not when it was relatively brief (e.g., Dickman, 1985; Matthews et al., 1989; Smith et al., 1991).

Tasks involving motor performance are also primarily SIT tasks. The data from these tasks also fail to consistently support Revelle's theory. On motor tasks requiring a brief response, there were no differences in performance between high and low impulsives (Cohen & Horn, 1974; Dickman & Meyer, 1988; Matthews et al., 1989). However, when the task required repeated responding, low impulsives were superior, as Revelle would predict (Amelang & Breit, 1983; Bachorowski & Newman, 1985; Barratt, 1967).

It is possible to reconcile these findings with Revelle's theory by assuming that the majority of tasks used in these studies were not resource limited. If these tasks could not benefit from additional SIT resources beyond those that the less aroused high impulsives had, high and low impulsives would not be expected to differ in performance. The data are therefore ambiguous.

Revelle's theory also predicts that high impulsives will be superior in STM because the relatively high levels of arousal characteristic of low impulsives will have negative effects on their STM. One implication of this is that on visual search tasks, increasing the load on STM should hurt low impulsives more than high impulsives. This was not the case (Anderson & Revelle, 1983; Smith et al., 1991). It was suggested earlier that the design of the visual search tasks reviewed here made it possible for subjects to minimize their use of memory if they chose. Because this possibility cannot be ruled out, the data from these visual search tasks are not necessarily inconsistent with Revelle's theory.

More troublesome for Revelle's theory are those studies that used tasks whose primary purpose was to measure STM. These studies have generally failed to find differences in performance between high and low impulsives (Anderson et al., 1989; Arnold et al., 1987; Barratt et al., 1981; Bowyer et al., 1983; Craig et al., 1979; Erikson et al., 1985; Matthews et al., 1989). It might be possible to account for these negative findings on the grounds that these STM tasks were not resource limited. However, on the more demanding STM tasks, low impulsives actually performed better than high impulsives—the opposite of what Revelle would predict (Barratt, 1967; Smith et al., 1991).

Similarly, for Anderson and Revelle's (1982) proofreading task, low impulsives were superior in the condition that was most demanding in terms of STM (i.e., searching for interword errors) and inferior in the condition requiring mainly SIT (i.e., searching for intraword errors). These findings are again the opposite of what Revelle would predict.

Another type of prediction made by Revelle's theory has to do with differences in the way increases in arousal will affect the performance of high and low impulsives. For SIT tasks, Revelle's theory predicts that increases in arousal are more likely to help high impulsives. The reasoning is that if high impulsives are less aroused than low impulsives, their SIT resources are more likely to be below the minimum level necessary for adequate task performance, and they are therefore more likely to benefit from the additional SIT resources produced by arousal.

Two studies using motor tasks that primarily involved SIT have examined the effects of arousal on performance. Neither of these studies found an effect of arousal on the performance of either high or low impulsives (Matthews et al., 1989; Smith et al., 1991). Again, this finding can be reconciled with Revelle's theory by assuming that these tasks were not demanding enough in terms of SIT resources for arousal increases to have a significant affect on performance. However, Matthews et al.'s (1989) finding that on a demanding SIT task involving simple encoding, higher levels of arousal were associated with greater accuracy for low impulsives and reduced accuracy for high impulsives is the opposite of what Revelle would predict.

For tasks that mainly require STM, Revelle's theory predicts that increases in arousal are more likely to hurt low impulsives than high impulsives. The reasoning is that the high arousal level of low impulsives reduces the resources available for STM. This increases the likelihood that further reductions in these resources, produced by increases in arousal, will cause them to fall below the level necessary for adequate task performance.

The studies reviewed here that have examined the manner in which arousal affects the STM of high and low impulsives have provided conflicting findings. Matthews et al. (1989) failed to find any effect of arousal on a STM task, whereas Smith et al.'s (1991) findings were consistent with Revelle's theory for one of their STM tasks but not for another. Again, it might be argued that the negative findings came from tasks that were resource-limited. If these tasks were not very demanding in terms of STM, increases in arousal could have negative effects on STM without significantly impairing performance. This explanation cannot be ruled out, although the error rates in these studies were not sufficiently low to provide strong support for such an explanation (for example, the overall error rate on Matthews et al.'s (1989) digit span task was .45).

Studies on visual search performance have found that arousal tends to hurt the performance of low impulsives (Anderson & Revelle, 1983; Smith et al., 1991). This is consistent with Revelle's theory, assuming that low impulsives used a strategy on this task that was demanding in terms of STM. As noted earlier, the visual search data for high impulsives are too inconsistent to interpret.

Because Revelle's theory of impulsivity is similar to Eysenck's in regarding arousal differences as the source of performance differences between high and low impulsives, Revelle must make the same general prediction that Eysenck makes: When it has been empirically demonstrated that arousal affects performance on a particular task, high and low impulsives will differ on that task. Data that fail to confirm this prediction were noted in the discussion of Eysenck's theory.

Another prediction that Revelle's theory makes is that on any task for which high and low impulsives differ in performance, the relative performance of the two groups should reverse from morning to evening. The reason is that, according to Revelle, the relative arousal levels of the two groups reverse between morning and evening. It is these arousal levels that are responsible for any performance differences that are found between them. Unfortunately, none of the studies reviewed here provide a test of this prediction. Those studies that compared morning and evening sessions did not find differences between high and low impulsives for either the morning or the evening sessions (Matthews et al., 1989; Revelle et al., 1980; Smith et al., 1991).

Because Revelle's theory assumes that the relationship between high and low impulsives' arousal levels reverses itself between morning and evening, that theory also predicts that if increases in arousal affect the two groups differently in the morning, the effects of arousal increases on the two groups will be reversed in the evening. Revelle et al. (1980) obtained data on a variety of complex tasks that tended to support Revelle's theory. In the morning, increases in arousal hurt the performance of low impulsives and helped the performance of high impulsives on these tasks. In the evening, the effects of arousal on the two groups tended to be the opposite of the morning, although the evening data were less consistent than the morning data. Matthews (1987), it should be noted, did not find this clear-cut reversal from morning to evening when examining the effects of arousal on a verbal IQ test similar to the tasks that Revelle et al. (1980) used.

Matthews et al. (1989) did find a time-of-day reversal for an attention-demanding simple encoding task. Matthews et al. found that in the morning, arousal hurt the performance of low impulsives on this task more than it did the performance of high impulsives, and the reverse was true for the evening. Matthews et al. (1989) data were not, however, entirely consistent with Revelle's theory. According to Revelle's theory, on a pure SIT task such as the one Matthews et al. used, higher levels of arousal should be associated with better performance, the opposite of what Matthews et al. found.

Overall, then, although some of the data reviewed here are consistent with Revelle's theory of impulsivity, a number of the findings of these studies fail to support that theory. It should be noted that because Revelle's theory of impulsivity incorporates a theory of arousal, it is possible that these negative findings reflect limitations in the theory of arousal rather than the theory of impulsivity.

Barratt

Barratt has proposed that individual differences in impulsivity are related to differences in the ability to maintain a cognitive tempo (Barratt, 1983, 1987; Barratt & Patton, 1983). According to Barratt, when a task requires that subjects maintain a certain tempo, or rate, of information processing, high impulsives have greater difficulty in maintaining such a tempo than do low impulsives, and therefore are likely to be less efficient in carrying out the task. Barratt has argued that the tasks most sensitive to these difficulties in maintaining a cognitive tempo are reaction time tasks (in which the subject must respond to stimuli presented at intervals) and tasks requiring rhythmic responses (e.g., paced tapping).

The data from studies reviewed here in which subjects had to respond to stimuli presented at regular intervals do not consistently support Barratt's theory. On tasks requiring simple encoding of stimuli presented at intervals, high impulsives performed more poorly than did low impulsives, as Barratt would predict (Barratt, 1967; Matthews et al., 1989; Robinson & Zahn, 1988; Roessler, 1973; Smith et al., 1991; Thackray et al., 1974). However, when more complex stimuli (or pairs of stimuli) were presented at regular intervals, the two groups failed to differ in performance (Anderson et al., 1989; Matthews et al., 1989; Nichols & Newman, 1986; Smith et al., 1991).

The data for tasks requiring only rhythmic motor responses are more consistent with Barratt's theory. Although high and low impulsives have not been found to differ in their ability to make brief motor responses, when they have been required to make motor responses rhythmically over a period of time, high impulsives performed more poorly (Amelang & Breit, 1983; Bachorowski & Newman, 1985; Barratt, 1967, 1981). Only Smith et al. (1991) failed to find differences between the two groups in this type of motor performance.

One limitation of Barratt's theory is that it does not seem to be able to account for all of the differences found between high and low impulsives. In particular, it does not appear to account for differences between high and low impulsives on tasks that do not involve either repetitive stimuli or repetitive responses. For example, high and low impulsives have been found to differ on the EFT (Loo & Townsend, 1977) and on proofreading tasks (Anderson & Revelle, 1982).

Another limitation of Barratt's theory, at least as it is currently stated, is that it does not account for the body of data on the differential effects of arousal on high and low impulsives. Thus, although Barratt's theory does account well for some of the data reviewed here, in its present form it leaves unexplained a number of important findings.

An Attentional Theory of Impulsivity

The data reviewed here seem to pose significant problems for each of the three theories of impulsivity discussed so far. It does appear possible to account for these data by means of a theory of impulsivity that views this trait in terms of individual differences in the mechanisms that allocate attention, rather than in terms of arousal or cognitive tempo.

More specifically, the data reviewed here are consistent with the view that individual differences in impulsivity reflect differences in the degree to which attention tends to remain fixed once it is directed to a particular source of information. To use the "flashlight" metaphor that is often applied to attention, high and low impulsives appear to differ in the ease with which various factors can cause the beam of the flashlight to move from its current location. The attention of low impulsives has a tendency to remain fixed, whereas the attention of high impulsives is readily shifted. The data seem best explained by assuming that such attentional differences are due to differences in the attentional mechanisms themselves, rather than to differences in arousal.

One implication of this attentional theory of impulsivity is that high and low impulsives will only differ on tasks that are especially demanding in terms of either focusing attention or shifting attention rapidly between multiple sources of information. Low impulsives would be expected to be superior on tasks that were especially demanding in terms of focusing attention, whereas high impulsives would be expected to be superior on tasks that were especially demanding in terms of shifting attention.

The tasks used in the studies reviewed here seem to place more of an emphasis on fixing attention on a single source of information than on shifting attention rapidly between multiple sources of information. A task might be especially demanding in terms of "attentional fixity" for at least two reasons: (a) Because good performance requires that attention be focused on the task continuously, without even momentary lapses, or (b) because good performance requires that some degree of attention be devoted to the task over an extended period of time. The former type of attentional demand will be referred to here as intensive attention, and the latter as sustained attention.

Many of the findings reviewed here are consistent with the view that low impulsives are superior on tasks that make demands in terms of focusing attention on a single source of information. A number of studies suggest the importance of the demand for sustained attention in producing differences in performance between high and low impulsives. Studies on simple encoding have found that high and low impulsives do not differ in their ability to carry out this type of encoding except when the task also demands that they maintain attention to such encoding over an extended period of time; under these circumstances, low impulsives are superior (Matthews et al., 1989; Roessler, 1973; Smith et al., 1991; Thackray et al., 1974). Similarly, studies on spatial ability have found that high and low impulsives do not differ except when the task requires them to maintain the image of the rotated object in visual memory for an extended period of time; under these circumstances, low impulsives have been found to be superior (Smith et al., 1991). In addition, it has been found that high and low impulsives do not differ in STM except when they have to use attention-demanding processes to maintain information in STM over an extended period of time; under these circumstances, low impulsives have again been found to be superior (Anderson et al., 1989; Arnold et al., 1987; Barratt, 1967; Barratt et al., 1981; Bowyer et al., 1983; Craig et al., 1979; Erikson et al., 1985; Matthews et al., 1989; Smith et al., 1991).

The research reviewed here also suggests the importance of the demand for intensive attention in producing differences in performance between high and low impulsives. Studies on encoding have found that even though high and low impulsives do not differ in simple, largely automatic encoding, low impulsives are superior on tasks requiring the sort of complex encoding that Treisman and others (Treisman & Gelade, 1980; Treisman & Schmidt, 1982) have found to be sensitive to even momentary lapses of attention (Dickman, 1985; Loo, 1979; Loo & Townsend, 1977). Similarly, studies on motor performance have found that high and low impulsives do not differ when the response required is a relatively brief one (e.g., Cohen & Horn, 1974; Dickman & Meyer, 1988; Matthews et al., 1989) but that when the two groups are required to respond repeatedly over a period of time, and their responding is continually monitored so that even momentary lapses in attention affect their score on the task, low impulsives are generally superior (Amelang & Breit, 1983; Bachorowski & Newman, 1985; Barratt, 1967, 1981). The Smith et al. (1991) study was the only one requiring such repeated responding that failed to find impulsivity-related differences in performance.

Even though the attentional fixity model of impulsivity proposed here does not view arousal differences as the source of differences in impulsivity, it can provide an explanation for the data on the differential effects of arousal increases on the performance of high and low impulsives. As was noted earlier, one of the effects of arousal (specifically, the type of arousal that peaks in the morning) is to increase the ability to maintain attention to a task. Arousal increases would therefore be expected to help performance on tasks that were demanding in terms of attentional fixity.

Such increases in arousal would only be expected to hurt performance if arousal became great enough to significantly interfere with those components of task performance that required the shifting of attention. As noted earlier, most of the tasks reviewed here require subjects to process a single source of information, and therefore do not appear to be particularly demanding in terms of rapid attention-shifting. For such tasks, arousal would have to be quite high before it significantly impaired performance.

Arousal increases would be expected to be of more benefit to high impulsives than to low impulsives, because high impulsives' initial level of attentional fixity was lower. Low impulsives' higher initial levels of attentional fixity would render them more vulnerable to having their performance impaired by increases in arousal; it is more likely that such increases would cause their level of attentional fixity to become great enough to interfere with whatever demands the task made in terms of attention-shifting.

The data for both mental rotation and STM tasks indicate that under conditions in which those tasks were especially attention demanding, arousal did help the performance of high impulsives (Smith et al., 1991). Increases in arousal also helped high impulsives on complex problem-solving tasks, at least in the morning (Revelle et al., 1980).

There were two studies in which increases in arousal hurt the performance of high impulsives. Revelle et al. (1980) found that increases in arousal hurt the performance of high impulsives on complex problem-solving tasks in the evening, although this finding was not consistent across all of the studies they reported. In addition, Matthews et al. (1989) found that on their attention-demanding simple encoding task, higher levels of arousal were associated with poorer performance for high-impulsive subjects in both the morning and the evening. One way that an attentional theory of impulsivity could account for findings such as these is to assume that under some circumstances, high impulsives will use strategies that capitalize on their ability to switch attention rapidly; such strategies would be especially likely to be impaired by the effects of arousal-induced increases in attentional fixity. Such an explanation must be considered speculative, because the data provided by these studies do not make it possible to directly confirm that such strategy effects accounted for their findings.

The data on the effects of arousal on low impulsives on attention-demanding tasks are consistent with an attentional theory of impulsivity. Increases in arousal would be expected to help the performance of low impulsives by increasing attentional fixity, unless arousal became so great that it significantly interfered with those components of task performance that required the shifting of attention. This would be most likely to happen in the morning because the circadian rhythm of the type of arousal that increases attentional fixity is at its peak at this time of day. Consistent with this reasoning, both Revelle et al. (1980) and Matthews (1987) found that increases in arousal helped the performance of low impulsives on complex tasks in the evening, but hurt their performance in the morning.

On tasks that were not demanding in terms of attention, arousal might still affect performance by affecting some other aspect of cognitive functioning besides the allocation of attention. However, because this aspect of cognitive functioning would not be one in which high and low impulsives differed, increases in arousal would not be expected to affect the two groups differently. This is how an attentional model of impulsivity would account for the fact that arousal affects performance on visual comparison tasks, measures of logical reasoning, and tasks requiring the retrieval of information from semantic memory, but does not affect the performance of high and low impulsives differently (Anderson et al., 1989; Smith et al., 1991). As was noted earlier, those theories of impulsivity that assume that arousal differences underlie differences in impulsivity (i.e., Eysenck, Revelle) have difficulty accounting for findings of this sort, because they must predict that high and low impulsives will differ on tasks that are affected by arousal.

CONCLUSION

It appears that by viewing individual differences in impulsivity as a function of differences in attention rather than in arousal or cognitive tempo, it is possible to account for most of the data reviewed in this chapter. This model of impulsivity also has the advantage of providing an explanation for a seeming paradox that is not well explained by the other theories discussed here.

Questionnaire measures of impulsivity such as those used in the studies reviewed here identify impulsive individuals on the basis of their report that they tend to act with less forethought than other individuals. Yet, the individuals identified as impulsive on the basis of these questionnaires sometimes respond more slowly than do other individuals on cognitive tasks. From the perspective of an attention theory of impulsivity, the reason that high impulsives report that they act with less forethought than others do is that even when they take longer than others to respond, they actually devote less time to considering their response because they have difficulty keeping their attention focused on the task during the time when they are preparing that response.

Finally, the attentional model of impulsivity proposed here may have the potential to provide a bridge between work on impulsivity in adults and work on impulsivity in children. There is a growing body of research on attention-deficit hyperactivity disorder (ADHD) that suggests that the attentional difficulties that these children have are closely related to their impulsive behavior (e.g., Douglas, 1983). It may be that the attentional characteristics that account for the performance of high-impulsive adults in the studies reviewed here also account for the impulsive behavior of these children.

REFERENCES

Akerstedt, T. (1977). Inversion of the sleep-wakefulness pattern: Effects on circadian variations in psychophysiological activation. *Ergonomics, 20, 459–474.*

Amelang, M., & Breit, C. (1983). Extraversion and rapid tapping: Reactive inhibition or general cortical activation as determinants of performance differences. *Personality and Individual Differences, 4, 103–105.*

Anderson, K., & Revelle, W. (1982). Impulsivity, caffeine, and proofreading: A test of the Easterbrook hypothesis. *Journal of Experimental Psychology: Human Perception and Performance, 8, 614–624.*

Anderson, K. J., & Revelle, W. (1983). The interactive effects of caffeine, impulsivity, and task demands on a visual search task. *Personality and Individual Differences, 4, 127–134.*

Anderson, K. J., Revelle, W., & Lynch, M. J. (1989). Caffeine, impulsivity, and memory scanning: A comparison of two explanations for the Yerkes-Dodson effect. *Motivation and Emotion, 13, 1–20.*

Arnold, M. E., Petros, T. V., Beckwith, B. E., Coon, G., & Gorman, N. (1987). The effects of caffeine, impulsivity, and sex on memory for word lists. *Physiology and Behavior, 41, 25–30.*

Bachorowski, J., & Newman, J. P. (1985). Impulsivity in adults and time-interval estimation. *Personality and Individual Differences, 6, 133–136.*

Bakan, P. (1959). Extraversion-introversion and improvement on an auditory vigilance task. *British Journal of Psychology, 50, 325–332.*

Baker, W. J., & Theologus, G. C. (1972). Effects of caffeine on visual monitoring. *Journal of Applied Psychology, 56, 422–427.*

Barratt, E. (1959). Anxiety and impulsiveness relative to psychomotor efficiency. *Perceptual and Motor Skills, 9, 191–198.*

Barratt, E. S. (1965). Factor analysis of some psychometric measures of impulsiveness and anxiety. *Psychological Reports, 16, 547–554.*

Barratt, E. (1967). Perceptual-motor performance relative to impulsivity and anxiety. *Perceptual and Motor Skills, 25, 485–492.*

Barratt, E. S. (1981). Time perception, cortical evoked potentials and impulsiveness among three groups of adolescents. In J. K. Hays, T. K. Roberts, & K. S. Solway (Eds.), *Violence and the violent individual* (pp. 87–95). New York: Spectrum Publications.

Barratt, E. S. (1983). The biological basis of impulsiveness: The significance of timing and rhythm disorders. *Personality and Individual Differences, 4,* 387–391.

Barratt, E. S. (1987). Impulsivity and anxiety: Information processing and electroencephalographic topography. *Journal of Research in Personality, 21,* 453–463.

Barratt, E. S., & Patton, J. H. (1983). Impulsivity: Cognitive, behavioral, and psychophysiological correlates. In M. Zuckerman (Ed.), *Biological bases of sensation seeking, impulsivity, and anxiety* (pp. 77–116). Hillsdale, NJ: Erlbaum.

Barratt, E., Patton, J., Olsson, N., & Zuker, G. (1981). Impulsivity and paced tapping. *Journal of Motor Behavior, 13,* 286–300.

Bentler, P. M., & McClain, J. (1976). A multitrait-multimethod analysis of reflection-impulsivity. *Child Development, 47,* 218–226.

Blake, M. J. (1967). Time of day effects on performance on a range of tasks. *Psychonomic Science, 9,* 345–350.

Blake, M. J. (1971). Temperament and time of day. In W. P. Colquhoun (Ed.), *Biological rhythms and human performance* (pp. 109–147). San Diego, CA: Academic Press.

Bowyer, P., Humphreys, M., & Revelle, W. (1983). Arousal and recognition memory: The effects of impulsivity, caffeine, and time on task. *Personality and Individual Differences, 4,* 41–49.

Cohen, D. B., & Horn, J. M. (1974). Extraversion and performance: A test of the theory of cortical inhibition. *Journal of Abnormal Psychology, 83,* 304–307.

Colquhoun, W. P. (1971). Circadian variation in mental efficiency. In W. P. Colquhoun (Ed.), *Biological rhythms and human performance* (pp. 9–107). San Diego, CA: Academic Press.

Craig, N., Humphreys, M., Rocklin, T., & Revelle, W. (1979). Impulsivity, neuroticism, and caffeine: Do they have additive effects on arousal? *Journal of Research in Personality, 13,* 104–119.

Davidson, W. B. (1982). Multimethod examination of field-independence and impulsivity. *Psychological Reports, 50,* 655–661.

Dickman, S. (1985). Impulsivity and perception: Individual differences in the processing of the local and global dimensions of stimuli. *Journal of Personality and Social Psychology, 48,* 133–149.

Dickman, S. (1990). Functional and dysfunctional impulsivity: Personality and cognitive correlates. *Journal of Personality and Social Psychology, 58,* 95–102.

Dickman, S., & Meyer, D. E. (1988). Impulsivity and speed–accuracy tradeoffs in information processing. *Journal of Personality and Social Psychology, 54,* 274–290.

Douglas, V. I. (1983). Attentional and cognitive problems. In M. Rutter (Ed.), *Developmental neuropsychiatry* (pp. 280–329). New York: Guilford Press.

Edman, G., Schalling, D. T., & Levander, S. E. (1985). Impulsivity and speed and errors in a reaction time task: A contribution to the construct validity of the concept of impulsivity. *Acta Psychologica, 53*, 1–8.

Erikson, G. C., Hager, L. B., Houseworth, C., Dungan, J., Petros, T., & Beckwith, B. E. (1985). The effects of caffeine on memory for word lists. *Physiology and Behavior, 35*, 47–51.

Eysenck, H. J. (1957). *The dynamics of anxiety and hysteria*. London: Routledge & Kegan Paul.

Eysenck, H. J., & Eysenck, M. W. (1985). *Personality and individual differences*. New York: Plenum Press.

Eysenck, H. J., & Eysenck, S. B. G. (1965). *Manual of the Eysenck Personality Inventory*. London: Hodder & Stoughton.

Eysenck, H. J., & Eysenck, S. B. G. (1975). *Manual of the Eysenck Personality Questionnaire*. London: Hodder & Stoughton.

Eysenck, M. W. (1974). Extraversion, arousal, and retrieval from semantic memory. *Journal of Personality, 42*, 319–331.

Eysenck, S. B. G., & Eysenck, H. J. (1977). The place of impulsiveness in a dimensional system of personality description. *British Journal of Social and Clinical Psychology, 16*, 57–68.

Eysenck, S. B. G., & Eysenck, H. J. (1978). Impulsiveness and venturesomeness: Their position in a dimensional system of personality description. *Psychological Reports, 43*, 1247–1253.

Folkard, S., & Monk, T. H. (1980). Circadian rhythms in human memory. *British Journal of Psychology, 71*, 295–307.

Gilpin, A. R., & Larsen, W. (1981). Conceptual and motoric impulsivity in college students. *Journal of General Psychology, 105*, 207–214.

Glanzer, M. (1977). Storage mechanisms in recall. In G. Bower (Ed.), *Human memory: Basic processes* (pp. 125–189). San Diego, CA: Academic Press.

Glanzer, M., & Cunitz, A. R. (1966). Two storage mechanisms in free recall. *Journal of Verbal Learning and Verbal Behavior, 16*, 21–28.

Glow, R. A., Lange, R. V., Glow, P. H., & Barnett, J. A. (1983). Cognitive and self-report impulsivity: Comparison of Kagan's MFFT and Eysenck's EPQ impulsivity measures. *Personality and Individual Differences, 4*, 179–187.

Hauty, G. T., & Payne, R. B. (1955). Mitigation of work decrement. *Journal of Experimental Psychology, 49*, 60–67.

Hockey, G. R. J., & Colquhoun, W. P. (1972). Diurnal variations in human performance: A review. In W. P. Colquhoun (Ed.), *Aspects of human efficiency: Diurnal rhythm and loss of sleep* (pp. 1–23). London: English Universities Press.

Humphreys, M. S., & Revelle, W. (1984). Personality, motivation, and performance: A theory of the relationship between individual differences and information processing. *Psychological Review, 91*, 153–184.

Kagan, J. (1966). Reflection-impulsivity: The generality and dynamics of conceptual tempo. *Journal of Abnormal Psychology, 71*, 17–24.

Kagan, J., Rosman, B. L., Day, D., Albert, J., & Phillips, W. (1964). Information processing in the child: Significance of analytic and reflective attitudes. *Psychological Monographs, 78*(1, Whole No. 578).

Klein, K. E., Hermann, R., Kuklinski, P., & Wegmann, H.-M. (1977). Circadian performance rhythms: Experimental studies in air operations. In R. R. Mackie (Ed.), *Vigilance: Theory, operational performance, and physiological correlates* (pp. 57–83). New York: Plenum Press.

Kleitman, N. (1963). *Sleep and wakefulness.* Chicago: University of Chicago Press.

Laird, D. A. (1925). Relative performance of college students as conditioned by time of day and day of week. *Journal of Experimental Psychology, 8,* 10–63.

Loo, R. (1979). Role of primary personality factors in the perception of traffic signs and driver violations and accidents. *Accident Analysis and Prevention, 11,* 125–127.

Loo, R., & Townsend, P. J. (1977). Components underlying the relationship between field dependence and extraversion. *Perceptual and Motor Skills, 45,* 528–530.

Malle, B. F., & Neubauer, A. C. (1991). Impulsivity, reflection, and questionnaire response latencies: No evidence for a broad impulsivity trait. *Personality and Individual Differences, 12,* 865–871.

Matthews, G. (1987). Personality and multidimensional arousal: A study of two dimensions of extraversion. *Personality and Individual Differences, 8,* 9–16.

Matthews, G., Jones, D. M., & Chamberlain, A. G. (1989). Interactive effects of extraversion and arousal on attentional task performance: Multiple resources or encoding processes? *Journal of Personality and Social Psychology, 56,* 629–639.

McCrae, R. R., & Costa, P. T., Jr. (1987). Validation of the five-factor model of personality across instruments and observers. *Journal of Personality and Social Psychology, 52,* 81–90.

Messer, S. B. (1976). Reflection–impulsivity: A review. *Psychological Bulletin, 83,* 1026–1052.

Millar, K., Styles, B., & Wastell, D. G. (1980). Time of day and retrieval from long-term memory. *British Journal of Psychology, 71,* 407–414.

Minors, D. S., & Waterhouse, J. M. (1981). *Circadian rhythms and the human.* Bristol, England: Wright PSG.

Newman, J. P. (1987). Reaction to punishment in extraverts and psychopaths: Implications for the impulsive behavior of disinhibited individuals. *Journal of Research in Personality, 21,* 464–480.

Newman, J. P., Widom, C. S., & Nathan, S. (1985). Passive avoidance in syndromes of disinhibition: Psychopathy and extraversion. *Journal of Personality and Social Psychology, 48,* 1316–1327.

Nichols, S. L., & Newman, J. P. (1986). Effects of punishment on response latency in extraverts. *Journal of Personality and Social Psychology, 50,* 624–630.

Norman, D. A., & Bobrow, D. G. (1975). On data-limited and resource-limited processes. *Cognitive Psychology, 7,* 44–64.

Plomin, R. (1976). Extraversion: Sociability and impulsivity? *Journal of Personality Assessment, 40,* 24–30.

Revelle, W., Humphreys, M. S., Simon, L., & Gilliland, K. (1980). The interactive effect of personality, time of day, and caffeine: A test of the arousal model. *Journal of Experimental Psychology: General, 109,* 1–31.

Robinson, T. N., & Zahn, T. P. (1988). Preparatory interval effects on the reaction time performance of introverts and extraverts. *Personality and Individual Differences, 4,* 749–761.

Rocklin, T., & Revelle, W. (1981). The measurement of extraversion: A comparison of the Eysenck Personality Inventory and the Eysenck Personality Questionnaire. *British Journal of Social Psychology, 20,* 279–284.

Roessler, R. (1973). Personality, psychophysiology, and performance. *Psychophysiology, 10,* 315–327.

Smith, A. P., Rusted, J. M., Savory, M., Eaton-Williams, P., & Hall, S. R. (1991). The effects of caffeine, impulsivity and time of day on performance, mood and cardiovascular function. *Journal of Psychopharmacology, 5,* 120–128.

Thackray, R. I., Jones, K. N., & Touchstone, R. M. (1974). Personality and physiological correlates of performance decrement on a monotonous task requiring sustained attention. *British Journal of Psychology, 65,* 351–358.

Thayer, R. E. (1967). Measurement of activation through self-report. *Psychological Reports, 20,* 663–678.

Thayer, R. E. (1978). Toward a psychological theory of multidimensional activation (arousal). *Motivation and Emotion, 2,* 1–34.

Treisman, A. M., & Gelade, G. (1980). A feature-integration theory of attention. *Cognitive Psychology, 12,* 97–136.

Treisman, A. M., & Schmidt, H. (1982). Illusory conjunctions in the perception of objects. *Cognitive Psychology, 14,* 141–149.

Walker, E. L. (1958). Action decrement and its relation to learning. *Psychological Review, 65,* 129–142.

Wallace, J. F., & Newman, J. P. (1990). Differential effects of reward and punishment cues on response speed in anxious and impulsive individuals. *Personality and Individual Differences, 11,* 999–1009.

Witkin, H. A., Dyk, R. B., Faterson, H. F., Goodenough, D. R., & Karp, S. A. (1962). *Psychological differentiation: Studies of development.* New York: Wiley.

Yerkes, R. M., & Dodson, J. D. (1908). The relation of strength of stimuli to rapidity of habit formation. *Journal of Comparative Neurology and Psychology, 18,* 459–482.

10

THE ROLE OF IMPULSIVENESS IN NORMAL AND DISORDERED EATING

MICHAEL R. LOWE and KATHLEEN L. ELDREDGE

This chapter will explore impulsivity in relation to normal eating, the moderately disordered eating of chronic dieters, and the markedly abnormal eating seen in anorexia and bulimia nervosa. In the first half of the chapter, we will review the history of attempts to apply psychological concepts to eating and weight problems. We will then consider how impulsivity—and related constructs such as sensation seeking, arousability, and external responsiveness—might determine eating patterns. The effect of dieting on eating control and the possibility that impulsivity and dieting combine synergistically to contribute to disordered eating are discussed. Finally, we will review studies on the relationship between impulsivity and eating disorders.

THE PSYCHOLOGY OF EATING BEHAVIOR: A HISTORICAL OVERVIEW

Past attempts to understand the role of psychological factors in normal and abnormal eating have gone through three main stages. The first stage, extending from the 1940s through the early 1970s, produced the so-called psychosomatic theory of obesity (Bruch, 1973; Freed, 1947; Kaplan & Kaplan, 1957). This theory emphasized the role of unconscious drives in producing overeating and obesity and suggested that eating served an anxiety-reducing role in vulnerable individuals.

The second stage, occurring from the mid 1960s to the mid 1970s, involved development of the "internal–external hypothesis" of obesity (Schachter & Rodin, 1974). This theory held that overweight people, compared with those of normal weight, were more responsive to external cues in the environment and less responsive to internal physiological cues. This viewpoint suggested that a combination of overresponsiveness to food cues and underresponsiveness to internal cues signaling hunger and satiety could explain tendencies toward overeating and obesity.

The third stage, dominating the past 20 years, emphasized the role of weight and dieting status in overeating. This stage began with Nisbett's (1972) suggestion that behavioral characteristics shown by overweight people stemmed from the suppression of weight below its biologically appropriate level. Herman and Polivy (1980, 1984) extended Nisbett's thesis in three major ways. First, rather than assessing weight suppression directly (in terms of weight or weight changes), they developed a self-report measure reflecting chronic concerns about eating and weight control. This measure, called the Restraint Scale, has been used in scores of studies during the last 20 years (for reviews, see Heatherton, Herman, Polivy, King, & McGree, 1988; Lowe, 1987). Second, these authors developed a cognitive explanation for why chronic concern with weight and eating could create a vulnerability to disinhibitory eating (Polivy & Herman, 1985). Third, they demonstrated that normal-weight individuals with high scores on the Restraint Scale showed anomalies in their eating similar to those previously observed among the overweight (Herman & Mack, 1975; Herman & Polivy, 1975).

These first two stages just described reflected a search for psychological *causes* of overeating and overweight, whereas the third stage reflected an interest in psychological *consequences* of chronic dieting and weight suppression. Unfortunately, as interest in the psychological consequences of dieting has increased, interest in psychological causes of eating and weight problems has decreased. We hope the present chapter takes a step toward redressing this imbalance. We begin by discussing definitional issues concerning the relation between impulsiveness and eating.

A useful way of viewing impulsivity and its relation to eating is to adapt the generic meaning of impulsivity ("inclined to act on impulse rather than thought," *American Heritage*, 1976, p. 662) to the specific act of eating. According to this viewpoint, "impulsive eating" would be defined simply as unplanned eating (i.e., eating that occurred outside of regular meals or snacks). Because for many people instances of such eating are in no way problematic, a further distinction needs to be made to identify impulsive eating that has adverse psychological or health-related consequences (viz., "dysfunctional impulsive eating").

It should be pointed out that whereas general agreement might be obtained on whether an instance of eating should be considered "impulsive," judgments of its dysfunctional nature will often depend on who is doing the judging. Thus, a bulimic may consider all unplanned eating as dysfunctional, whereas a therapist might consider some such instances as appropriate responses to a negative energy balance or overly rigid dietary rules.

DYSFUNCTIONAL IMPULSIVE EATING: AROUSABILITY AND EXTERNAL RESPONSIVENESS

There is a dearth of research on the relationship between eating and impulsivity in normal (non-eating-disordered) populations. Much more research and theory exists on the relationship between eating and two other constructs related to impulsivity: arousability and external responsiveness. Of course, as has been pointed out before (Rodin, 1981), these two constructs may constitute different ways of representing a common underlying process.

Arousability and Eating

L. Spitzer and Rodin (1983) provided an excellent overview of the topic of arousal-induced eating. They noted that although the data are not perfectly consistent, a variety of arousing stimuli have been associated with increased eating in both humans and rats. Of particular interest is the authors' theoretical discussion of why arousal might be linked to eating behavior.

Spitzer and Rodin started by noting that hunger and the desire to eat are governed not only by an organism's deprivation level, but by past stimuli associated with eating. A number of physiological responses (e.g., salivation, insulin release) related to food consumption are elicited by stimuli that have been regularly paired with eating. These physiological responses are thought to prepare the body to metabolize the food to be ingested, but they may also increase the amount that can be eaten (Powley, 1977). Thus, any influence intensifying the magnitude of these so-called cephalic phase responses should increase the amount of food eaten. Holding other relevant factors constant, individuals who tend to acquire the strongest cephalic phase responses would be most vulnerable to overeating (i.e., to being in a positive energy balance) and, over time, to gaining weight.

Spitzer and Rodin (1983) suggested that arousal could affect the magnitude of conditioned physiological responses in at least two ways:

> First, given that the association between various physiological responses and environmental or cognitive cues has already been learned, the presence of an optimal level of arousal may serve to energize, and thus intensify, these learned responses. Second, a more subtle role that arousal may play in determining the strength of these learned physiological responses is that it may impact on the conditioning process itself. If optimal levels of arousal enhance and strengthen conditioning of responses to particular environmental stimuli, food consumption in a given situation may be dependent not only on the individual's present level of arousal but also on arousal levels present in similar past situations. (p. 575)

These authors cited Revusky (1967) as one of several studies supporting the idea that arousal facilitates learning. Revusky showed that pairing the consumption of particular foods with food deprivation led to a subsequent increase in the ingestion of those foods. L. Spitzer and Rodin suggested that because food deprivation increases arousal, Revusky's results might be due to a deprivation-induced heightening of arousal that intensified the strength of cephalic phase responses and led to increased eating.

Spitzer and Rodin (1983) also reported the results of an interesting study whose design and results will only be summarized here. The authors attempted to develop conditioned responses to a novel food (mango sherbet) by repeatedly pairing the sight of the food with its ingestion. They measured arousal level by recording subjects' electrodermal activity (EDA) during conditioning. Because a bell was used to signal the start of each conditioning trial and the subjects showed clear EDA responses to the bell, the size of the subjects' EDA to the bell preceding the first conditioning trial was used as a measure of arousability. Subjects rated their hunger before the conditioning trials and their hunger and liking for the sherbet after the conditioning trials. Subjects returned on a second day to assess the strength of conditioning that occurred on Day 1 and rated their hunger when they arrived and again after being presented with a bowl of mango sherbet. Finally, they were also given lime sherbet and allowed to eat as much as they wished of either sherbet.

The researchers first examined the correlation between changes in hunger ratings on Day 1 (from before to after the conditioning trials) and arousability. The correlation between increases in hunger and arousability was sizable ($r = .65$, $p < .01$). Of even greater interest was the fact that arousability on Day 1 was strongly correlated ($r = .76$, $p < .005$) with increases in hunger on Day 2 (when the mango sherbet was presented but not yet tasted). Because there is reason to believe that hunger ratings may reflect the magnitude of underlying metabolic processes, these correlations suggest that more arousable subjects may have developed stronger metabolic reactions to the food conditioning procedure.

The authors reported two other findings of interest. One was a marginally significant correlation between arousability and amount of mango sherbet consumed ($r = .39, p < .10$). The other was that their measure of arousability correlated significantly with Eysenck's (Eysenck & Eysenck, 1975) measure of introversion–extroversion, $r = .59, p < .025$). Because there is independent evidence that extroverts are more arousable (Cacioppo & Petty, 1983), this correlation supports Spitzer and Rodin's use of the EDA as a measure of arousability.

Spitzer and Rodin (1983) suggested that because food

> is one of the more salient stimuli that people are often exposed to throughout their lifetime, one might expect that those people who are very arousable to environmental stimuli will develop strong conditioned responses to food, thereby showing an arousal-induced prepotent response. In this instance arousal may influence both the acquisition of a conditioned response of a given magnitude and its subsequent expression. (p. 585)

Rodin (1991) recently replicated these findings. Her study added further evidence that heightened arousal independently strengthens both the acquisition of conditioned responses to food and the amount of food eaten under conditions of elevated arousal.

External Responsiveness and Eating

In its original form, the so-called "internal–external" theory of obesity suggested that overweight people, relative to those of normal weight, were overresponsive to external (especially food) cues and underresponsive to internal cues signaling hunger and satiety (Schachter & Rodin, 1974). These characteristics presumably led to overeating and were therefore viewed as responsible for weight gain. However, a number of conceptual, methodological, and empirical considerations led Rodin (1981) to reject this hypothesis as overly simplistic.

Although responsiveness to internal and external cues may be insufficient as predictors of body weight, cue responsiveness could still be a direct determinant of food intake and thus contribute to body weight. According to Rodin (1981), cue responsiveness might affect food intake in at least two ways. First, external food cues are themselves arousing, in terms of both energizing a specific motivational state (i.e., a readiness to eat) and producing more generalized arousal (e.g., autonomic arousal).

Thus, individual differences in arousability could contribute to overeating and obesity in some hyperarousable individuals.

Second, arousability may modulate the magnitude of metabolic responses to food. These metabolic responses, which begin well before any food is ingested, prepare an organism to consume and digest more food than would otherwise be possible (Rodin, 1985). Thus, if arousability influences the extent of metabolic responses to food, the most arousable individuals should experience the largest preparatory metabolic reactions and be most prone to excessive eating. As Rodin (1985) noted in regard to individuals who are hyperresponsive to external and sensory cues, "rather than secreting quantities of hormones and digestive enzymes appropriate for effective utilization of the ingested material, such a person might oversecrete in response to external cues, and then ingest more calories in order to balance the hormonal and metabolic adjustments" (p. 6).

Rodin (1985) supported this reasoning by examining insulin release among normal-weight and overweight subjects who were categorized as "external" or "non-external" on a standardized set of measures assessing eating behavior, emotional responsiveness, and cognitive processing (Rodin & Slochower, 1976). She showed that externally responsive subjects in both weight groups had greater insulin release than nonexternal subjects when presented with a highly palatable food stimulus (a steak being grilled). Since hyperinsulinimia produces greater hunger and increased food intake (Rodin, Wack, Ferrannini, & DeFronzo, 1985), externally responsive subjects would presumably be susceptible to increased caloric consumption because of the vast number of foods and eating opportunities that people in food-abundant environments typically encounter. Along these lines, Rodin and Slochower (1976) showed that externally responsive children gained more weight during summer camp than did less external children.

The preceding discussion suggests that the concepts of arousability and externality are largely interchangeable ways of discussing the same phenomenon. Arousability refers to a variety of internal events (e.g., autonomic activation, insulin release) that occur following exposure to relevant external stimuli (e.g., food or cues that have been associated with eating). Also, as Rodin (1981) pointed out, internal and external signals interact. Thus, exposure to a palatable food (an external stimulus) may elicit insulin release (an internal stimulus), which further heightens the attention paid to, or the appeal of, the palatable food. Finally, a person's arousal level during eating may intensify the strength of conditioning to contiguous stimuli (Rodin, 1991), providing a final pathway through which arousability may be relevant in understanding eating behavior.

IMPULSIVITY AND DYSFUNCTIONAL IMPULSIVE EATING

The preceding review raises the central question addressed in this chapter: How is impulsivity relevant in understanding normal and disordered eating? Before addressing this question, it is important to distinguish between impulsiveness as a *causal* factor in eating and impulsiveness as a *description* of certain types of eating behavior. The most straightforward way in which impulsivity may be relevant is as a cause of the frequency or extent of eating. That is, impulsivity may cause certain individuals to eat more frequently or to consume more food.

However, impulsiveness is also relevant at a purely descriptive level in characterizing a form of eating behavior that occurs on the spur of the moment and without forethought. Such impulsive eating is of interest in its own right, whatever its cause. Thus, impulsive eating could be causally related to preexisting impulsiveness or to any of a variety of other influences (medications, dieting, etc.). As noted earlier, when this type of eating is viewed as contrary to an individual's interests because of its psychological or biological consequences, it is referred to as dysfunctional impulsive eating.

We turn now to a consideration of how impulsivity as an individual difference variable might be relevant to the development of "excessive" eating. In doing so, we will also refer to literature on sensation seeking, under the assumption that impulsivity and sensation seeking are closely intertwined constructs (see Zuckerman, chapter 5, this volume). Also, because impulsive and sensation-seeking individuals are thought to be similar on the dimension of arousal and arousability, impulsiveness and sensation seeking will be treated as interchangeable constructs here. To make this assumption explicit, we will use Zuckerman's term "ImpSS" to discuss the impulsiveness/sensation-seeking construct.

At a purely descriptive level, it is perhaps logical to expect that high-ImpSS individuals would be vulnerable to developing a wide variety of problems related to immediate need gratification. This may occur, at least in part, because impulsive individuals are especially sensitive to reward (Gray, 1987). In addition to the possibility of developing problems because of an overresponsiveness to the immediate rewards of eating, high-impulsive people might be expected to develop other problems involving immediate need gratification (e.g. alcohol and drug abuse, sexual promiscuity). Thus, one might expect that if impulsivity contributes to the development of eating disorders, eating-disordered individuals would be more likely to show generalized problems involving impulse control. This possibility will be considered in the second half of this chapter.

Of greater immediate interest is exploring how the ImpSS dimension may be relevant to the previous discussion of the relationships among arousability, external responsiveness, and eating. As already noted, the work of Rodin and others has suggested that arousability[1] may play a role in determining the strength of the conditioned responses that people develop to food and food-related stimuli. Furthermore, Rodin (1991) has suggested that "there may be prewired determinants that distinguish individuals in their degree of arousability" (p. 142). For instance, Rodin reviewed data based on a study by Milstein (1980) that indicated that newborns of overweight parents showed greater visual and gustatory responsiveness than did newborns of normal-weight parents. Furthermore, although the two groups of newborns did not differ in weight at birth, follow-up data indicated that both measures of responsiveness significantly predicted weight when the children were 3 years old.

It therefore appears possible to link the area of arousal-induced eating to impulsive eating if it can be shown that impulsivity and arousability are positively related. In fact, such a relationship does exist. A general characterization of the high-ImpSS individual is someone whose basal arousal level is low (S. B. G. Eysenck, chapter 8, this volume; King, Jones, Scheuer, Curtis, & Zarcone, 1990; O'Gorman & Lloyd, 1987; Zuckerman, 1991) but whose arousability to external stimulation is high. For example, Barratt, Pritchard, Faulk, and Brandt (1987) found that impulsiveness was associated with increased amplitude of the visual evoked potential with increasing intensities of external stimulation. Zuckerman, Simons, and Como (1988) found that high sensation seekers had augmented evoked potentials and stronger heart rate responses to visual and auditory stimuli. Smith, Davidson, Smith, Goldstein, and Perlstein (1989) found that high sensation seekers showed enhanced phasic electrodermal responsiveness, and noted that "clearly, high sensation seekers are the more aroused or arousable group, and this positive correlation between sensation seeking and psychophysiological arousal is enhanced at the higher intensities of stimulation" (p. 677).

Although these studies suggest a connection between impulsivity and arousability, it is unknown whether highly arousable individuals actually experience more impulse control problems related to eating. However, if highly impulsive individuals are lower in basal arousal but more arousable to salient environmental stimulation, these characteristics might influence their eating behavior in several ways. First, because eating itself may increase arousal, high-ImpSS individuals might tend to use eating to raise their arousal to a more optimal level. In this case, eating would be used in much the same way as drugs or highly stimulating activities to regulate arousal level (Zuckerman, 1991). This type of eating could produce a positive energy balance and weight gain over time.

---

[1]Because Rodin (1981) has suggested that the effect of external responsiveness may by mediated by the activation of general (e.g., electrodermal) and metabolic (e.g., insulin) arousability, the term *arousability* will henceforth be used to refer to both the literature on arousal-induced eating (L. Spitzer & Rodin, 1983) and eating associated with external responsiveness (Rodin, 1981, 1985).

Second, stress-inducing life events may have different appetitive effects on individuals differing in ImpSS. For low-ImpSS individuals, whose characteristic arousal level is relatively high, external stressors might create excessive autonomic arousal, thereby suppressing appetite (Schachter, Goldman, & Gordon, 1968). High-ImpSS individuals, whose characteristic arousal level is lower, would presumably not experience a stress-induced increase in autonomic arousal sufficient to suppress eating.

Third, if high-ImpSSs are more arousable than low-ImpSSs, Rodin's theory of arousal-induced eating (Rodin, 1991; L. Spitzer & Rodin, 1983) would predict stronger associations between eating cues and the metabolic responses that precede and help drive eating behavior. Thus, if low- and high-ImpSS individuals had identical conditioning histories of eating paired with particular stimuli, future exposure to such stimuli would be more likely to produce eating in high-ImpSS individuals.

It must be acknowledged, however, that although these considerations point to a possible connection between impulsivity and eating behavior, this link is based on theoretical considerations, not on empirical findings. Little research exists on the relationship between impulsiveness (or sensation seeking) and eating behavior (for exceptions, see studies by Israel, Stolmaker, & Prince, 1983; Shaye, 1989). By providing a rationale for a possible link between impulsivity and eating, we hope that more research will be conducted on these important topics.

DIETING AND IMPULSIVE EATING

Impulsive eating can occur for a number of other reasons besides an individual's underlying impulsiveness, arousability, or external responsiveness. One influence that may create a vulnerability to impulse eating is dieting behavior. Such a possibility is of interest in its own right and may also be important in understanding how impulsivity and rigorous dieting may combine to produce disordered eating of clinical proportions.

The theory of restrained eating suggests that chronic dieting may make restrained eaters vulnerable to disinhibitory eating. This occurs when negative affect or consumption of a "forbidden" food undermines a restrained eater's "diet boundary," thereby permitting the behavioral expression of underlying hunger (Polivy & Herman, 1985). Restrained eaters' affect- or preload-induced overeating can be viewed as dysfunctional impulsive eating because it is spontaneous, largely uncontrolled, and runs counter to the restrained eater's long-term goals.

Lowe (1993) recently developed a model of dieting behavior that suggests that dieting could produce excessive eating for two different reasons. Essentially, Lowe argued that "restrained eating" and "dieting" are separate phenomena that have different effects on appetite. For example, restrained eaters show an enhanced salivary response to food (Klajner, Herman, Polivy, & Chhabra, 1981), whereas current dieters show a reduced salivary output to food (Durrant, 1981; Rosen, 1981). Also, whereas restrained eaters tend to eat more with than without a preload (Herman & Mack, 1975), current dieters do the opposite (Lowe, Whitlow, & Bellwoar, 1991).

Lowe (1993) suggested that restrained eaters' vulnerability to negative-affect eating and preload-induced eating stemmed not from a current state of cognitive or dietary restraint, but from an extensive history of dieting and overeating. In the area of emotional eating, Lowe noted that restrained eaters have not only dieted frequently in the past, but have also repeatedly "gone off" their diets when stressed or upset. Thus, restrained eaters presumably have an extensive conditioning history of overeating when distressed, and this conditioning history could explain their susceptibility to negative-affect eating (Booth, 1988). Lowe (1993) also suggested that frequent dieting and overeating could make restrained eaters, compared with unrestrained eaters, less sensitive to internal cues and more sensitive to external cues to determine how much they should eat. Thus, restrained eaters may overeat following the required consumption of a milkshake preload because the forced preload creates expectations for the sort of eating that "should" occur in the eating test. Restrained eaters may be more sensitive than unrestrained eaters to such expectations because their history of ignoring or overriding internal eating controls produces an overreliance on external guides for eating (see Heatherton, Polivy, & Herman, 1989; Lowe, 1993).

Lowe (1993) also argued that the vulnerability of restrained eaters to disinhibitory eating could not stem from their current dieting behavior because dieters respond to disinhibitory challenges by decreasing, rather than increasing, their eating (Eldredge, 1993; Lowe, 1993; Lowe et al., 1991). Furthermore, when not preloaded, individuals who say they are "currently on a diet to lose weight" eat significantly more than restrained eaters who are not dieting. These findings suggest that current dieters may be vulnerable to overeating when they are merely exposed to palatable food and are invited to eat it. Because these dieters were trying to lose weight, this overeating response could be characterized as "impulsive."

Another recent study (Logue & King, 1991) found results consistent with this conclusion. This study used a self-control paradigm in which subjects had to repeatedly choose between the immediate consumption of a small amount of a preferred beverage or the delayed consumption of a larger amount. Subjects showed wide individual differences in their level of self-control, and only one variable—current dieting status—predicted self-control. Current dieters showed an "impulsive" consummatory pattern; that is, they choose the immediate over the delayed reward relatively more often than did the nondieters.

Overall, the research described in this section suggests that frequent dieter/overeaters and current dieters are both vulnerable to overeating, but under different circumstances. Frequent dieter/overeaters appear to be most vulnerable when experiencing negative affect or when exposed to social influences promoting eating. Current dieters seem more vulnerable when simply exposed to palatable food in the absence of any compelling environmental provocation (see also Grilo, Shiffman, & Wing, 1989). However, each of these types of diet-related eating are—from the perspective of the weight-conscious individual—unplanned, unwelcome, and counterproductive to long-term weight control. Thus, there is considerable evidence that dieting itself may predispose to the development of problems with "dysfunctional impulsive eating."

IMPULSIVITY, DIETING, AND DYSFUNCTIONAL IMPULSIVE EATING: POSSIBLE INTERACTIONS

We have considered the possibility that impulsivity or sensation seeking may contribute to the development of overeating. We also considered the hypothesis that a second influence—dieting—may independently contribute to the development of overeating problems. Also of considerable interest is the possibility that these two influences could combine synergistically to create disordered eating, perhaps of clinical proportions.

As noted previously, one way in which impulsivity and dieting could be related is through the effect of impulsiveness on eating. If it is true that impulsive people learn to eat in response to more varied stimuli, or to consume more calories overall, then impulsive individuals in our weight-conscious society would be most likely to develop concerns regarding weight. Along these lines, Jansen, Klaver, Merckelbach, and van den Hout (1989) noted similarities between eating disorders and addictions and suggested that restrained eaters, like addicts, would show rapid psychophysiological habituation and be high in sensation seeking. Restrained eaters did indeed differ from unrestrained eaters on these variables. Jansen et al. explained these findings by suggesting that sensation seekers have a greater risk of being "excessive consumers" of various stimulating substances, including food. Thus, individuals may become restrained in their eating as a means of protecting themselves from undesirable consequences (e.g., guilt, weight gain) of such overconsumption. According to this viewpoint, it is rapid habituation, rather than chronic low arousal, that motivates sensation-seeking behavior and the subsequent need for restraint.

Although the possibility of a link among sensation seeking, restraint, and overeating should be pursued, it is unlikely that the relationship between sensation seeking (or impulsivity) and disordered eating is sufficient to explain the development of eating and weight problems. For instance, it may be that only individuals whose impulsivity exceeds some elevated threshold become vulnerable to using food to modulate arousal. (A similar argument has been made concerning the relationship between sensation seeking and psychophysiological responsiveness; see Smith et al., 1989.) Also, it is important to recognize that increased caloric intake will not always translate into weight-control problems, because metabolic and other biological factors determine the propensity to gain weight (Rodin, 1981).

To the extent that highly impulsive individuals are susceptible to overeating and weight gain, they may turn to dieting to prevent weight gain or to lose weight. However, dieting is unlikely to produce a long-term solution to weight control problems; on the contrary, both acute and chronic dieting may create new problems with dysfunctional impulsive eating. These problems could be exacerbated among highly impulsive individuals. For instance, if dietary control tends to deteriorate under emotional stress (LaPorte, 1990; Lowe, 1993), frequent dieters will have numerous "conditioning trials" that pair negative affect with loss-of-control eating. If it is true that impulsive (or highly arousable) individuals tend to develop stronger conditioned reactions, then such conditioning might be especially powerful among individuals who are dispositionally high in impulsivity.

There are two additional ways in which dieting and impulsivity may combine to cause or maintain an eating disorder. One is that highly impulsive individuals may find dieting particularly difficult. Successful dieting involves repeatedly denying oneself a powerful source of gratification, an ability notably lacking in impulsive individuals. Thus, over time impulsive individuals may experience more dieting failures, thereby creating further problems in their ability to regulate eating (Lowe, 1993). The second is that if dieting is unsuccessful, the impulsive individual may go to more extreme lengths—such as purging—to control his or her weight.

In sum, we are suggesting that there are several different pathways through which impulsivity is relevant to eating behavior. However, it is critical to emphasize that the role of impulsivity among psychologically healthy people may be quite different than that among the relatively small subgroup of individuals with eating disorders. Among the general population, impulsivity may be one of a host of factors that determine susceptibility to developing eating or weight problems. Among those with such problems, however, impulsivity may be more relevant in understanding eating-disorder symptoms (e.g., whether the disorder involves binging or not), the co-occurrence of other impulsive problems, or the prognosis of individual cases. With this caveat in mind, we now turn to a consideration of the role of impulsivity in eating-disordered individuals.

Impulsivity as a Risk Factor for the Development of Eating Disorders

The relation between impulsivity and clinical eating disorders is multifaceted. Psychologists have begun to recognize the impulsive nature of some eating-disordered individuals to present with a wide constellation of impulsive behaviors. These observations, coupled with the growing recognition that not all eating-disordered patients respond equally well to focused interventions, have prompted investigations into the role of impulsivity in the development, manifestation, and treatment of eating disorders. Research conducted to date on each of these related topics will be the focus of the remainder of this chapter.

It should be noted that although no standard definition of impulsivity has been commonly adopted by eating-disorder researchers, a variety of behaviors have been frequently identified as impulsive in nature. These behaviors include substance abuse, interpersonal violence, impulsive shopping, stealing, promiscuity, and repetitive suicidal gestures or attempts. These behaviors have been considered impulsive because of the inherent disregard for long-term consequences associated with each. Although many other types of behavior might also fit this definition, they have not as yet been a primary focus of research in connection with eating disorders and will therefore not be addressed in this overview.

If impulsivity is a risk factor for the development of eating disorders, then it should be possible to demonstrate empirically that possessing a personal or family history of impulsivity would increase the probability of developing an eating disorder. The question of the predisposing nature of a personal history of impulsivity could best be addressed in longitudinal studies in which baseline measures of impulsivity are examined as predictors of subsequent eating-disorder symptomatology. To date, this research strategy has not been used. A sizable body of research has been conducted, however, on the possible link between a family history of impulsivity and the development of eating disorders. Such studies have sought to determine the prevalence of eating disorders and other impulsive behaviors in the relatives of impulsive individuals, as well as the prevalence of impulse control problems in the relatives of eating-disordered individuals compared with appropriate control groups. Results from studies using each of these research strategies will be addressed in turn.

If a family history of impulsivity predisposes to the development of eating disorders, then it could be predicted that a higher prevalence of eating disorders and other impulsive behaviors would be found among individuals with a positive family history of impulsivity, compared with individuals with no such family history. In a survey of 141 female and 72 male college students, Leon and colleagues (Leon, Carroll, Chernyk, & Finn, 1985) found support for this hypothesis among female but not male students. Forty-three percent of women who engaged in binge-eating episodes at some time during the past few years, as compared with 17% of the female nonbingers, reported having at least one first-degree relative with a history of substance abuse. Among men, bingers and nonbingers did not differ in the percentage reporting a family history of substance abuse (26% each).

An investigation by Mitchell, Hatsukami, Pyle, and Eckert (1988) also lends tentative support to this hypothesis. Mitchell and colleagues compared the substance abuse histories of *DSM–III* (American Psychiatric Association, 1980) diagnosed bulimic outpatients both with ($n =$ 102) and without ($n = 173$) self-reported family histories of drug abuse. Although no difference between groups was found for a self-report of alcoholism, drug abuse, or problems with alcohol or drugs, bulimics with a positive family history were significantly more likely to have received prior treatment for drug usage problems (27% with a positive family history vs. 12.3% with a negative family history). The authors rightly cautioned, however, that these self-report results may not represent a differing prevalence rate of drug problems between the two groups, but rather a higher tendency to seek treatment among individuals more highly sensitive to the issue of drug abuse.

The hypothesis that impulsivity is a risk factor for the development of eating disorders suggests that impulsivity may be transmitted along familial lines. This predisposition may operate such that individuals from impulsive families are at heightened risk for the development of eating disorders in general, or specific eating-disordered behavior in particular (e.g., binge eating). It is possible, however, that a family history of impulsivity serves as a risk factor for the development of psychopathology, rather than posing a particular risk factor for the development of eating disorders. Thus, a more specific hypothesis is that a family history of impulsivity is more likely to be associated with the development of eating disorders than some other (nonimpulsive) form of psychopathology.

A preliminary investigation of this first hypothesis was conducted by Hatsukami, Mitchell, Eckert, and Pyle (1986). These authors elicited self-report information about substance abuse in the parents of 46 women with bulimia only, 34 bulimic women with a history of affective disorder, and 34 bulimic women with a history of substance abuse (all subjects were referrals to an outpatient eating disorders treatment program). Hatsukami and colleagues found that 18% to 32% of the subjects' fathers and 8% to 12% of the subjects' mothers were described as having alcohol problems (these statistics were not broken down by bulimic subgroups). The authors compared these rates with the general population rates for alcoholism of 15% in men and 6% in women, and suggested that a parental history of alcoholism serves as a risk factor for the development of bulimia. A study by Eckert and colleagues (Eckert, Goldberg, Halmi,

Casper, & Davis, 1979) with a sample of 103 anorexic inpatients, however, did not indicate an elevated risk of alcoholism among first-degree relatives. These researchers found a 16.5% prevalence among fathers and a 1.9% prevalence among mothers. Unfortunately, these researchers did not report on whether there was a higher risk of familial alcoholism among bulimic than restricting anorexics; if this were the case, then it would suggest that substance abuse is a specific risk factor for binge eating.

In research with binge eaters, Leon and colleagues (Leon et al., 1985) conducted another study in which they administered a self-report inventory to assess the personal and familial history of substance abuse among 37 female bulimic (by DSM–III diagnosis) outpatients and their first-degree relatives. The results indicated that 51% of the bulimics reported having at least one first-degree relative who had been diagnosed as chemically dependent by a health professional. Interestingly, a positive family history of chemical dependency was related neither to the bulimics' own self-reported history of chemical dependency or drug abuse nor to their patterns of shoplifting or laxative use or the frequency of their binge eating or vomiting. These results are consistent with the hypothesis that chemical dependency in the family is a risk factor for the development of bulimia. However, they also indicate that having a family history of chemical dependency is apparently unrelated to the presence or extent of additional forms of impulsive behavior in diagnosed bulimics. Because this study did not include male bulimics, it cannot be concluded that such a proposed link is gender specific. However, the fact that only a small fraction of bulimics are male would support gender specificity. The results of all of the studies reported thus far must be interpreted cautiously because of the primary reliance on self-reported diagnoses by patients rather than on structured diagnostic interviews with their relatives.

Hudson and colleagues (Hudson, Pope, Jonas, Yurgelun-Todd, & Frankenburg, 1987) improved on the aforementioned methodology in a study comparing the family history of psychiatric disorders among 69 bulimic probands (9 of whom had a prior history of anorexia nervosa), 24 probands with major depression, and 28 nonpsychiatric control probands. These researchers conducted diagnostic interviews with probands to determine the risk of psychiatric disorders (using DSM–III criteria) among their first-degree relatives. The risk of substance abuse disorders (including alcohol and other substances) was significantly greater among the relatives of bulimic probands (19.1% of a sample of 283 relatives) compared with those of depressed (9.1% of a sample of 104 relatives) or nonpsychiatric control (6.5% of a sample of 149 relatives) probands. Once again, however, it must be cautioned that these findings are not based on structured diagnostic interviews with the relatives themselves, nor is any distinction made between bulimics with and without a prior history of anorexia nervosa (two groups that possibly do not share the same etiology).

A study by Stern and colleagues (Stern et al., 1984) addressed some of the methodological limitations of the previous studies. These researchers interviewed one parent (in all but two instances, the mother) of 27 women who met DSM–III criteria for bulimia (with no prior history of anorexia nervosa) and of 27 nonpsychiatric controls matched for age, sex, and race. On the basis of this interview, lifetime prevalence of psychiatric disorders for all first- and second-degree relatives, excluding children, of each proband was determined using Research Diagnostic Criteria (RDC; R. L. Spitzer, Endicott, & Robins, 1975). Stern found no difference in the prevalence of substance use disorders among relatives of bulimic (18.2% of a sample of 368) and control (6.5% of a sample of 384) probands.

A recent study conducted at the National Institute of Mental Health laboratories, however, did demonstrate such a difference. Kassett and colleagues (Kassett et al., 1989) used RDC to establish psychiatric diagnoses for first-degree relatives of 40 female bulimics (DSM–III–R criteria; American Psychiatric Association, 1987) with no prior history of anorexia nervosa and 24 nonpsychiatric control probands. Of the 303 first-degree relatives identified, diagnoses for 62% were based on structured diagnostic interviews, with the remaining 38% diagnosed on the basis of information provided by interviewed probands and relatives. A significantly higher rate of alcoholism was found among relatives of bulimics (27.6%) than among those of nonpsychiatric controls (13.6%). In addition, the rate of drug abuse among relatives of bulimics with a concurrent major affective disorder (15%) was significantly higher than that of relatives of controls (5.9%).

An early study conducted by Strober (1983), which sought to develop a typology of anorexia nervosa subtypes through multivariate cluster analyses on personality variables, addressed this issue. Strober's analyses of 130 female anorexia nervosa inpatients yielded three distinct personality subtypes. Of particular interest to the present review, one subtype (which comprised 15% of the patients and was labeled "Type 3" in the study) was characterized by Minnesota Multiphasic Personality Inventory (MMPI) profiles suggestive of significantly greater tendencies toward impulsivity, emotional lability, and moodiness than the other two subtypes. Strober conducted structured interviews (RDC) with each of the patients' parents to establish clinical ratings of the presence or absence of a history of alcoholism or depressive disorder prior to the patient's development of her eating disorder. Interestingly, Strober found a significantly higher parental history of alcoholism or depression among Type 3 anorexics than among the other two subtypes. Unfortunately, Strober did not report the separate prevalence rates for depression and alcoholism, making it impossible to determine if familial alcoholism, as a form of impulsive behavior, is a specific risk factor for probands' own impulsivity.

The Link Between Familial Impulsivity and Binge Eating

The symptom of binge eating usually occurs as an unplanned eating episode involving a feeling of loss of control and is thus experienced to have an impulsive quality. The impulsive nature of binge eating raises the issue of whether eating-disordered individuals who engage in binge eating might not be at increased risk of developing this particular symptom because of a family history of impulsivity. In research investigations with anorexia nervosa patients, this issue has been addressed through comparisons of the family histories of restricting anorexics without a history of bulimia with those of bulimic anorexics or anorexics with a past history of bulimia nervosa. Three studies will be reported that have made such comparisons.

Hudson and colleagues (Hudson, Pope, Jonas, & Yurgelun-Todd, 1983a) compared the rates of psychiatric disturbance in first-degree relatives of 14 restricting anorexics without a history of bulimia nervosa, 20 anorexics with either a current or a past history of bulimia, and 55 bulimics. The authors made DSM–III diagnoses on the basis of interviews with probands (interviewers were not unaware of the probands' diagnoses) and, when available, their family members. The results suggested group differences in the rate of substance abuse disorders among first-degree relatives: 8.6% of anorexia nervosa versus 10.8% of bulimia and 18.2% of anorexia nervosa with bulimia. Unfortunately, no statistical tests were conducted on these rates to determine whether these differences are significant.

Another study, conducted by Piran and colleagues (Piran, Lerner, Garfinkel, Kennedy, & Brouillette, 1988), compared the family histories of 30 restricting and 38 bulimic anorexics (all were inpatients or awaiting inpatient admission at the time of assessment, and received DSM–III diagnoses). On a self-report questionnaire, bulimic anorexics reported significantly higher rates than did restricting anorexics of parental substance abuse (34.2% vs. 6.7%), sibling substance abuse (26.3% vs. 3.3%), and interpersonal violence both between parents (34.2% vs. 6.7%) and among siblings (26.3% vs. 3.3%). These results must be viewed cautiously, however, because they are based on self-reports rather than on clinical diagnostic interviews.

The most methodologically sophisticated study on this issue was conducted by Strober, Salkin, Burroughs, and Morrell (1982). These authors matched 35 bulimic anorexics to 35 restricting anorexics on age, duration of illness, percentage weight loss, and social class. All anorexics were selected from a pool of consecutive admissions to a university hospital. At the time of admission, MMPIs were administered to the patients' parents; shortly thereafter, Schedule for Affective Disorders and Schizophrenia interviews were conducted on all first-degree relatives over the age of 15, with additional information solicited from parents concerning second-degree relatives, using the Family History Interview (Endicott, Spitzer, & Andreasen, 1975). Strober and colleagues found evidence for greater impulsivity among family members of bulimic anorexics. Fathers but not mothers of bulimic anorexics scored significantly higher on the MMPI 4 scale (a marker of impulsivity), whereas 83% of bulimic anorexics, compared with 49% of restrictors, had a positive family history of alcoholism. Once again, fathers but not mothers of bulimic anorexics evidenced a significantly higher rate of alcoholism (29%) than did their restrictor counterparts (9%). No differences among groups were found for all relatives on the prevalence of drug use disorders.

Finally, only one study to date has addressed the question of whether a family history of impulsivity is associated with binge eating in the obese. Wilson, Nonas, and Rosenblum (1993) classified as binge eaters or non-binge eaters 71 men and 99 women in a weight control program for moderate to severe obesity. On the basis of self-reports of family history, no difference was found between binge eaters and non-binge eaters on the rate of alcohol abuse in parents or siblings.

Summary of Findings Regarding the Role of Impulsivity in the Development of Eating Disorders

Although the findings of the studies cited above vary, overall they lend tentative support to the hypothesis that a family history of impulsivity predisposes to the development of specific eating-disorder symptoms. In particular, a family history of substance abuse or some other impulsive behavior appears to be associated with an elevated risk of the development of binge eating, whether seen in bulimia or in bulimic anorexia. The data are inconclusive as to whether this link with familial impulsivity is specifically for binge eating or for more generalized difficulties with impulse control. Furthermore, many of the these studies did not perform diagnostic interviews with probands and their relatives, instead relying heavily on self-report for family diagnoses. Such methodological limitations introduce the possibility of bias in the results presented. Nonetheless, the data warrant further exploration.

EVIDENCE FOR AN ASSOCIATION BETWEEN EATING DISORDERS AND OTHER IMPULSIVE DISORDERS

In the previous section, evidence was cited that suggests that impulsivity plays a role in the development of at least certain forms of eating-disordered behavior (i.e., binge eating). If such eating-disordered behavior stems from a more generalized trait of impulsivity, then it is reasonable to expect an elevated rate of additional forms of impulsive behavior among individuals with eating disorders. A survey of the literature shows that research on this issue has been conducted from three complementary perspectives: (a) examinations of the prevalence of eating disorders among women with other impulsive disorders, (b) examinations of the prevalence of impulse control problems among eating-disordered individuals (at times in comparison with that of select normal samples), and (c) comparisons of impulse control problems among subsets of eating-disordered individuals. Research findings from each of these perspectives will be considered in turn.

The Prevalence of Eating Disorders Among Individuals With Other Impulsive Disorders

Much of the research on this topic has focused on female alcoholics. Beary, Lacey, and Merry (1986) examined the rate of previous eating disorders in a sample of 20 consecutive female admissions for alcohol dependence to an inpatient alcohol unit. On the basis of a semistructured clinical interview by a psychiatrist experienced in the treatment of eating disorders, 7 of these women (35%) were diagnosed with a history of bulimia (4 patients), anorexia nervosa (2 patients), or massive obesity (1 patient). In a survey of 27 female alcohol clinic attenders (both inpatients and outpatients), Lacey and Moureli (1986) found similar results. All of the these women, who had diagnoses of mild to moderate alcohol dependence, were administered a structured interview to determine their rate of eating-disordered behavior. Eleven of the women, or 40%, reported a pattern of either current (7 patients) or past (4 patients) binge eating; each of these women admitted the use of some method of compensation for their binge eating (vomiting, food avoidance, or purging).

A final survey of alcoholics, conducted by Peveler and Fairburn (1990) with consecutive female first-attenders of an alcohol unit (inpatients and outpatients), supported the preceding findings. These authors administered the EDE-Q (Beglin & Fairburn, 1992), a self-report diagnostic interview for eating disorders, to 31 women whose alcohol dependence ranged in severity from mild to severe (52% met the latter criterion). Thirty-six percent of the women sampled admitted to having recent difficulties with binge eating, whereas 26% met diagnostic criteria for a probable current eating disorder. Additionally, 19% of the sample met diagnostic criteria for a probable history of anorexia nervosa. Peveler and Fairburn pointed out that the co-occurrence of alcohol dependence and eating disorders found in this clinic sample (and that of Lacey and Moureli) was higher than might be expected by chance, given the significantly lower rate of eating disorders in the general population. However, these authors also caution that these results may reflect a selection bias in that patients with dual diagnoses may be more likely to seek treatment.

These results with female alcohol abusers suggest the possibility of an elevated rate of eating disorders among other substance abusers (for a more thorough review of the literature on the connection between eating and addictive disorders, see Wilson, 1991). The only study that has addressed this possibility was conducted with cocaine abusers. Jonas and colleagues (Jonas, Gold, Sweeney, & Pottash, 1987) administered by telephone a structured clinical interview to diagnose the lifetime prevalence of eating disorders among 259 consecutive callers to the National Cocaine Hotline (all callers met DSM–III criteria for cocaine abuse). Twenty-two percent of the callers met diagnostic criteria for bulimia without self-induced vomiting, whereas 7% met criteria for both anorexia nervosa and bulimia. The one caveat to these results, however, is the unusually high percentage of men (44%) in the identified sample of bulimics. Because men usually compose only a small percentage of the bulimic population (roughly 5%), these results suggest something unusual about the sample of callers, or possibly the diagnostic criteria used for bulimia (only 9% of the callers met criteria for bulimia with compensatory purging).

Two studies in this area that did not focus on drug abusers are noteworthy. The first study, conducted by Myers and Burket (1989), examined the eating-disorder symptoms, by self-report questionnaire, of 41 female adolescent residents of juvenile delinquent facilities (typical offenses warranting commitment included stealing, truancy, substance abuse, and assault). Experience with binge eating was reported by 73% of the subjects, with 38% reporting two or more binges weekly. Vomiting was also reported by 37%, again with a smaller percentage (30%) meeting the diagnostic criterion of two or more times per week. Overall, 22% of the residents were diagnosed with probable eating disorders: 7.5% with bulimia, 2.5% with bulimic anorexia, and 12% with eating disorders not otherwise specified. The second study was a review by McElroy and colleagues (McElroy, Hudson, Pope, & Keck, 1991) of studies concerning the characteristics and associated pathology of kleptomaniacs. These authors noted that across studies, 11% of kleptomaniacs reported "eating disturbances" ranging from cravings for sweets to binge or purge episodes. In a review of studies with eating-disorder patients, the authors collapsed data to yield an overall average rate of 21% displaying kleptomania or impulsive stealing behavior. They further noted that eating-disorder patients who stole were more likely than their nonstealing counterparts to be bulimics or bulimic anorexics than to be restricting anorexics.

The Prevalence of Impulse Control Problems Among Eating-Disordered Individuals

The conceptualization of binge eating as a form of impulsive behavior has prompted a large number of investigations into the comorbidity of eating disorders and other impulsive behaviors. The bulk of this work has been conducted with bulimics. A number of researchers have reported on the current or past history of alcohol or chemical dependency among their outpatient or inpatient bulimics. In one of the earliest published surveys, researchers at the University of Minnesota (Pyle, Mitchell, & Eckert, 1981) evaluated 34 outpatient bulimics (with *DSM–III* diagnoses) for past or current comorbid psychopathology. Eight of the bulimics surveyed (23.5%) had received prior treatment for chemical dependency, with 1 additional patient admitting to an untreated history of alcohol abuse. In a subsequent larger survey of 108 consecutive bulimic outpatients (*DSM–III* diagnoses), researchers from this clinic found 18.5% of their patients to meet *DSM–III* criteria for alcohol or drug abuse (by current or past history), with 13.9% of the sample reporting prior treatment for their substance abuse (Hatsukami, Eckert, Mitchell, & Pyle, 1984). The largest scale survey published by this group of researchers reported diagnoses based on the self-reports of 275 bulimic outpatients (Mitchell, Hatsukami, Eckert, & Pyle, 1985). In this sample, 34.4% admitted to a history of problems with alcohol or other drugs, whereas 23% indicated a history of alcohol abuse and 17.7% reported a prior history of treatment for chemical dependency.

Lacey and Moureli (1986) noted an 18% prevalence rate of alcoholism or alcohol abuse among their bulimic clinic patients, which is comparable to the prevalence rates found in the Minnesota group's studies. Unfortunately, information was not provided on the sample size or diagnostic criteria used to obtain this statistic. In a nonconsecutive sample of 20 bulimic clinic attenders, Beary et al. (1986) found 40% of the bulimics to meet diagnostic criteria for alcohol abuse, with an additional 10% classified as engaging in "excessive" alcohol use. A further study, included in the same report, of 112 consecutive bulimic patients demonstrated that rates of alcohol abuse and excessive use rise with age, such that 50% of these patients met one of these criteria by the age of 35.

It should be noted that the majority of the results reported thus far come from a single research site in Minnesota. The consistency of these findings may therefore reflect the particular make-up of the patient population. Two surveys conducted in different settings yielded somewhat lower prevalence rates. Leon and colleagues (Leon et al., 1985) conducted a survey of 37 bulimic outpatients (*DSM–III* diagnoses). A large percentage of the sample (61.1%) admitted to having "engaged in excessive use of alcohol," whereas current drug use was reported by 21.2% of the bulimics. These rates may be misleading, however, because it cannot be determined whether such alcohol or drug use could be classified as abuse or dependence. Indeed, a far smaller percentage of bulimics, 6.7%, admitted to having been diagnosed by a health professional as chemically dependent. All of the results from this study may be biased, however, because the sample consisted of only 57% of a larger pool of 65 subjects asked to participate in the study. Johnson and colleagues (Johnson, Stuckey, Lewis, & Schwartz, 1982) also found a low prevalence rate of alcohol or street drug use (< 10%) in a self-report mail survey of 316 bulimic women (all with probable *DSM–III* diagnoses). Note that this statistic reflects a current rather than a past history of substance use/abuse and may therefore significantly underrepresent the extent of substance use problems. Also, these results are based on a 68% response rate to the initial mailing and may represent a select subset of bulimic individuals.

One other form of impulsive behavior that is commonly believed to be present among bulimics is stealing. Clinicians often note the tendency of their bulimic patients to steal food, usually for their binges. Fifty percent of the bulimics surveyed by Leon et al. (1985) and 44% of those surveyed by Pyle et al. (1981) admitted to stealing binge food. The question remains whether this represents a problem with impulse control in general or financial difficulties in meeting bingeing requirements. However, 9 bulimics (26%) in Pyle et al.'s sample reported a history of stealing predating the development of their eating disorder, and 21% admitted to currently stealing items other than food.

A number of studies have sought to compare the prevalence of impulsivity among bulimics with that found in normal, nonpsychiatric populations. The Minnesota group (Pyle et al., 1983) compared the responses to a self-report questionnaire by 1,355 college freshmen and 37 bulimic outpatients (*DSM–III* diagnoses). Among the student sample, probable diagnoses of *DSM–III* bulimia were derived on the basis of diagnostic items on the questionnaire. Pyle et al. found a significant difference in the self-reported prevalence of alcohol or drug abuse between nonbulimic and bulimic students[2] (3.6% vs. 13.3%, respectively), as well as a significant difference in the self-reported history of stealing (3.1% of nonbulimic vs. 13.3% of bulimic students). Compared with their student counterparts, bulimic outpatients did not differ in their prevalence of a history of alcohol or drug abuse (27% of bulimic outpatients) but were significantly more likely to have a history of stealing behavior (56.8% of the sample).

---

[2]It should be noted that one study failed to find differences between bulimics and nonbulimics on a measure of sensation seeking (Schumaker, Groth-Marnat, Small, & Macaruso, 1986).

Bulik conducted two separate studies that investigated the comparative prevalence rates of substance abuse among bulimics and normal controls. In one study, 12 bulimics (*DSM–III* diagnoses) currently involved in outpatient eating-disorder support groups were compared with 12 normal controls (solicited by advertisement) matched for age, gender, and socioeconomic status (Bulik, 1987a). No differences were found between groups on the Michigan Alcoholism Screening Test (Selzer, 1971), a standard self-report screening instrument for current and past alcohol abuse problems. In a larger scale survey, Bulik (1987b) conducted structured clinical interviews to derive *DSM–III* diagnoses with 35 bulimic (*DSM–III* diagnoses) women and 35 normal female controls (all subjects were solicited by radio, TV, and print announcements). Bulimics had significantly higher rates of current alcohol and drug abuse and dependence (from 22.9% to 48.6%) than did control subjects (from 0% to 8.6%). Although the prevalence rates of substance abuse among bulimics are informative, Bulik's comparison with those of normals are suspect because she screened the normal sample for, among other things, alcoholism and major depression.

A less biased study was conducted by Stern et al. (1984) with 27 outpatient bulimics (*DSM–III*) with no prior history of anorexia nervosa and 27 age-, race-, and gender-matched controls (subjects were screened out only for a prior history of anorexia, bulimia, or nonaffective psychosis, or for current evidence of a probable eating disorder). Lifetime *DSM–III* diagnoses (by structured clinical interview) of substance use disorder were significantly more prevalent among bulimics than controls (30% vs. 4%, respectively), with marijuana and amphetamines the drugs most commonly abused by bulimics. Weiss and Ebert (1983) likewise found greater evidence of impulsivity among bulimics than normal controls. These authors conducted diagnostic interviews with 15 normal-weight bulimics (*DSM–III* diagnoses) and 15 normal controls matched for age, gender, IQ, and socioeconomic status. Bulimics reported a significantly higher prevalence of suicide attempts, stealing, and impulse buying (40%, 67%, and 67%, respectively) than did normal controls (0%, 14%, and 26%). Bulimics also had a higher prevalence of drug use (including cocaine, amphetamines, marijuana, and barbiturates, but excluding alcohol or cigarettes) than did controls. Because the authors did not make distinctions between use and abuse or dependence and between current and past histories of such, it is unclear if this finding represents problematic use, a greater tendency among bulimics toward sensation seeking, or both.

Using a subject population distinct from that thus far presented, Schmidt and Telch (1990) evaluated the behavior of undergraduate volunteers (23 per group) who were classified as bulimics, subclinical binge eaters, or non-binge eaters on the basis of their responses to a 15–item screening questionnaire. Bulimics scored significantly higher than binge eaters, who in turn scored higher than non-binge eaters on two self-report measures of impulsivity. The first of these measures, the Global Impulsivity Scale, included 4 items assessing global self-perceptions of tendencies to behave impulsively; the second measure, the Dickman Scale, consisted of 7 items related to impulsive eating behavior. Interestingly, no differences were found among groups on history of suicide attempts, shoplifting/stealing, cigarette use, or drug abuse. A significantly greater proportion of bulimics admitted, however, to having threatened suicide (27.3%) than did the binge eaters (0%) and non-binge eaters (4.3%).

An article by Dykens and Gerrard (1986) reported on two related studies that they conducted with female students who met diagnostic criteria for bulimia or were classified as repeat dieters or nondieters. In the first study, the bulimics ($n = 29$) were found to score significantly higher on the MMPI 4 scale (a marker of impulsive tendencies) than the repeat dieters ($n = 27$) and the nondieters ($n = 27$). However, none of the three groups scored beyond the normal range on this scale. The second study compared 39 bulimics (12 of whom had a past history of purging but were currently only engaging in binge eating) with 31 repeat dieters and 20 nondieters on, among other things, their substance use and MMPI scores. Nonpurging bulimics scored significantly higher on the MMPI 4 scale than did repeat dieters and nondieters, although all scores were again within the normal range. Both groups of bulimics currently consumed significantly more alcohol than did repeat dieters and nondieters, in addition to having begun consuming alcohol at a significantly younger age.

Three studies have reported on the prevalence of impulse control problems among anorexics. Eckert et al. (1979) found that in an inpatient sample of 105 female anorexics, 6.7% of the patients had current or historical significant problems with alcohol. Interestingly, compared with those patients without such problems, anorexics with alcohol problems were significantly more likely to engage in bulimia and kleptomania, with the onset of the former consistently preceding the onset of the latter. The researchers unfortunately did not report on what types of items were typically stolen (i.e., foodstuffs vs. nonfoodstuffs). Crisp, Hsu, and Harding (1980a, 1980b) conducted two studies involving retrospective chart reviews of the medical records of female outpatient anorexics (with 102 patients in each sample). In one study (1980b), 10 patients (9.8%) admitted to stealing a variety of items (including food); this stealing was reported to occur only during the bulimic phase of their anorexia. In the second study (1980a), these authors found that 14, or 13.7%, of the anorexics admitted to stealing. Thirteen of these 14 patients reported this behavior to occur during bulimic phases of their anorexia. It is noteworthy in this second study that only 4 individuals stole items other than food or money (6 of the remaining anorexics reported stealing food only, whereas 4 admitted to stealing money and food). In a study of 105 female anorexic inpatients, Casper and colleagues (Casper, Eckert, Halmi, Goldberg, & Davis, 1980) similarly reported a higher prevalence of compulsive stealing and higher MMPI 4 scale scores among bulimic anorexics than among restricting anorexics (these findings were based on structured clinical interviews). Unfortunately, these researchers did not report actual prevalence rates, the types of items stolen, or MMPI scale scores to allow for comparison with Crisp et al.'s results. Furthermore, the question remains whether a pattern of stealing money and food reveals underlying difficulties with impulsivity, reflects a manifestation of starvation effects or denial (e.g., of the need for food or the tendency to binge eat), or all of the above.

The Prevalence of Impulse Control Problems Among Subsets of Eating-Disordered Individuals

A number of investigations have been conducted into the comparative rates of impulse control problems among specific subsets of eating-disordered individuals. These investigations have included examinations of the prevalence of impulse control problems among eating-disordered individuals with additional forms of psychopathology (either personality disorders or affective disorders), as well as comparisons between individuals with different types of eating disorders. It should be noted that a large number of studies have been conducted that have attempted to demonstrate a higher-than-chance comorbidity of borderline personality disorder (the diagnosis of which may include binge eating as a form of impulsive behavior) and bulimia. Although quite interesting, rarely do these studies specifically report on the criteria met for the diagnosis of borderline personality disorder (i.e., the type of impulsive behavior, such as binge eating, stealing, or sexual acting out), making it difficult to decipher whether such multiply diagnosed bulimics actually have more than one form of impulsive behavior. Thus, these types of studies will not be covered in this overview (for representative studies, see Johnson, Tobin, & Enright, 1989; Yates, Sieleni, & Bowers, 1989).

Impulse Control Problems Among Eating-Disordered Individuals With Additional Forms of Psychopathology

Two studies have been reported that suggest that eating-disordered individuals with multiple diagnoses may be more prone to multiple forms of impulsive behavior. Lacey and Evans (1986) conducted a literature review of studies that reported on the prevalence rates of impulsive behavior among individuals with substance abuse disorders, eating disorders, personality disorders, and histories of self-harm and impulse control problems. These authors found that within each of these clinical populations, a subset of individuals could be identified who exhibited more than one form of impulsive behavior (e.g., eating-disordered individuals who also abuse alcohol; arsonists or shoplifters who also abuse drugs). On the basis of this review, Lacey and Evans argued for the existence of a "multi-impulsive" personality group that warrants treatment for impulsivity per se. Thus, any eating-disordered individuals who exhibit multiple forms of impulsive behavior should not be diagnosed as eating disordered, but rather as possessing a multi-impulsive personality disorder. The implication of this argument is that the connection between impulsivity and eating disorders exists only for those individuals who are prone to be impulsive, and not the other way around (e.g., possessing an eating disorder predisposes to the development of additional impulsive behaviors).

The second study of this type suggested that elevated levels of impulsive behavior among bulimics may be linked to a history of affective disorder. Hatsukami et al. (1986) compared 46 outpatients with bulimia only, 34 bulimics with a history of affective disorder, and 34 bulimics with a history of substance abuse (all subjects were outpatients) on their rates of many forms of behavior that might be considered impulsive. Bulimics with a history of substance abuse were found to have significantly higher rates of diuretic use, financial and work problems, stealing (both before and after the onset of their bulimia), and drinking (after the onset of bulimia) than did the other two groups, whereas both the substance abuse and affective disorder groups demonstrated a significantly higher prevalence of attempted suicide than did the bulimia-only group. These results appear consistent with Lacey and Evans's (1986) suggestion that only particular eating-disordered individuals will demonstrate multiple forms of impulsive behavior.

Impulse Control Problems Among Purging Versus Nonpurging Bulimics

Although binge eating, with its requisite experience of a sense of loss of control, has been considered by many to be impulsive in nature, it is likewise possible to characterize purging (i.e., vomiting) as a form of impulsive behavior. The act of purging is engaged in not only in the hope of preventing weight gain, but also for the more immediate purpose of eliminating feelings of anxiety and depression associated with the fear of weight gain. For this reason, it might be hypothesized that, holding binge frequency constant, bulimics who purge are more likely to engage in additional forms of impulsive behavior than nonpurging bulimics. Another argument as to why purging bulimics may be more impulsive than their nonpurging counterparts is related to a predisposition in the former group toward impulsivity. As suggested earlier, it is possible to speculate that one reason why specific dieters become purging bulimics in the first place is related to underlying problems with impulsivity. Successful dieting requires an ability to withstand feelings of deprivation and to frequently delay gratification. Thus, it may be that those women highest in impulsivity have a more difficult time dieting and are therefore more likely to resort to purging in attempts to meet their dieting goals. It is surprising to note that no studies to date have reported any data to investigate the hypothesis that purging bulimics are more impulsive than their non-purging counterparts.

Impulse Control Problems Among Restricting and Bulimic Anorexics

There are a variety of reasons why bulimic anorexics may be hypothesized to show more impulsive behavior than their restricting counterparts. These reasons include that both binge eating and purging may be conceptualized as impulsive in nature and that bulimic anorexics may have greater difficulties with deprivation, delay of gratification, or tolerance of negative affect.

A number of studies have compared the impulse control problems of bulimic and restricting anorexics. Strober (1982) conducted a study that compared the clinical symptomatology and premorbid functioning of 22 restricting and 22 bulimic anorexics matched for age and duration of illness. Strober found bulimic anorexics to have a greater prevalence of premorbid obesity, as well as a higher weight at the onset of dieting. Bulimic anorexics also evidenced a higher prevalence of depression and were more often described by their parents as premorbidly affectively unstable than were restrictors. Furthermore, only bulimic anorexics reported alcohol use.

In a retrospective review of data from intake interviews with 68 bulimic and 73 restricting anorexics, Garfinkel, Moldofsky, and Garner (1980) found evidence of greater impulsivity among bulimic anorexics. Compared with their restricting counterparts, bulimic anorexics used significantly more alcohol and street drugs (both at the time of interview and by history), reported more frequent stealing, had more prior suicide attempts and higher rates of self-mutilation, and were more frequently deemed to have labile moods, on the basis of a mental status examination. Casper et al. (1980) also compared the prevalence of impulsive behavior among restricting ($n = 56$) and bulimic ($n = 49$) anorexics presenting for inpatient treatment. These authors reported that compulsive stealing was found almost exclusively among bulimic anorexics, who additionally evidenced a higher rate of alcoholism and received higher 4 scale scores on the MMPI. In a comparison of MMPI profiles of 14 bulimics, 15 bulimic anorexics, and 10 restricting anorexics (all inpatients), Norman and Herzog (1983) similarly found bulimics and bulimic anorexics to score significantly higher on the 4 scale (both groups at clinical levels) than restricting anorexics. However, in a comparison of 30 restricting and 38 bulimic anorexics, Piran et al. (1988) did not find any between-groups differences on the MMPI 4 scale. These researchers did nonetheless find higher prevalence rates of self-harm, drug or alcohol abuse, and promiscuous behavior among bulimic anorexics.

Woznica (1990) conducted one of two studies comparing bulimic and restricting anorexics that also included a psychiatric control group. In this study, Woznica administered the Self-Report Test of Impulse Control (Lazzaro, 1969) to 12 restricting anorexics, 12 bulimic anorexics, and 24 psychiatric controls (female outpatients diagnosed with either dysthymic disorder, major depression, or generalized anxiety disorder). Restricting anorexics were found to score higher in impulse control than did bulimic anorexics, who in turn scored higher than the psychiatric controls. DeSilva and Eysenck (1987) conducted a study comparing 59 restricting anorexics, 122 bulimics (60 with "major weight loss," 18 without weight loss, and 44 whose weight status was left undescribed), 66 female drug addicts, and 1,546 normal female controls on the addiction scale of the Eysenck Personality Questionnaire (Eysenck & Eysenck, 1975). Bulimics were found to have higher addiction scale scores than did anorexics (who scored "relatively low"); bulimics' scores were also noted to approach the high scores of the addicts (no tests of statistical significance were reported).

A number of studies comparing impulsivity rates among restricting and bulimic anorexics have also included comparison groups of normal-weight bulimics. Using a self-report personality inventory that includes a control/impulsivity subscale (the Multidimensional Personality Inventory; Tellegen, 1985), Casper and colleagues (Casper, Hedeker, & McClough, 1992) compared impulsivity scale scores for 12 restricting anorexics, 19 bulimic anorexics, and 19 bulimics (subjects were consecutive female patients given DSM–III–R diagnoses). Restricting anorexics scored significantly lower on impulsivity than did bulimics, who scored in the high normal range. Bulimic anorexics, who scored in the low normal range on impulsivity, did not significantly differ from the other two groups. Laessle, Wittchen, Fichter, and Pirke (1989) compared the lifetime prevalence of substance use disorders among a sample of 91 consecutive female patients (21 restricting anorexics, 20 bulimic anorexics, 23 bulimics with a history of anorexia, and 27 bulimics with no such history). All three bulimic groups had significantly higher lifetime prevalence rates of substance abuse disorders than did the restricting anorexic group. These authors also pointed out that the prevalence rates of substance abuse among the bulimic groups (20% for the bulimic anorexics, 30.4% for the bulimics with a history of anorexia, and 37% for the bulimics) were higher than the lifetime risk for substance use disorders in their native German population (6.1%).

Hudson and colleagues (Hudson, Pope, Jonas, & Yurgelun-Todd, 1983b) also found elevated rates of substance abuse and impulsivity among bulimics and bulimic anorexics. On the basis of structured diagnostic interviews for lifetime diagnoses with 90 consecutive referrals (inpatients, outpatients, and respondents to research advertisements), bulimic anorexics had higher lifetime rates of alcohol abuse or dependence (36%) than did bulimics (22%) or restricting anorexics (6%). Bulimic anorexics also had significantly higher rates of kleptomania (44%) than did restricting anorexics (6%), although once again, no mention was made of the types of items stolen. Finally, Garner, Garfinkel, and O'Shaughnessy (1985) compared three samples of consecutive clinic referrals, including restricting anorexics ($n = 59$), bulimic anorexics ($n = 59$), and individuals with bulimia nervosa ($n = 59$). On the basis of clinical interviews and responses to questionnaires, no differences were found among the groups on history of self-mutilation or suicide attempts. Bulimic anorexics also did not differ from normal-weight bulimics on alcohol or drug use or history of stealing. Restricting anorexics reported significantly less alcohol use than did normal-weight bulimics, and significantly lower use of street drugs and history of stealing than did both bulimic groups. These and the other previously reported data are consistent with a recent review by DaCosta and Halmi (1992) that concluded that bulimic anorexics evidence more impulsive behavior than do restricting anorexics.

Impulse Control Problems Among Obese Binge Eaters and Non-Binge Eaters

Finally, a number of studies have compared the prevalence of impulse control problems among normal-weight bulimics, obese binge eaters, obese non-binge eaters, and normal controls, or some subsets of these groups. In a comparison of 15 bulimic (DSM–III), 15 normal-weight, and 15 obese women (all groups were matched for age and height; normals and bulimics were also matched for weight), Williamson and colleagues (Williamson, Kelley, Davis, Ruggiero, & Blouin, 1985) found bulimics to score significantly higher than normals (though not than the obese) on the MMPI 4 scale, although no group scored in the clinical range. The number of bulimics who did score in the clinical range on this scale was marginally significantly higher than the number of obese or normals.

Hudson and colleagues (Hudson et al., 1988) compared lifetime prevalence rates of substance abuse among 47 normal-weight bulimics, 47 obese nonbulimics, and 23 obese bulimics (all diagnoses were by DSM–III, based on structured clinical interviews). Normal-weight bulimics had a significantly higher prevalence of alcohol abuse or dependency (43%) than did obese nonbulimics (9%), although not higher than that of obese bulimics (22%). Normal-weight bulimics were significantly higher than both groups on a combined measure of the prevalence of alcohol and other substance abuse or dependency (normal-weight bulimics = 55%, obese bulimics = 21%, obese nonbulimics = 15%). The Minnesota group (Mitchell, Pyle, Eckert, Hatsukami, & Soll, 1990) conducted a similar comparison between 25 normal-weight and 25 obese bulimics (*DSM–III–R* criteria; data for all subjects came from a retrospective chart review of evaluations conducted at their Eating Disorders Clinic). Normal-weight and obese bulimics did not differ in their self-reported prevalence of alcohol abuse (37.5% vs. 44%, respectively), stealing (50% vs. 60%), or family history of alcohol abuse (37.5% vs. 62.5%). Obese bulimics, however, had significantly higher prevalence rates of suicide attempts (56% vs. 18.2%) and self-injurious behavior (62.5% vs. 30.8%) than did normal-weight bulimics.

Two further studies have reported on differences between obese binge eaters and non-binge eaters. Marcus and colleagues (Marcus et al., 1990) used structured clinical interviews to determine lifetime prevalence rates of substance abuse disorders among 25 obese binge eaters and 25 obese non-binge eaters matched for similar age and weight. No difference was found between groups, with 4% of non-binge eaters versus 12% of binge eaters meeting diagnostic criteria for current or past substance abuse disorder. In a survey of 71 male and 99 female attenders of a weight control program for the moderately to severely obese, Wilson et al. (1993) similarly found no differences between binge eaters and non-binge eaters in their self-reports of current personal alcohol problems or a family history of such.

Summary of Findings on the Association Between Eating Disorders and Other Impulsive Disorders

Once again, the data on the comorbidity of eating disorders and other impulse control problems indicate a specific link between impulsivity and binge eating. Studies with substance abusers, juvenile delinquents, and kleptomaniacs have all strongly suggested a higher than expected prevalence of binge eating, whereas studies with eating-disorder patients have indicated a higher prevalence of impulsive behavior among binge eaters (i.e., normal-weight bulimics, bulimic anorexics, and obese binge eaters). Also noteworthy are the studies that have indicated that only a select subset of binge eaters are multiply impulsive. Further research into this latter possibility may be particularly helpful in guiding treatment.

The methodological limitations of this research, however, must be considered. Specifically, studies that draw from inpatient or outpatient settings may overestimate rates of comorbidity, because of the possibility that dually diagnosed people are more frequent treatment seekers. On the other hand, studies using very young subject pools may underestimate comorbidity, because some substance abuse problems, such as alcoholism, usually surface later in life. Future research accounting for these biases may shed further light on the association between impulsivity and eating-disorder symptomatology.

IMPULSIVITY AND TREATMENT OUTCOME

Much of the research reviewed thus far indicates that impulsivity may be related to the development or manifestation of specific eating-disorder symptomatology. An important question remains whether individuals with more impulsive forms of eating-disorder symptomatology (i.e., binge eaters) or with multiple forms of impulsivity fare more poorly in eating-disorder treatment programs. Although few studies have addressed this question directly, a number of studies have suggested a relationship between impulsivity and poor treatment outcome.

Lacey (1983) reported on pretreatment indicators of poor outcome among a sample of 30 bulimics who participated in a 10–week combined individual and group treatment program for bulimia. Although only 1 of the 20 patients who were asymptomatic at the end of treatment had a history of alcohol abuse, half of the bulimics who reported mild occasional bulimic episodes at the end of treatment had a history of alcohol abuse (4 of 8 patients). Three of these 4 former alcohol abusers also had a history of previous anorexia nervosa.

Two additional studies have suggested that affective and behavioral instability, although not impulsivity per se, are poor prognostic indicators in the treatment of bulimia. Johnson, Tobin, and Dennis (1990) provided a combined cognitive behavioral and psychodynamic treatment program to 55 women who met DSM–III–R diagnostic criteria for bulimia nervosa. At pretreatment, 21 bulimics were classified as having a probable additional diagnosis of borderline personality organization (on the basis of the Borderline Syndrome Index, BSI; Conte, Plutchik, Karasu, & Jerrett, 1980), and 19 were classified as clearly nonborderline. Whereas only 21% of the nonborderlines were still symptomatic at 1 year after entry into treatment, 62% of the probable borderlines remained symptomatic. Those borderline patients who were not symptomatic after 1 year no longer scored within the borderline range on the BSI.

In a study of predictor variables in the treatment of 71 bulimic outpatients, Rossiter and colleagues (Rossiter, Agras, Telch, & Schneider, 1992) similarly found a relationship between personality pathology and poor treatment outcome. Bulimics who at pretreatment scored high on

Cluster B personality pathology (combined scores on the borderline, narcissistic, histrionic, and antisocial subscales of the Personality Disorder Examination; Loranger, Susman, Oldham, & Russakoff, 1987) had significantly poorer treatment outcome after 16 weeks of therapy than did low Cluster B bulimics. Unlike the results of Johnson et al. (1990), a diagnosis of borderline personality disorder at pretreatment did not predict outcome. At 1-year follow-up, there remained a trend toward Cluster B predicting poorer outcome, although this trend was nonsignificant.

Two studies have compared treatment outcome between restricting and bulimic anorexics. In an early report, Beaumont, George, and Smart (1976) conducted a retrospective chart review of 31 female inpatients, 17 diagnosed as restricting anorexics and 14 diagnosed as bulimic anorexics. Bulimic anorexics were significantly more likely than restricting anorexics to have symptoms persisting for more than 3 years (79% vs. 29%, respectively) and less likely to have recovered completely during the period of assessment (21% vs. 41%, respectively). Vandereycken and Pierloot (1983) also found poorer treatment response among bulimic anorexics. These authors studied the symptomatology at follow-up to treatment (mean follow-up time was 4.2 years) of 141 consecutive inpatients with anorexia nervosa (65 restricting anorexics, 21 binge-eating anorexics, and 55 vomit/purging anorexics). Significantly fewer vomit/purging anorexics (38%) were excellent or much improved at follow-up compared with restricting anorexics (66%) and binge-eating anorexics (66%). Furthermore, a higher proportion of vomit/purging anorexics (8 of 55), compared with restricting anorexics (1 of 65) and binge-eating anorexics (1 of 21), were dead at follow-up. Although neither of these two reports were based on controlled treatment studies, the data strongly suggest poorer treatment outcome among bulimic (or perhaps only purging) anorexics.

Sohlberg and colleagues (Sohlberg, Norring, Holmgren, & Rosmark, 1989) provided one of the only studies that has directly measured the influence of impulsivity on treatment outcome. (It should be noted that neither the length nor the content of treatment was strictly controlled in this study.) Impulsivity was defined on a 0-to-4 continuum based on the presence or absence of binge eating, stealing, alcohol/drug abuse, and recent suicide attempts. In a combined sample of 35 anorexics and bulimics, impulsivity was found to account for 25% of anorexic symptoms at 2- to 3-year follow-up, and 14% at 4- to 6-year follow-up.

In summary, although there have been very few controlled research studies on this topic, the evidence published to date indicates a trend toward poorer outcome among binge eaters, substance abusers, and patients who might be characterized as behaviorally or affectively unstable (i.e., borderlines or individuals with Cluster B diagnoses). Methodologically, future research on this topic could be strengthened by controlled studies that measure the predictive value of a wide range of impulsive behaviors at pretreatment.

SUGGESTIONS FOR FUTURE RESEARCH

The foregoing theoretical and empirical reviews suggest the value of further studies of the relationship between impulsivity and eating disorders. Clearly, studies using more sophisticated methodologies are needed. For example, the routine use of appropriate control groups (whether normal or psychiatric, or matched for gender, age, race, socioeconomic status, and body mass index), structured clinical interviews rather than self-report inventories, and community samples will all significantly reduce the types of bias that were common in the research reviewed. Furthermore, the development of a more uniform construct of impulsivity and impulsive behavior would allow for greater comparison of findings across samples and studies.

To date, no longitudinal research has been conducted on the manifestation of impulsive behavior over the life span. It is of interest to determine not only the stability of impulsive behavior over time (i.e., whether impulsive tendencies as a child predict impulsivity later in life), but also whether various intervening factors (i.e., a history of dieting/weight loss or gain, the development of a more differentiated cognitive style, or the successful treatment of one form of impulsive behavior) alter the later display of impulsivity. Such longitudinal research may enrich our understanding of the nature of both impulsivity and eating disorders.

Finally, it is important to comment on a disparity that exists between the first and second halves of this chapter. The first half of the chapter, which discussed the possible relationship between impulsiveness/sensation seeking and eating, said little of a definitive nature about this relationship because so little data exist in this area. The second half of the chapter presented many data-based studies that examined the role of impulsivity in eating disorders. However, little could be said about the relevance of arousability or external responsiveness in eating disorders because little or no data exist on these topics. Clearly, it would be desirable to bridge this gap by directly examining the relevance of arousability and external responsiveness in understanding disordered eating. This could be done, for example, by comparing eating-disordered individuals with appropriate control groups on measures of arousability and external responsiveness, by comparing different subgroups of eating-disordered individuals (e.g., bulimic vs. restricting anorexics), and by examining what impact successful treatment had on measures of these constructs.

REFERENCES

American Heritage Dictionary of the English Language. (1976). New York: Houghton Mifflin.

American Psychiatric Association. (1980). *Diagnostic and statistical manual of mental disorders* (3rd ed.). Washington, DC: Author.

American Psychiatric Association. (1987). *Diagnostic and statistical manual of mental disorders* (3rd. ed., rev.). Washington, DC: Author.

Barratt, E. S., Pritchard, W. S., Faulk, D. M., & Brandt, M. E. (1987). The relationship between impulsiveness sub-traits, trait anxiety, and normal N100-augmenting-reducing: A typographic analysis. *Personality and Individual Differences, 8,* 43–51.

Beary, M. D., Lacey, J. H., & Merry, J. (1986). Alcoholism and eating disorders in women of fertile age. *British Journal of Addiction, 81,* 685–689.

Beaumont, P. J. V., George, G. C. W., & Smart, D. E. (1976). Dieters and vomiters and purgers in anorexia nervosa. *Psychological Medicine, 6,* 617–622.

Beglin. S. J., & Fairburn, C. G. (1992). The evaluation of a new instrument for detecting eating disorders in community samples. *Psychiatry Research, 44,* 191–201.

Booth, D. A. (1988). Culturally corralled into food abuse: The eating disorders as physiologically reinforced excessive appetites. In K. M. Pirke, W. Vandereycken, & D. Ploog (Eds.), *The psychobiology of bulimia nervosa* (pp. 18–32). Berlin: Springer-Verlag.

Bruch, H. (1973). *Eating disorders.* New York: Basic Books.

Bulik, C. M. (1987a). Alcohol use and depression in women with bulimia. *American Journal of Drug and Alcohol Abuse, 13,* 343–355.

Bulik, C. M. (1987b). Drug and alcohol abuse by bulimic women and their families. *American Journal of Psychiatry, 144,* 1604–1606.

Cacioppo, J. T., & Petty, R. E. (1983). *Social psychophysiology: A sourcebook.* New York: Guilford Press.

Casper, R. C., Eckert, E. D., Halmi, K. A., Goldberg, S. C., & Davis, J. M. (1980). Bulimia: Its incidence and clinical importance in patients with anorexia nervosa. *Archives of General Psychiatry, 37,* 1030–1035.

Casper, R. C., Hedeker, D., & McClough, J. F. (1992). Personality dimensions in eating disorders and their relevance for subtyping. *Journal of the American Academy of Child and Adolescent Psychiatry, 31,* 830–840.

Conte, H. R., Plutchik, R., Karasu, T. B., & Jerrett, I. (1980). A self-report borderline scale: Discriminative validity and preliminary norms. *Journal of Nervous and Mental Disease, 168,* 428–435.

Crisp, A. H., Hsu, L. K. G., & Harding, B. (1980a). Clinical features of anorexia nervosa: A study of a consecutive series of 102 female patients. *Journal of Psychosomatic Research, 24,* 179–191.

Crisp, A. H., Hsu, L. K. G., & Harding, B. (1980b). The starving hoarder and voracious spender: Stealing in anorexia nervosa. *Journal of Psychosomatic Research, 24,* 225–231.

DaCosta, M., & Halmi, K. A. (1992). Classifications of anorexia nervosa: Questions of subtypes. *International Journal of Eating Disorders, 11,* 305–313.

deSilva, P., & Eysenck, S. (1987). Personality and addictiveness in anorexic and bulimic patients. *Personality and Individual Differences, 8,* 749–751.

Durrant, M. (1981). Salivation: A useful research tool? *Appetite, 2,* 362–365.

Dykens, E. M., & Gerrard, M. (1986). Psychological profiles of purging bulimics, repeat dieters, and controls. *Journal of Consulting and Clinical Psychology, 54,* 283–288.

Eckert, E. D., Goldberg, S. C., Halmi, K. A., Casper, R. C., & Davis, J. M. (1979). Alcoholism in anorexia nervosa. In R. W. Pickens & L. L. Heston (Eds.), *Psychiatric factors in drug abuse* (pp. 267–283). New York: Grune & Stratton.

Eldredge, K. L. (1993). An investigation of the influence of dieting and self-esteem on dietary inhibition. *International Journal of Eating Disorders, 13,* 57–67.

Endicott, J., Spitzer, R. L., & Andreasen, N. (1975). *The family history—Research diagnostic criteria.* New York: New York State Psychiatric Institute, Biometric Research Division.

Eysenck, H. J., & Eysenck, S. B. G. (1975). *Manual of the Eysenck Personality Questionnaire.* London: Hodder & Stoughton.

Freed, S. O. (1947). Psychic factors in the development and treatment of obesity. *Journal of the American Medical Association, 133,* 369–373.

Garfinkel, P. E., Moldofsky, H., & Garner, D. M. (1980). The heterogeneity of anorexia nervosa: Bulimia as a distinct subgroup. *Archives of General Psychiatry, 37,* 1036–1040.

Garner, D. M., Garfinkel, P. E., & O'Shaughnessy, M. (1985). The validity of the distinction between bulimia with and without anorexia nervosa. *American Journal of Psychiatry, 142,* 581–587.

Gray, J. A. (1987). *The psychology of fear and stress* (2nd ed.). New York: Cambridge University Press.

Grilo, C. M., Shiffman, S., & Wing, R. R. (1989). Relapse crises and coping among dieters. *Journal of Consulting and Clinical Psychology, 57,* 488–495.

Hatsukami, D., Eckert, E., Mitchell, J. E., & Pyle, R. (1984). Affective disorder and substance abuse in women with bulimia. *Psychological Medicine, 14,* 701–704.

Hatsukami, D., Mitchell, J. E., Eckert, E. D., & Pyle, R. (1986). Characteristics of patients with bulimia only, bulimia with affective disorder, and bulimia with substance abuse problems. *Addictive Behaviors, 11,* 399–406.

Heatherton, T. F., Herman, C. P., Polivy, J., King, G. A., & McGree, S. T. (1988). The (mis)measurement of restraint: An analysis of conceptual and psychometric issues. *Journal of Abnormal Psychology, 97,* 19–28.

Heatherton, T. F., Polivy, J., & Herman, C. P. (1989). Restraint and internal responsiveness: Effects of placebo manipulations of hunger state on eating. *Journal of Abnormal Psychology, 98,* 89–92.

Herman, C. P., & Mack, D. (1975). Restrained and unrestrained eating. *Journal of Personality, 43,* 647–660.

Herman, C. P., & Polivy, J. (1975). Anxiety, restraint and eating behavior. *Journal of Abnormal Psychology, 84,* 666–672.

Herman, C. P., & Polivy, J. (1980). Restrained eating. In A. J. Stunkard (Ed.), *Obesity* (pp. 208–225). Philadelphia: Saunders.

Herman, C. P., & Polivy, J. (1984). A boundary model for the regulation of eating. In A. J. Stunkard & E. Stellar (Eds.), *Eating and its disorders* (pp. 141–156). New York: Raven.

Hudson, J. I., Pope, H. G., Jr., Jonas, J. M., & Yurgelun-Todd, D. (1983a). Family history study of anorexia nervosa and bulimia. *British Journal of Psychiatry, 142,* 133–138.

Hudson, J. I., Pope, H. G., Jr., Jonas, J. M., & Yurgelun-Todd, D. (1983b). Phenomenologic relationship of eating disorders to major affective disorder. *Psychiatry Research, 9,* 345–354.

Hudson, J. I., Pope, H. G., Jonas, J. M., Yurgelun-Todd, D., & Frankenburg, F. R. (1987). A controlled family history study of bulimia. *Psychological Medicine, 17,* 883–890.

Hudson, J. I., Pope, H. G., Wurtman, J., Yurgelun-Todd, D., Mark, S., & Rosenthal, N. E. (1988). Bulimia in obese individuals: Relationship to normal-weight bulimia. *Journal of Nervous and Mental Disease, 176,* 144–152.

Israel, A. C., Stolmaker, L., & Prince, B. (1983). The relationship between impulsivity and eating behavior in children. *Child and Family Behavior Therapy, 5,* 71–75.

Jansen, A., Klaver, J., Merckelbach, H., & van den Hout, M. (1989). Restrained eaters are rapidly habituating sensation seekers. *Behaviour Research and Therapy, 27,* 247–252.

Johnson, C., Stuckey, M. K., Lewis, L. D., & Schwartz, D. M. (1982). Bulimia: A descriptive survey of 316 cases. *International Journal of Eating Disorders, 2,* 3–15.

Johnson, C., Tobin, D. L., & Dennis, A. (1990). Differences in treatment outcome between borderline and nonborderline bulimics at one-year follow-up. *International Journal of Eating Disorders, 9,* 617–627.

Johnson, C., Tobin, D., & Enright, A. (1989). Prevalence and clinical characteristics of borderline patients in an eating-disordered population. *Journal of Clinical Psychiatry, 50,* 9–15.

Jonas, J. J., Gold, M. S., Sweeney, D., & Pottash, A. L. C. (1987). Eating disorders and cocaine abuse: A survey of 259 cocaine abusers. *Journal of Clinical Psychiatry, 48,* 47–50.

Kaplan, H. I., & Kaplan, H. S. (1957). The psychosomatic concept of obesity. *Journal of Nervous and Mental Disease, 125,* 181–189.

Kassett, J. A., Gershon, E. S., Maxwell, M. E., Guroff, J. J., Kazuba, D. A., Smith, A. L., Brandt, H. A., & Jimerson, D. C. (1989). Psychiatric disorders in the first-degree relatives of probands with bulimia nervosa. *American Journal of Psychiatry, 146*, 1468–1471.

King, R. J., Jones, J., Scheuer, J. W., Curtis, D., & Zarcone, V. P. (1990). Plasma cortisol correlates of impulsivity and substance abuse. *Personality and Individual Differences, 11*, 287–291.

Klajner, F., Herman, C. P., Polivy, J., & Chhabra, R. (1981). Human obesity, dieting and the anticipatory salivation to food. *Physiology and Behavior, 27*, 195–198.

Lacey, J. H. (1983). Bulimia nervosa, binge eating, and psychogenic vomiting: A controlled treatment study and long-term outcome. *British Medical Journal, 286*, 1609–1613.

Lacey, J. H., & Evans, C. D. H. (1986). The impulsivist: A multi-impulsive personality disorder. *British Journal of Addiction, 81*, 641–649.

Lacey, J. H., & Moureli, E. (1986). Bulimic alcoholics: Some features of a clinical sub-group. *British Journal of Addiction, 81*, 389–393.

Laessle, R. G., Wittchen, H. U., Fichter, M. M., & Pirke, K. M. (1989). The significance of subgroups of bulimia and anorexia nervosa: Lifetime frequency of psychiatric disorders. *International Journal of Eating Disorders, 8*, 569–574.

LaPorte, D. J. (1990). A fatiguing effect in obese patients during partial fasting: Increase in vulnerability to emotion-related events and anxiety. *International Journal of Eating Disorders, 9*, 345–355.

Lazzaro, T. A. (1969). The development and validation of the Self-Report Test of Impulse Control (Doctoral dissertation, Southern Illinois University, 1968). *Dissertation Abstracts International, 29*, 2620B.

Leon, G. R., Carroll, K., Chernyk, B., & Finn, S. (1985). Binge eating and associated habit patterns within college student and identified bulimic populations. *International Journal of Eating Disorders, 4*, 43–57.

Logue, A. W., & King, G. R. (1991). Self-control and impulsiveness in adult humans when food is the reinforcer. *Appetite, 17*, 105–120.

Loranger, A. W., Susman, V. L., Oldham, J. M., & Russakoff, L. M. (1987). The personality disorder examination: A preliminary report. *Journal of Personality Disorders, 1*, 1–13.

Lowe, M. R. (1987). Set point, restraint, and the limits of weight loss: A critical analysis. In W. Johnson (Ed.), *Advances in Eating Disorders: Vol. I. Treating and preventing obesity* (pp. 1–37). Greenwich, CT: JCI Press.

Lowe, M. R. (1993). The effects of dieting on eating behavior: A three-factor model. *Psychological Bulletin, 114*, 100–121.

Lowe, M. R., Whitlow, J. W., & Bellwoar, V. (1991). Eating regulation: The role of restraint, dieting, and weight. *International Journal of Eating Disorders, 10*, 461–471.

Marcus, M. D., Wing, R. R., Ewing, L., Kern, E., Gooding, W., & McDermott, M. (1990). Psychiatric disorders among obese binge eaters. *International Journal of Eating Disorders, 9,* 69–77.

McElroy, S. L., Hudson, J. I., Pope, H. G., & Keck, P. E. (1991). Kleptomania: Clinical characteristics and associated psychopathology. *Psychological Medicine, 21,* 93–108.

Milstein, R. M. (1980). Responsiveness in newborn infants of overweight and normal weight parents. *Appetite, 1,* 65–74.

Mitchell, J. E., Hatsukami, D., Eckert, E. D., & Pyle, R. L. (1985). Characteristics of 275 patients with bulimia. *American Journal of Psychiatry, 142,* 482–485.

Mitchell, J. E., Hatsukami, D., Pyle, R., & Eckert, E. (1988). Bulimia with and without a family history of drug abuse. *Addictive Behaviors, 13,* 245–251.

Mitchell, J. E., Pyle, R. L., Eckert, E. D, Hatsukami, D., & Soll, E. (1990). Bulimia nervosa in overweight individuals. *Journal of Nervous and Mental Disease, 178,* 324–327.

Myers, W. C., & Burket, R. C. (1989). Eating attitudes, behaviors, and disorders in female juvenile delinquents. *Psychosomatics, 30,* 428–432.

Nisbett, R. E. (1972). Hunger, obesity, and the ventromedial hypothalamus. *Psychological Review, 79,* 433–453.

Norman, D. K., & Herzog, D. B. (1983). Bulimia, anorexia nervosa, and anorexia nervosa with bulimia: A comparative analysis of MMPI profiles. *International Journal of Eating Disorders, 2,* 43–52.

O'Gorman, J. G., & Lloyd, J. E. M. (1987). Extraversion, impulsiveness, and EEG alpha activity. *Personality and Individual Differences, 8,* 169–174.

Peveler, R., & Fairburn, C. (1990). Eating disorders in women who abuse alcohol. *British Journal of Addiction, 85,* 1633–1638.

Piran, N., Lerner, P., Garfinkel, P. E., Kennedy, S. H., & Brouillette, C. (1988). Personality disorders in anorexic patients. *International Journal of Eating Disorders, 7,* 589–599.

Polivy, J., & Herman, C. P. (1985). Dieting and binging: A causal analysis. *American Psychologist, 40,* 193–201.

Powley, T. L. (1977). The ventromedial hypothalamic syndrome, satiety, and a cephalic phase hypothesis. *Psychological Review, 84,* 89–126.

Pyle, R. L., Mitchell, J. E., & Eckert, E. D. (1981). Bulimia: A report of 34 cases. *Journal of Clinical Psychiatry, 42,* 60–64.

Pyle, R. L., Mitchell, J. E., Eckert, E. D., Halvorson, P. A., Neuman, P. A., & Goff, G. M. (1983). The incidence of bulimia in freshman college students. *International Journal of Eating Disorders, 2,* 75–85.

Revusky, S. H. (1967). Hunger level during food consumption: Effect on subsequent preference. *Psychonomic Science, 7,* 109–110.

Rodin, J. (1981). The current status of the internal–external obesity hypothesis: What went wrong. *American Psychologist, 36,* 361–372.

Rodin, J. (1985). Insulin levels, hunger and food intake: An example of feedback loops in body weight regulation. *Health Psychology, 4*, 1–18.

Rodin, J. (1991). Stress-induced eating: Implications for diabetes. In P. M. McCabe, N. Schneiderman, T. M. Field, & J. S. Skyler (Eds.), *Stress, coping and disease* (pp. 135–146). Hillsdale, NJ: Erlbaum.

Rodin, J., & Slochower, J. (1976). Externality in the non-obese: The effects of environmental responsiveness on weight. *Journal of Personality and Social Psychology, 29*, 557–565.

Rodin, J., Wack, J., Ferrannini, E., & DeFronzo, R. A. (1985). Effect of insulin and glucose on feeding behavior. *Metabolism, 34*, 826–831.

Rosen, J. C. (1981). Effects of low-calorie dieting and exposure to diet-prohibited food on appetite and anxiety. *Appetite: Journal for Intake Research, 2*, 366–369.

Rossiter, E. M., Agras, W. S., Telch, C. F., & Schneider, J. A. (1992). *Cluster B personality disorder characteristics predict outcome in the treatment of bulimia nervosa.* Unpublished manuscript, Stanford University, Stanford, CA.

Schachter, S., Goldman, R., & Gordon, A. (1968). Effects of fear, food deprivation, and obesity on eating. *Journal of Personality and Social Psychology, 10*, 91–97.

Schachter, S., & Rodin, J. (1974). *Obese humans and rats.* Washington, DC: Erlbaum/Halsted.

Schmidt, N. B., & Telch, M. J. (1990). Prevalence of personality disorders among bulimics, nonbulimic binge eaters, and normal controls. *Journal of Psychopathology and Behavioral Assessment, 12*, 169–185.

Schumaker, J. F., Groth-Marnat, G., Small, L., & Macaruso, P. A. (1986). Sensation seeking in a female bulimic population. *Psychological Reports, 59*, 1151–1154.

Selzer, M. L. (1971). The Michigan Alcoholism Screening Test: The quest for a new diagnostic instrument. *American Journal of Psychiatry, 127*, 89–94.

Shaye, R. (1989). Relationships between impulsivity and eating behaviour under varying conditions of stress and food deprivation. *Personality and Individual Differences, 7*, 805–808.

Smith, B. D., Davidson, R. A., Smith, D. L., Goldstein, H., & Perlstein, W. (1989). Sensation seeking and arousal: Effects of strong stimulation on electrodermal activation and memory task performance. *Personality and Individual Differences, 10*, 671–679.

Sohlberg, S., Norring, C., Holmgren, S., & Rosmark, B. (1989). Impulsivity and long-term prognosis of psychiatric patients with anorexia nervosa/bulimia nervosa. *Journal of Nervous and Mental Disease, 177*, 249–258.

Spitzer, L., & Rodin, J. (1983). Arousal induced eating: Conventional wisdom or empirical findings? In J. Cacioppo & R. Petty (Eds.), *Social psychophysiology* (pp. 565–591). New York: Guilford Press.

Spitzer, R. L., Endicott, J., & Robins, E. (1975). *Research diagnostic criteria.* New York: Biometric Research Division, New York State Psychiatric Institute.

Stern, S. L., Dixon, K. N., Nemzer, E., Lake, M. D., Sansone, R. A., Smeltzer, D. J., Lantz, S., & Schrier, S. S. (1984). Affective disorder in the families of women with normal weight bulimia. *American Journal of Psychiatry, 141,* 1224–1227.

Strober, M. (1982). The significance of bulimia in juvenile anorexia nervosa: An exploration of possible etiologic factors. *International Journal of Eating Disorders, 1,* 28–43.

Strober, M. (1983). An empirically derived typology of anorexia nervosa. In P. L. Darby, P. E. Garfinkel, D. M. Garner, & D. V. Coscina (Eds.), *Anorexia nervosa: Recent developments in research* (pp. 185–196). New York: Liss.

Strober, M., Salkin, B., Burroughs, J., & Morrell, W. (1982). Validity of the bulimia–restricter distinction in anorexia nervosa: Parental personality characteristics and family psychiatric morbidity. *Journal of Nervous and Mental Disease, 170,* 345–351.

Tellegen, A. (1985). Structures of mood and personality and their relevance to assessing anxiety, with an emphasis on self-report. In A. H. Tuma & J. D. Maser (Eds.), *Anxiety and anxiety disorders* (pp. 681–706). Hillsdale, NJ: Erlbaum.

Vandereycken, W., & Pierloot, R. (1983). The significance of subclassification in anorexia nervosa: A comparative study of clinical features in 141 patients. *Psychological Medicine, 13,* 543–549.

Weiss, S. R., & Ebert, M. H. (1983). Psychological and behavioral characteristics of normal-weight bulimics and normal-weight controls. *Psychosomatic Medicine, 45,* 293–303.

Williamson, D. A., Kelley, M. L., Davis, C. J., Ruggiero, L., & Blouin, D. C. (1985). Psychopathology of eating disorders: A controlled comparison of bulimic, obese, and normal subjects. *Journal of Consulting and Clinical Psychology, 53,* 161–166.

Wilson, G. T. (1991). The addiction model of eating disorders: A critical analysis. *Advances in Behaviour Research and Therapy, 13,* 27–72.

Wilson, G. T., Nonas, C. A., & Rosenblum, G. D. (1993). Assessment of binge-eating in obese patients. *International Journal of Eating Disorders, 13,* 25–33.

Woznica, J. G. (1990). Delay of gratification in bulimic and restricting anorexia nervosa patients. *Journal of Clinical Psychology, 46,* 706–713.

Yates, W. R., Sieleni, B., & Bowers, W. A. (1989). Clinical correlates of personality disorder in bulimia nervosa. *International Journal of Eating Disorders, 8,* 473–477.

Zuckerman, M. (1991). *Psychobiology of personality.* New York: Cambridge University Press.

Zuckerman, M., Simons, R. F., & Como, P. G. (1988). Sensation seeking and stimulus intensity as modulators of cortical, cardiovascular, and electrodermal response: A cross-modality study. *Personality and Individual Differences, 9,* 361–372.

11

IMPULSIVE BEHAVIOR AND SUBSTANCE ABUSE

WILLARD L. JOHNSON, ROBERT M. MALOW,
SHEILA A. CORRIGAN, and JEFFERY A. WEST

Impulsivity and disorders of impulse control have traditionally been a central etiological concept in many theoretical models of substance abuse, whether involving alcohol, drugs, or combinations of both (e.g., Khantzian, 1979; McKenna, 1979; Wishnie, 1977). Given this theoretical attention, it would seem critical for substance abuse treatment programs to incorporate components specifically targeting impulsivity (McCown, 1990). Furthermore, understanding the pathophysiology of impulsivity is valuable in developing psychopharmacological interventions for substance abusers.

As discussed in earlier chapters, the operational definition of impulsivity has varied across contexts. For example, impulsivity has been defined as a cognitive style by some authors (e.g., Kagan, 1966; Kendall,

We gratefully acknowledge Stacey Cunningham for her assistance with the interpretation of selected articles reviewed in the section on psychobiology and for her comments on earlier drafts of this chapter.

Moses, & Finch, 1980; Messer, 1976). In this context, then, impulsivity is contrasted with reflectivity as a problem-solving approach. Other authors (e.g., Eysenck & Eysenck, 1978; Eysenck, Pearson, Easting, & Allsopp, 1985) have viewed impulsivity as a personality trait. Perhaps the most detailed analysis of impulsivity has been provided by Shapiro (1965), who described various factors related to impulsivity, including (a) the subjective experience of impulse (i.e., a feeling of detachment from the act); (b) the quality of impulsive action (i.e., quick, abrupt, or discontinuous with previous activity, with a lack of planning); (c) the impulsive mode of cognition (i.e., concrete and egocentric with poor judgment and lack of planning, concentration, logical objectivity, and reflectiveness); and (d) relationships among the foregoing factors and affective functioning. In contrast with this multidimensional view, researchers in this area do not typically operationalize the construct of impulsiveness; rather, the terms *impulsivity* and *impulsiveness* are used as if their meanings were self-evident or as if the reader could assume that impulsiveness is defined in terms of the construct measured by the Impulsivity (I5 or I7) scale of the Eysenck Personality Inventory.

In this chapter we discuss the prevalence of impulsivity among substance abusers and the interrelationships among impulsivity, psychopathology, and substance abuse. The psychobiology of impulsive behavior as it relates to substance abuse, treatment issues, and suggested research directions in this area are also addressed. We use the terms *impulsivity* and *impulsiveness* interchangeably and do not attempt to exclude or differentiate between studies on the basis of definition. Rather, we survey a representative sample of studies relevant to the topic of impulsivity among substance abusers.

PREVALENCE OF IMPULSIVENESS AMONG ABUSERS

Although the prevalence of the characteristic of impulsiveness among substance abusers has been addressed by numerous authors, many such discussions have been more subjective and impressionistic than data based. For example, Khantzian (1979) described a predominance of impulse control problems among abusers of various classes of drugs. More empirically based was the initial study in a series of reports examining impulsivity among substance abusers by McCown (1988). In this study, impulsivity scores for abusers of a single substance were compared with those of polydrug abusers. Polydrug abusers were found to score significantly higher on a measure of impulsivity.

One limitation of the McCown (1988) study was the lack of a comparison group of nondrug users. However, King, Jones, Scheuer, Curtis, and Zarcone (1990) compared inpatient substance abusers with a nonpatient control group on impulsivity scores from the Eysenck Personality Inventory. Substance abusers were found to score significantly higher than controls, further supporting the notion of increased impulsivity as a factor potentially contributing to substance abuse.

Thus, although not definitive, there is empirical evidence of increased levels of impulsivity within groups of substance abusers. In addition to the evidence of impulsiveness among substance abusers presented here, further support for the existence of this trait among substance abusers may be found in the literature relating substance abuse to other mental disorders.

CO-OCCURRENCE OF SUBSTANCE ABUSE WITH OTHER MENTAL DISORDERS

According to the Alcohol, Drug Abuse, and Mental Health Administration (National Institute of Mental Health), studies have indicated that as many as 28% of patients meeting diagnostic criteria for a primary mental disorder also abuse drugs or alcohol, whereas 45% of the individuals diagnosed as abusing alcohol and 71% of those abusing drugs have been found to have a coexisting mental disorder (Goodwin, 1989). The co-occurrence of a psychoactive substance use disorder and another Axis I or Axis II disorder from the revised third edition of the *Diagnostic and Statistical Manual of Mental Disorders* (DSM–III–R; American Psychiatric Association, 1987) has been referred to by various labels, including mentally ill chemical abuse (M. P. Carey, Carey, & Meisler, 1990), psychiatrically impaired substance abuse, and dual diagnosis (K. B. Carey, 1991; K. B. Carey & Carey, 1990; Evans & Sullivan, 1990). As noted by Uddo, Malow, and Sutker (1993), the relation between substance abuse and other psychiatric disorders may be conceptualized in several ways. For example, psychopathology may serve as an antecedent or predisposing factor for substance abuse, psychopathology may occur concomitantly with or consequent to substance abuse, or substance abuse and psychopathology may coexist without significant interaction. In addition, there has been much debate as to whether the "addictive personality" should be recognized as a distinct diagnostic entity (Sutker & Allain, 1988). The co-occurrence of substance abuse and psychopathology has been reviewed in greater detail elsewhere (e.g., Brown, Ridgely, Pepper, Levine, & Ryglewicz, 1989; Bukstein, Brent, & Kaminer, 1989; Malow, Pintard, Sutker, & Allain, 1988; Mirin, 1984). Hence, in the following section we discuss the topic only insofar as impulsivity is suggested as a contributing or related factor.

Psychopathology in Childhood and Adolescence

Behavior that may be considered impulsive (e.g., difficulty waiting one's turn, blurting out answers before completion of the question, temper outbursts) are included among the *DSM-III-R* criteria for childhood diagnoses such as attention-deficit hyperactivity disorder (ADHD) and oppositional defiant disorder. Several authors have linked these disorders, subsequent development of antisocial personality disorder (ASPD), and the occurrence of substance abuse (e.g., Gittelman, Mannuzza, Shenker, & Bonagura, 1985; McCord & McCord, 1960; McCown, 1988; Shedler & Block, 1990; Tomas, Vlahov, & Anthony, 1990).

In a study representative of this perspective, Robins (1974) found that almost half of the boys referred for treatment of antisocial behaviors later displayed social problems related to excess alcohol intake, even though only 8% were noted to abuse alcohol as minors. Further findings of relevance cited in this study include the following: Of all the children who were referred for antisocial behavior, 28% were diagnosed as "sociopathic personality" as adults, and 35% of these were characterized as impulsive as children. Finally, severe problems with alcohol were found in 62% of those diagnosed as sociopathic as adults. These findings, although supportive of relationships among impulsivity, development of ASPD, and alcoholism, also suggest that the correlations among these factors may be less than perfect.

Other authors have argued that impulsivity, although frequently present in the childhood histories or current behaviors of adolescents and adults with substance abuse disorders, is not a relevant factor. These authors emphasize that impulsive behaviors are not predictive of later substance abuse, whereas antisocial types of acting out are related (cf. Bukstein et al., 1989). One particularly good example of a study demonstrating this finding was that of Hesselbrock (1986), who, by separating factors such as attention, hyperactivity, impulsivity, and conduct problems, found that hyperactivity, conduct problems, and aggressiveness were predictive of antisocial personality and substance abuse in adults but that impulsivity was not.

Therefore, the contribution of impulsivity during childhood to the later development of substance abuse is open to question. Currently, the available evidence disputes the hypothesis that childhood impulsivity leads to later substance abuse. However, there will likely be further studies exploring this relationship.

Axis I Disorders

Several authors have addressed the co-occurrence of substance abuse and other Axis I symptomatology, such as mood disorders (e.g., Dackis

& Gold, 1984; Malow, Corrigan, Peña, Calkins, & Bannister, 1992; Mirin, Weiss, Sollogub, & Michael, 1984), anxiety (e.g., Malow, West, Corrigan, Peña, & Lott, 1992), and schizophrenic or psychotic behavior (e.g., Brady et al., 1990; McLellan, Woody, & O'Brien, 1979; Mueser et al., 1990). Other investigators have examined the co-occurrence of substance abuse and specific impulsive behaviors, such as suicide attempts, which are often displayed by individuals with Axis I diagnoses (e.g., Craig & Olson, 1990). Tentative links between impulsivity, substance abuse, and some other Axis I disorders also have been suggested in that substances may be abused in conjunction with the impulsive characteristics involved in disorders characterized by mania (e.g., Gawin & Kleber, 1986; Weiss, Mirin, Michael, & Sollogub, 1986). Furthermore, McLellan and Druley (1977) noted the links between decreased levels of serotonin, which may contribute to impulsivity (see the section on psychobiology) and mania, hyperactivity, and possibly schizophrenia.

To the extent that impulsivity and Axis I disorders share a genetic or biochemical basis, arguments for consideration of the importance of biological vulnerability in treatment selection seem particularly relevant (e.g., Bigelow, Brooner, McCaul, & Svikis, 1988; Rounsaville, 1988; Rounsaville, Dolinsky, Babor, & Meyer, 1987). However, the importance of the role that impulsivity plays in linking Axis I disorders with substance abuse has yet to be clearly demonstrated.

Axis II Disorders

Interest in the personality characteristics of substance abusers has a long-standing and controversial history (Butcher, 1988). Indeed, it was not until 1980 that substance abuse disorders were separated from personality disorders in the DSM (Nace, 1990). As is evident in the current literature, the debate over the significance of personality factors in understanding and treating substance abuse continues. For example, in summarizing the results of a recent study, Nace, Davis, and Gaspari (1991) concluded that (a) personality disorders were present in more than half of their sample of substance abusers, indicating the clinical heterogeneity of such groups; (b) substance abusers with personality disorders appear to have greater involvement with illegal drugs, differ in patterns of alcohol abuse, and show greater psychopathology (including more impulsivity); and (c) individualized treatment approaches related to differences in dysfunction between patients were indicated. On the other hand, the author of another recent study stated that "successful [substance abuse] treatment outcomes can occur regardless of the personality disorder classification of the person" (Lennings, 1990, p. 211).

The extensive literature on personality factors and substance abuse has been critically reviewed by Sutker and Allain (1988), who suggested that work in this area, although potentially important, has been characterized by inadequate experimental designs and oversimplified conceptualizations. In light of these observations, we attempt only a limited overview of the substance abuse literature involving impulsiveness as a personality trait or characteristic of specific personality disorders.

The relationship of substance abuse to crime has been well documented (e.g., De La Rosa, Lambert, & Gropper, 1990; Hammersley, Forsyth, Morrison, & Davies, 1989; Shaffer, Nurco, & Kinlock, 1984). For example, several authors have noted connections between ASPD (psychopathic, sociopathic) and substance abuse (e.g., Adler, 1990; Craig, 1986; Grande, Wolf, Schubert, Patterson, & Brocco, 1984; Shaffer et al., 1984; Stabenau & Hesselbrock, 1984; Weddington et al., 1990; Woody, McLellan, Luborsky, & O'Brien, 1985). In fact, reviewers have concluded that "the form of psychopathology which has been most consistently linked to both alcoholism and substance abuse is antisocial personality disorder" (Meyer & Hesselbrock, 1984, p. 2).

In relating ASPD specifically to substance abuse, authors have argued the importance of recognizing the heterogeneity among substance abusers with ASPD (Alterman & Cacciola, 1991; Lykken, 1957; Whitters, Cadoret, & McCalley-Whitters, 1987). In this regard, the characteristic of impulsivity, consistently included in general descriptions of ASPD (Brantley & Sutker, 1984; Sutker, Archer, & Kilpatrick, 1983; Sutker & King, 1985), has also been emphasized as an important factor in the personality structure of substance abusers (e.g., Craig, 1986; Forgays, 1986; Grande et al., 1984). However, again stressing the importance of recognizing heterogeneity among abusers, Cloninger (1988) described his scheme for subgrouping alcoholics that differentiated one subgroup ("Type 2") with a constellation of behaviors that included impulsivity from other subgroups of alcoholics that did not.

In the DSM–III–R, ASPD is grouped with borderline personality disorder (BPD), histrionic personality disorder, and narcissistic personality disorder in Cluster B. At least one study has documented a preponderance of substance-abusing subjects with personality disorders (30 of 57 subjects) displaying one of these four personality types (Nace et al., 1991). Along this line, although not specifically addressing substance abuse, recent position papers have encouraged replacement of the current categorical approach to personality classification with one that is dimensional (Nurnburg et al., 1991). At least one such article emphasized the dimension of impulsivity as linking disorders such as ASPD and BPD (Siever & Davis, 1991).

A number of other researchers have examined impulsivity, substance abuse, and personality. For example, Malow, West, Williams, and Sutker (1989) used the Structured Clinical Interview for DSM–III–R to examine the extent to which personality disorders and associated symptom criteria would be found among cocaine and opiate addicts on a Veterans Affairs drug dependence treatment unit. They obtained base rates for the impulsiveness criterion for BPD of 47% and 56%, respectively, for cocaine addicts (6% of the total sample met BPD criteria) and opiate addicts (35% met BPD criteria). Among the subjects meeting criteria for adult ASPD, 10% of cocaine addicts (14% met ASPD criteria) and 20% of opiate addicts (44% met ASPD criteria) met the criteria for impulsivity and poor planning.

Jensen, Pettinati, Evans, Myers, and Valliere (1990) studied impulsive characteristics among groups of substance abusers with and without a dual diagnosis. Subjects with dual diagnoses were found to be significantly more impulsive than others, and those with an Axis II personality disorder in Cluster B or C (*DSM–III–R*) were significantly more impulsive than were subjects without such diagnoses. However, because impulsive behaviors are among the criteria for many Cluster B and C disorders, this latter finding is not surprising.

Arguing in support of the heuristic value of impulsivity as a personality type, Lacey and Evans (1986) reviewed the literature on impulsiveness as related to the following disorders: substance abuse, eating disorders, self-harm, and BPD. They found subgroups of individuals within each of these diagnostic categories who presented symptoms of one or more of the other categories. Those authors argued for the creation of a "multi-impulsive personality disorder" diagnostic category (Lacey & Evans, 1986, p. 646) and suggested that treatment for individuals falling into this category, regardless of other diagnoses, should emphasize "management of delayed gratification and control of impulsivity" (Lacey & Evans, 1986, p. 646).

However, in a recent empirical study, Kennedy and Grubin (1990) specifically attempted to test the Lacey and Evans (1986) position that some individuals may have a multi-impulsive personality disorder. A number of interview scales and personality measures were administered to a group of imprisoned sex offenders. Kennedy and Grubin reasoned that if multi-impulsive personality disorder does indeed exist, it would be found more frequently among the prisoner population. A high score on a measure of impulsiveness was significantly correlated with the overall number of types of impulsive behavior reported by each subject (i.e., alcoholism, sedative dependence, other drug abuse, pathological gambling, repeated aggression, self-harm). In addition, certain specific types of problems were found to occur together significantly more often than would be expected by chance: (a) other drug abuse with other categories of substance abuse and with repeated aggression and (b) alcoholism and aggression. Gambling and self-harm were not significantly related to the other types of problems. Despite these findings, Kennedy and Grubin (1990) concluded that their results "do not support the concept of a multi-impulsive personality disorder, but simply demonstrate the truism that impulsive people do impulsive things." These authors argued in favor of a conceptualization of impulsiveness as a continuum, with individuals who display multiple types of impulsive behavior as falling at one extreme rather than as manifesting a syndrome that differentiates them as having a specific personality disorder.

Summary of Co-Occurrence Issues

Impulsive behavioral features are among the defining criteria of several *DSM–III–R* diagnoses on both Axis I and Axis II, including ADHD; ASPD and BPD; organic personality syndrome; bipolar disorder; and manic and impulse control disorders not classified elsewhere, such as intermittent explosive disorder and pathological gambling. To the extent that impulsivity proves to be a common link between substance abuse and a second mental disorder, it may be possible to identify different subgroups of abusers, the members of which manifest common etiologies or respond differentially (from those of other subgroups) to treatment. One example of the important issues related to such co-occurrence of substance abuse and psychopathology is the question of whether one's preferred substance of abuse serves a self-medicating function dependent on the nature of the psychopathology and the choice of substance. That is, it may be possible that abusers displaying one type of psychopathology prefer a particular substance over those preferred by another group displaying a different kind of psychopathology (e.g., Khantzian, 1990; Khantzian, Halliday, & McAuliffe, 1990).

However, the interrelationships among impulsivity, substance abuse, and other mental disorders are potentially complex and challenging to investigate. Sutker and her associates have discussed the various issues involved in conceptualizing and researching the relationship of psychopathology and substance abuse (Sutker & Allain, 1988; Sutker & Archer, 1984; Uddo et al., 1993). In these overviews of the complexities involved in such research, the authors offered the following conclusions: (a) Most researchers now realize that univariate, unidirectional models are not helpful in understanding the various dimensions of addictive behavior; (b) a greater understanding of possible origins, underlying mechanisms, and forms of addictive behaviors is needed; and (c) future research must include person and setting factors, biological risk markers, longitudinal measurement approaches, and relationships between substance abuse and other addictive behaviors (e.g., gambling, hypersexuality, and overeating).

PSYCHOBIOLOGICAL LINKS BETWEEN IMPULSIVENESS AND SUBSTANCE ABUSE

The ever-increasing confluence of the fields of neurophysiology and psychology provides a potentially rich research area for exploring relationships between impulsivity and substance abuse. Whether a precursor to or consequence of substance abuse, complex neurochemical factors that interact presumably covary with the expression of impulsivity. Investigation of these factors is likely to elucidate clinically relevant differences between substance abusers and nonabusers prior to or following drug use. Furthermore, an understanding of neurochemical correlates of impulsivity among substance abusers may be valuable in guiding pharmacological intervention (Dackis & Gold, 1985; Lewis, 1991).

The most thoroughly documented link between impulsivity and neurobiology involves the neurotransmitter serotonin (5-hydroxytryptamine [5-HT]). Goodwin (1989) noted that

> neurobiological research has yielded diverse strands of evidence that suggest linkages among: violence and impulse control disorders; subtypes of affective illness and alcoholism (i.e., "Type III") that are associated with genetic loading, early age of onset, and aggressive and impulsive behavior; and suicide. Interestingly, serotonin disturbances have been shown, independently, to be involved in each of these conditions. (p. 3517)

Klar and Siever (1985) conceptualized impulsivity as being one of three core psychobiological vulnerabilities (with affective instability and schizotypy representing the others) potentially affecting personality. These authors reviewed the research literature related to personality disorders characterized by impulsivity, with particular emphasis on psychobiological mechanisms, and the reader is referred to that discussion for additional details. We present selected examples of research in this area.

Reviewing central neurotransmitter system research with human and animal subjects, Fishbein, Lozovsky, and Jaffe (1989) concluded that reduced serotonergic activity may produce disinhibition, leading to aggressive or impulsive behavior. The administration of a 5-HT agonist was then used to confirm the hypothesis that the substance abusers who scored high on baseline measures of impulsivity would be found to have impaired 5-HT system functioning (as reflected in levels of the hormones prolactin and cortisol, which reflect 5-HT levels, but can be measured less invasively than the cerebral spinal fluid sampling required to measure 5-HT activity directly). The authors concluded that the obtained results support the role of serotonin in determining neurochemical vulnerability to drug abuse.

The linkage of neuroendocrine functioning, impulsivity, and substance abuse was also discussed by King et al. (1990), who reviewed several animal and human studies suggesting a link between low plasma cortisol levels and impulsivity. The researchers measured plasma cortisol levels and the impulsivity scores of substance abusers and nonabusers and concluded that substance abusers may show low baseline levels of hypothalamic–pituitary–adrenal axis arousal, resulting in a decreased responsiveness to aversive conditioning. In other words, the impulsivity of drug abusers may be the product of a failure to learn social rules normally learned to avoid punishment.

Examining the relation between serotonin levels and behavioral features such as aggression and impulsivity, particularly among alcohol abusers in treatment, Buydens-Branchey and colleagues (Branchey, Buydens-Branchey, & Lieber, 1988; Buydens-Branchey, Branchey, & Noumair, 1989; Buydens-Branchey, Branchey, Noumair, & Lieber, 1989) reported that impulsivity was a characteristic of a subgroup of alcoholics found to have a highly heritable form of alcoholism. These authors suggested that this subgroup of alcoholics differs from another, characterized by later alcohol involvement and greater environmental influences in the availability of the serotonin precursor tryptophan. They emphasized the need for research involving serotonin-directed agents (agonist or antagonist) to explore the feasibility of reducing impulsivity and other negative behaviors that appear to be associated with reduced serotonergic levels.

Reviewing the neurobiology of alcohol abuse, Wallace (1988) suggested that (a) the impact of drug abuse is "biphasic" (p. 208), having different acute versus chronic effects, and (b) studies have shown genetic bases for alcoholism, but what appears inherited are "biologic risk factors" (p. 209) and, consequently, the environment also plays a role. For example, Li, Lumeng, McBride, and Murphy (1987) found that alcohol-preferring rats were deficient in both 5-HT and serotonin metabolites (5-hydroxyindoleacetic acid; 5-HIAA), prior to exposure to alcohol. Other studies have shown that rats, not predisposed to alcohol consumption, increased drinking following ablation of certain brain structures (Zhukov, Varkov, & Burov, 1987) and following administration of a single alcohol dose (Burov, Zhukov, & Khodorova, 1986)—both manipulations that deplete serotonergic activity. Wallace (1988) reviewed other studies suggesting links between alcohol consumption and serotonin, stressing that although such relationships are apparent, the mechanisms involved are unclear.

Moss (1987) related such findings concerning 5-HT and alcoholism to forms of psychopathology characterized by disturbances in inhibition (including impulsive behavior). Therefore, it is reasonable to hypothesize that psychopathology characterized by low levels of 5-HT prior to alcohol exposure may lead to impulsive behavior. This may in turn prompt alcohol use or abuse, which may then exacerbate the depletion of 5-HT and 5-HIAA. Thus, continued alcohol use may enhance both impulsivity and alcohol abuse. Although hypothetical, such sequelae suggest directions for prevention and treatment research efforts.

Furthermore, disruptive effects on the serotonergic system are not limited to alcohol. LSD and phenethylamine hallucinogens (e.g., mescaline and 2,5-dimethoxy-4-methylamphetamine) have been shown to affect serotonin levels in animals (e.g., Aghajanian, Foote, & Sheard, 1970; Jacobs, 1984; Rasmussen & Aghajanian, 1986; Trulson, Heym, & Jacobs, 1981). Similarly, a number of researchers and reviewers (e.g., De Souza & Battaglia, 1989; Ellinwood & Lee, 1989; L. H. Gold, Geyer, & Koob, 1989; Molliver, Mamounas, & Wilson, 1989) have noted that amphetamine analogue substances alter serotonergic functioning. For example, several designer drugs and other amphetamine analogue substances, including methamphetamine, 3,4-methylenedioxymethamphetamine, 3,4-methylenedioxyamphetamine, p-chloroamphetamine, and fenfluramine have produced irreversible, toxic damage to cortical functioning in animals (Fuller, 1989; Zaczek, Hurt, Culp, & De Souza, 1989). Finally, although acute (i.e., initial) cocaine use increased the levels of 5-HT (Mule, 1984), the results of other studies have indicated eventual decreases in 5-HT and 5-HIAA resulting from chronic (i.e., repeated) cocaine exposure (e.g., M. S. Gold & Dackis, 1984).

In summary, then, the research literature suggests links among neurotransmitter substances (e.g., 5-HT and metabolites), neuroendocrine levels (e.g., prolactin, cortisol), impulsivity, and substance abuse. However, much of the work to date has involved animal subjects or is otherwise more suggestive than definitive. Thus, additional research is required to fully elucidate causal relationships between neurochemical function, impulsivity, and substance abuse (Roy, Virkkunen, & Linnoila, 1987). The potential for pharmacotherapy, guided by these neurobiological findings, is discussed shortly.

TREATMENT CONCERNS

McKenna (1979) argued for the importance of adjusting treatment to the individual needs of the patient, on the basis of the results of diagnostic assessment and formulation. For example, the treatment of an abuser with a history of depression might involve tricyclic antidepressants, whereas another abuser with a character disorder may require a confrontational, cognitively oriented psychotherapeutic approach. In previous sections of this chapter we have explored a number of variables that may be helpful in the development of effective treatment programs. In this section, previous work relating treatment efforts to impulsivity and substance abuse is reviewed. In addition, we indicate potentially important areas that have not yet been addressed with respect to treatment.

In a study that could be interpreted as contradicting the importance of considering impulsivity as a treatment concern, Kendall et al. (1980) examined adult male offenders (drug and alcohol abusers and sex offenders) and found no relationship between subjects' cognitive style, impulsivity, and persistence in a treatment program. Persistence in treatment was measured as the total days each subject remained in the treatment program prior to quitting or termination (discharge).

However, McCown (1989), in a study involving Twelve Step self-help group members, found significant negative correlations between impulsivity and months of abstinence, as well as between impulsivity and number of "slips" (i.e., relapses). Interestingly, significant positive correlations were found for impulsivity and total months in self-help throughout the individual's life. McCown speculated that impulsive substance abusers may come in contact with self-help groups at an early age, resulting in a longer history of involvement, despite difficulties with abstinence and slips.

In a subsequent study, McCown (1990) compared initial measures of impulsivity with self-reports of abstinence and slips collected at the end of 1 year and found impulsivity to be positively correlated with the number of slips and negatively correlated with the total weeks of abstinence. This study involved new members of four Twelve Step groups who were polysubstance abusers with less than 6 months of abstinence.

Studies with emotionally disturbed children have indicated that impulsivity can be modified through cognitive–behavioral treatment (e.g., Kendall & Finch, 1978). Such results raise the question of whether impulsivity can be modified through substance abuse treatment. Jensen et al. (1990) investigated the results of substance abuse treatment on impulsivity among substance abusers and reported an overall decrease in impulsivity following treatment. Although the details of treatment were not provided, these data suggest that impulsivity may be amenable to treatment and may not represent an entirely enduring personality trait.

Woody et al. (1985) compared the effectiveness of paraprofessional drug counseling alone and in combination with psychotherapy for four groups of opiate addicts, including opiate addiction only; opiate dependence plus depression; opiate addiction plus ASPD; and opiate addiction, depression, and ASPD. Regarding the effectiveness of psychotherapy, Woody et al. concluded that the dual diagnosis of opiate addiction and ASPD generally suggests a negative prognosis for psychotherapy. However, they also found that the presence of depression alone or in combination with ASPD may be related to a more positive psychotherapy outcome for opiate addicts.

In an attempt to integrate various findings such as those just described, Sandberg, Greenberg, and Birkman (1991) presented a clinical model for matching the treatment modality to the type of abuser. Sandberg et al. described four categories of abusers: (a) Group 1, self-medication of Axis I disorder; (b) Group 2, addictive disease; (c) Group 3, antisocial behavior or character; and (d) Group 4, combination of other, or severe borderline or narcissistic character. Their model specifies, for example, that Group 1 abusers should receive a combination of inpatient psychotherapy (supportive, exploratory, or psychodynamic), outpatient psychotherapy, and Alcoholics Anonymous or Narcotics Anonymous meetings. Accompanying the description of the model is a review of the literature supporting its design. However, impulsivity was not specifically addressed by those authors, and it remains to be seen whether such consideration would be helpful in regard to such a treatment classification scheme. At this time, then, the potential importance of addressing impulsivity as a treatment consideration has not been determined and awaits further research.

FUTURE RESEARCH

In previous sections of this chapter, we discussed a variety of empirical and nonempirical works relevant to the construct of impulsivity and the relationship between this construct and the phenomenon of substance abuse. Despite the wide variety of writings in this area, one is left with a sense of uncertainty regarding the importance of impulsivity as an important factor in either the understanding or treatment of substance abuse.

The conspicuous absence of a widely used and accepted definition of impulsivity contributes to the uncertainty. Another source of ambiguity involves the difficulty faced in interpreting research in which impulsivity appears to have been a factor in both independent and dependent variables. This appears to be the case in the studies cited earlier in which measures of impulsivity were obtained from samples diagnosed as displaying *DSM–III–R* disorders for which impulsive behavior is one of the defining criteria. Consequently, future researchers investigating impulsivity within groups of substance abusers would be advised to define impulsivity operationally and to include only subjects who would have met the criteria for diagnosis regardless of whether impulsive characteristics were considered.

Despite the shortcomings in the literature, there are a number of potentially fruitful directions for future researchers to explore. The first of these involves obtaining more direct documentation of substance abuse as an antecedent, consequence, or unrelated co-occurring symptom with impulsive behaviors (or a personality trait of impulsivity). For example, it may be possible to conduct further, more extensive longitudinal studies correlating individual levels of impulsivity prior to the use of substances with the later development of a substance abuse disorder. Previous studies (e.g., Robins, 1974) have suggested positive correlations in this regard, but are not definitive, and their conclusions have been challenged (Bukstein et al., 1989).

A related issue was initially raised by McCown (1988), who concluded that further research will be necessary to determine whether high impulsivity leads individuals to greater levels of exposure to situations encouraging the use of addictive agents or whether those with high levels of impulsiveness might have personality structures that are particularly vulnerable to addictions. Although difficult to conduct, research disentangling these factors would provide an important contribution to this literature.

It is anticipated that researchers will also continue to explore the psychobiological mechanisms underlying impulsivity and substance abuse. Further evidence is needed on the extent to which the apparent neurochemical (impaired 5-HT transmission) basis of impulsivity may be an important factor in the high prevalence of substance abuse problems among various *DSM–III–R* Axis I and Axis II disorders. In other words, multifactorial studies are needed to explore the relationships among impulsivity, 5-HT functioning, psychopathology, and substance abuse. To date, such studies have not been published.

Finally, the relevance of the impulsivity construct to the treatment of substance abuse remains to be demonstrated. At least two approaches seem possible. Further research is needed to determine the extent to which the tendency toward impulsivity can be controlled by an individual after psychotherapy or psychoeducational efforts specifically designed to teach behaviors such as reflectivity (i.e., consideration of the consequences of one's actions) and planning. A second potentially useful treatment approach awaiting research involves the use of psychopharmacological agents designed to target impaired 5-HT functioning, for example (e.g., Gorelick, 1989; Kosten, 1990).

CONCLUSION

In this review we have sampled the professional literature relevant to the construct of impulsivity or impulsiveness as it relates to understanding and treating substance abuse. We noted that impulsivity is a characteristic consistently cited when describing substance abusers and appears to occur among a significant percentage of such individuals. Furthermore, impulsivity has been shown, in numerous studies, to have heuristic value in understanding the relationship of substance abuse and various types of psychopathology. Psychobiological studies lend support to the notion that impulsivity may be a link between substance abuse and psychopathology. To the extent that such a link exists, it would be valuable to address the issue of impulsive behavior of substance abusers as part of the intervention provided in treatment programs. However, currently, the importance of impulsivity in the understanding and treatment of substance abuse would be better characterized as a hypothesis than a substantiated conclusion. For this reason, we provided suggestions about several research issues that should be addressed in the future.

REFERENCES

Adler, T. (1990, February). Drug abuse risk linked with antisocial disorder. *APA Monitor*, p. 12.

Aghajanian, G. K., Foote, W. E., & Sheard, M. H. (1970). Action of psychotogenic drugs on single midbrain raphe neurons. *Journal of Pharmacology Experimentation and Therapy, 171*, 178–187.

Alterman, A. I., & Cacciola, J. S. (1991). The antisocial personality disorder diagnosis in substance abusers: Problems and issues. *Journal of Nervous and Mental Disease, 179*, 401–409.

American Psychiatric Association. (1987). *Diagnostic and statistical manual of mental disorders* (3rd ed., rev.). Washington, DC: Author.

Bigelow, G. E., Brooner, R. K., McCaul, M. E., & Svikis, D. S. (1988). Biological vulnerability: Treatment implications/applications. In R. W. Pickens & D. S. Svikis (Eds.), *Biological vulnerability to drug abuse* (NIDA Research Monograph 89, Department of Health and Human Services, pp. 165–173). Washington, DC: U.S. Government Printing Office.

Brady, K., Anton, R., Ballenger, J. C., Lydiard, R. B., Adinoff, B., & Selander, J. (1990). Cocaine abuse among schizophrenic patients. *American Journal of Psychiatry, 147*, 1164–1167.

Branchey, M. H., Buydens-Branchey, L., & Lieber, C. S. (1988). P3 in alcoholics with disordered regulation of aggression. *Psychiatry Research, 25*, 49-58.

Brantley, P. J., & Sutker, P. B. (1984). Antisocial behavior disorders. In H. E. Adams & P. B. Sutker (Eds.), *Comprehensive handbook of psychopathology* (pp. 439–478). New York: Plenum Press.

Brown, V. B., Ridgely, M. S., Pepper, B., Levine, I. S., & Ryglewicz, H. (1989). The dual crisis: Mental illness and substance abuse. *American Psychologist, 44*, 565–569.

Bukstein, O. G., Brent, D. A., & Kaminer, Y. (1989). Comorbidity of substance abuse and other psychiatric disorders in adolescents. *American Journal of Psychiatry, 146*, 1131–1141.

Burov, Y. V., Zhukov, V. N., & Khodorova, N. A. (1986). Serotonin content in different brain areas and in the periphery: Effect of ethanol in rats predisposed and non-predisposed to alcohol intake. *Biogenic Amines, 4*, 205–209.

Butcher, J. N. (1988). Personality factors in drug addiction. In R. W. Pickens & D. S. Svikis (Eds.), *Biological vulnerability to drug abuse* (NIDA Research Monograph 89, Department of Health and Human Services, pp. 87–92). Washington, DC: U.S. Government Printing Office.

Buydens-Branchey, L., Branchey, M. H., & Noumair, D. (1989). Age of alcoholism onset: I. Relationship to psychopathology. *Archives of General Psychiatry, 46,* 225–230.

Buydens-Branchey, L., Branchey, M. H., Noumair, D., & Lieber, C. S. (1989). Age of alcoholism onset: II. Relationship to susceptibility to serotonin precursor availability. *Archives of General Psychiatry, 46,* 231–236.

Carey, K. B. (1991). Research with dual diagnosis patients: Challenges and recommendations. *The Behavior Therapist, 14,* 5–8.

Carey, K. B., & Carey, M. P. (1990). Social problem-solving in dual diagnosis patients. *Journal of Psychopathology and Behavioral Assessment, 12,* 247–254.

Carey, M. P., Carey, K. B., & Meisler, A. W. (1990). Training mentally ill chemical abusers in social problem solving. *Behavior Therapy, 21,* 515–518.

Cloninger, C. R. (1988). Etiologic factors in substance abuse: An adoption study perspective. In R. W. Pickens & D. S. Svikis (Eds.), *Biological vulnerability to drug abuse* (NIDA Research Monograph 89, Department of Health and Human Services, pp. 52–72). Washington, DC: U.S. Government Printing Office.

Craig, R. J. (1986). The personality structure of heroin addicts. In S. I. Sazara (Ed.), *Neurobiology of behavioral control in drug abuse* (NIDA Research Monograph 74, Department of Health and Human Services, pp. 25–36). Washington, DC: U.S. Government Printing Office.

Craig, R. J., & Olson, R. E. (1990). MMPI characteristics of drug abusers with and without histories of suicide attempts. *Journal of Personality Assessment, 55,* 717–728.

Dackis, C. A., & Gold, M. S. (1984). Depression in opiate addicts. In S. M. Mirin (Ed.), *Substance abuse and psychopathology* (pp. 19–40). Washington, DC: American Psychiatric Press.

Dackis, C. A., & Gold, M. S. (1985). Pharmacological approaches to cocaine addiction. *Journal of Substance Abuse Treatment, 2,* 139–145.

De La Rosa, M., Lambert, E. Y., & Gropper, B. (Eds.). (1990). *Drugs and violence: Causes, correlates, and consequences* (NIDA Research Monograph 103, Department of Health and Human Services, p. 127). Washington, DC: U.S. Government Printing Office.

De Souza, E. B., & Battaglia, G. (1989). Effects of MDMA and MDA on brain serotonin neurons: Evidence from neurochemical and autoradiographic studies. In K. Asghar & E. De Souza, (Eds.), *Pharmacology and toxicology of amphetamine and related designer drugs* (NIDA Research Monograph 94, Department of Health and Human Services, pp. 196–222). Washington, DC: U.S. Government Printing Office.

Ellinwood, E. H., & Lee, T. H. (1989). Dose and time-dependent effects of stimulants. In K. Asghar & E. De Souza (Eds.), *Pharmacology and toxicology of amphetamine and related designer drugs* (NIDA Research Monograph 94, Department of Health and Human Services, pp. 323–340). Washington, DC: U.S. Government Printing Office.

Evans, K., & Sullivan, J. M. (1990). *Dual diagnosis: Counseling the mentally ill substance abuser*. New York: Guilford Press.

Eysenck, S. B. G., & Eysenck, H. J. (1978). Impulsiveness and venturesomeness: Their position in a dimensional system of personality description. *Psychological Reports, 43*, 1247–1255.

Eysenck, S. B. G., Pearson, P. R., Easting, G., & Allsopp, J. F. (1985). *Personality and Individual Differences, 6*, 613–619.

Fishbein, D. H., Lozovsky, D., & Jaffe, J. H. (1989). Impulsivity, aggression, and neuroendocrine responses to serotonergic stimulation in substance abusers. *Biological Psychiatry, 25*, 1049–1066.

Forgays, D. G. (1986). Personality characteristics and self-abusive behavior. In S. I. Szara (Ed.), *Neurobiology of behavioral control in drug abuse* (NIDA Research Monograph 74, Department of Health and Human Services, pp. 45–58). Washington, DC: U.S. Government Printing Office.

Fuller, R. W. (1989). Recommendations for future research on amphetamines and related designer drugs. In K. Asghar & E. De Souza (Eds.), *Pharmacology and toxicology of amphetamine and related designer drugs* (NIDA Research Monograph 94, Department of Health and Human Services, pp. 341–357). Washington, DC: U.S. Government Printing Office.

Gawin, F. H., & Kleber, H. D. (1986). Abstinence symptomatology and psychiatric diagnosis in cocaine abusers: Clinical observations. *Archives of General Psychiatry, 43*, 107–113.

Gittelman, R., Mannuzza, S., Shenker, R., & Bonagura, N. (1985). Hyperactive boys almost grown up: I. Psychiatric status. *Archives of General Psychiatry, 42*, 937–948.

Gold, L. H., Geyer, M. A., & Koob, G. F. (1989). Neurochemical mechanisms involved in behavioral effects of amphetamines and related designer drugs. In K. Asghar & E. De Souza (Eds.), *Pharmacology and toxicology of amphetamine and related designer drugs* (NIDA Research Monograph 94, Department of Health and Human Services, pp. 101–126). Washington, DC: U.S. Government Printing Office.

Gold, M. S., & Dackis, C. A. (1984). New insights and treatments: Opiate withdrawal and cocaine addiction. *Clinical Therapeutics, 7*, 6–21.

Goodwin, F. K. (1989). From the Alcohol, Drug Abuse, and Mental Health Administration. *Journal of the American Medical Association, 261*, 3517.

Gorelick, D. A. (1989). Serotonin uptake blockers and the treatment of alcoholism. In M. Galanter (Ed.), *Recent developments in alcoholism: Vol. 7. Treatment research* (pp. 26–37). New York: Plenum Press.

Grande, T. P., Wolf, A. W., Schubert, D. S. P., Patterson, M. B., & Brocco, K. (1984). Associations among alcoholism, drug abuse, and antisocial personality: A review of the literature. *Psychological Reports, 55*, 455–474.

Hammersley, R., Forsyth, A., Morrison, V., & Davies, J. B. (1989). The relationship between crime and opioid use. *British Journal of Addiction, 84*, 1029–1043.

Hesselbrock, M. N. (1986). Childhood behavior problems and adult antisocial personality disorder in alcoholism. In R. Meyers (Ed.), *Psychopathology and addictive disorders* (pp. 78–95). New York: Guilford Press.

Jacobs, B. L. (1984). Postsynaptic serotonergic action of hallucinogens. In B. L. Jacobs (Ed.), *Hallucinogens: Neurochemical, behavioral, and clinical perspectives* (pp. 183–202). New York: Raven Press.

Jensen, J. M., Pettinati, H. M., Evans, B. D., Meyers, K., & Valliere, V. N. (1990). Impulsivity and substance abusers: Changes with treatment and recovery. In *Problems of drug dependence 1990: Proceedings of the 52nd Annual Scientific Meeting* (NIDA Research Monograph 105, Department of Health and Human Services, pp. 287–288). Washington, DC: U.S. Government Printing Office.

Kagan, J. (1966). Reflection–impulsivity: The generality and dynamics of conceptual tempo. *Journal of Abnormal Psychology, 71,* 17–24.

Kendall, P. C., & Finch, A. J. (1978). A cognitive–behavioral treatment for impulsivity: A group comparison study. *Journal of Consulting and Clinical Psychology, 46,* 110–118.

Kendall, P. C., Moses, J. A., & Finch, A. J. (1980). Impulsivity and persistence in adult inpatient "impulse" offenders. *Journal of Clinical Psychology, 36,* 363–365.

Kennedy, H. G., & Grubin, D. H. (1990). Hot-headed or impulsive? *British Journal of Addiction, 85,* 639–643.

Khantzian, E. J. (1979). Impulse problems and drug addiction: Cause and effect relationships. In H. A. Wishnie & J. Nevis-Olsen (Eds.), *Working with the impulsive person* (pp. 97–112). New York: Plenum Press.

Khantzian, E. J. (1990). Self-regulation and self-medication factors in alcoholism and the addictions: Similarities and differences. In M. Galanter (Ed.), *Recent developments in alcoholism* (pp. 255–271). New York: Plenum Press.

Khantzian, E. J., Halliday, K. S., & McAuliffe, W. E. (1990). *Addiction and the vulnerable self: Modified dynamic group therapy for substance abusers.* New York: Guilford Press.

King, R. J., Jones, J., Scheuer, J. W., Curtis, D., & Zarcone, V. P. (1990). Plasma cortisol correlates of impulsivity and substance abuse. *Personality and Individual Differences, 11,* 287–291.

Klar, H., & Siever, L. J. (Eds.). (1985). *Biologic response styles: Clinical impressions.* Washington, DC: American Psychiatric Press.

Kosten, T. R. (1990). Neurobiology of abused drugs: Opioids and stimulants. *Journal of Nervous and Mental Disease, 178,* 217–227.

Lacey, J. H., & Evans, C. D. H. (1986). The impulsivist: A multi-impulsive personality disorder. *British Journal of Addiction, 81,* 641–649.

Lennings, C. J. (1990). Drug dependence and personality disorder: Its relationship to the treatment of drug dependence. *Drug and Alcohol Dependence, 27,* 209–212.

Lewis, C. E. (1991). Neurochemical mechanisms of chronic antisocial behavior (psychopathy): A literature review. *Journal of Nervous and Mental Disease*, 179, 720–727.

Li, T. K., Lumeng, L., McBride, W. J., & Murphy, J. M. (1987). Rodent lines selected for factors affecting alcohol consumption. *Alcohol and Alcoholism, Suppl. 1*, 91–96.

Lykken, D. T. (1957). A study of anxiety in the sociopathic personality. *Journal of Abnormal Social Psychology*, 55, 6–10.

Malow, R. M., Corrigan, S. A., Peña, J. M., Calkins, A. M., & Bannister, T. M. (1992). Mood and HIV risk behavior among drug dependent veterans.*Psychology of Addictive Behaviors*, 2, 131–134.

Malow, R. M., Pintard, P. F., Sutker, P. B., & Allain, A. N. (1988). Psychopathology subtypes: Drug use motives and patterns. *Psychology of Addictive Behaviors*, 2, 1–13.

Malow, R. M., West, J. A., Corrigan, S. A., Peña, J. M., & Lott, W. C. (1992). Cocaine and speedball users: Psychopathological differences. *Journal of Substance Abuse Treatment*, 8 171–176.

Malow, R. M., West, J. A., Williams, J. A., & Sutker, P. B. (1989). Personality disorders classification and symptoms in cocaine and opioid addicts. *Journal of Consulting and Clinical Psychology*, 57, 765–767.

McCord, W., & McCord, J. (1960). *Origins of alcoholism*. Stanford, CA: Stanford University Press.

McCown, W. (1988). Multi-impulsive personality disorder and multiple substance abuse: Evidence from members of self-help groups. *British Journal of Addiction*, 83, 431–432.

McCown, W. (1989). The relationship between impulsivity, empathy, and involvement in Twelve Step self-help substance abuse treatment groups. *British Journal of Addictions*, 84, 391–393.

McCown, W. (1990). The effect of impulsivity and empathy on abstinence of polysubstance abusers: A prospective study. *British Journal of Addiction*, 85, 635–637.

McKenna, G. J. (1979). Fitting different treatment modes to patterns of drug use. In H. A. Wishnie & J. Nevie-Olsen (Eds.), *Working with the impulsive person* (pp. 113–123). New York: Plenum Press.

McLellan, A. T., & Druley, K. A. (1977). Non-random relation between drugs of abuse and psychiatric diagnosis. *Journal of Psychiatric Research*, 13, 179–184.

McLellan, A. T., Woody, G. E., & O'Brien, C. P. (1979). Development of psychiatric illness in drug abusers: Possible role of drug preference. *New England Journal of Medicine*, 301, 1310–1314.

Messer, S. B. (1976). Reflection–impulsivity: A review. *Psychological Bulletin*, 83, 1026–1052.

Meyer, R. E., & Hesselbrock, M. N. (1984). Psychopathology and addictive disorders revisited. In S. M. Mirin (Ed.), *Substance abuse and psychopathology* (pp. 1–18). Washington, DC: American Psychiatric Press.

Mirin, S. M. (Ed.). (1984). *Substance abuse and psychopathology*. Washington, DC: American Psychiatric Press.

Mirin, S. M., Weiss, R. D., Sollogub, A., & Michael, J. (1984). Affective illness in substance abusers. In S. M. Mirin (Ed.), *Substance abuse and psychopathology* (pp. 57–78). Washington, DC: American Psychiatric Press.

Molliver, M. E., Mamounas, L. A., & Wilson, M. A. (1989). Effects of neurotoxic amphetamines on serotonergic neurons: Immunocytochemical studies. In K. Asghar & E. De Souza (Eds.), *Pharmacology and toxicology of amphetamine and related designer drugs* (NIDA Research Monograph 94, Department of Health and Human Services, pp. 270–305). Washington, DC: U.S. Government Printing Office.

Moss, H. B. (1987). Serotonergic activity and disinhibitory psychopathology in alcoholism. *Medical Hypotheses, 23*, 353–361.

Mueser, K. T., Yarnold, P. R., Levinson, D. F., Singh, H., Bellack, A. S., Kee, K., Morrison, R. L., & Yadalam, K. G. (1990). Prevalence of substance abuse in schizophrenia: Demographic and clinical correlates. *Schizophrenia Bulletin, 16*, 31–56.

Mule, S. J. (1984). The pharmacodynamics of cocaine abuse. *Psychiatric Annals, 14*, 724–727.

Nace, E. P. (1990). Personality disorder in the alcoholic patient. *Psychiatric Annals, 19*, 256–260.

Nace, E. P., Davis, C. W., & Gaspari, J. P. (1991). Axis II comorbidity in substance abuse. *American Journal of Psychiatry, 148*, 118–120.

Nurnburg, H. G., Raskin, M., Levine, P. E., Pollack, S., Siegel, O., & Prince, R. (1991). The comorbidity of borderline personality disorder and other DSM-III-R Axis II personality disorders. *American Journal of Psychiatry, 148*, 1371–1377.

Rasmussen, K., & Aghajanian, G. K. (1986). Effect of hallucinogens on spontaneous and sensory-evoked locus coeruleus unit activity in the rat: Reversal by selective 5-HT antagonists. *Brain Research, 385*, 395–400.

Robins, L. N. (1974). *Deviant children grown up: A sociological and psychiatric study of sociopathic personality*. Huntington, NY: Robert E. Krieger.

Rounsaville, B. J. (1988). The role of psychopathology in the familial transmission of drug abuse. In R. W. Pickens & D. S. Svikis (Eds.), *Biological vulnerability to drug abuse* (NIDA Research Monograph 89, Department of Health and Human Services, pp. 108–119). Washington, DC: U.S. Government Printing Office.

Rounsaville, B. J., Dolinsky, Z. S., Babor, T. F., & Meyer, R. E. (1987). Psychopathology as a predictor of treatment outcome in alcoholics. *Archives of General Psychiatry, 44*, 505–513.

Roy, A., Virkkunen, M., & Linnoila, M. (1987). Reduced central serotonin turnover in a subgroup of alcoholics? *Progress in Neuropsychopharmacology and Biological Psychiatry, 11*, 173–177.

Sandberg, C., Greenberg, W. M., & Birkman, J. C. (1991). Drug-free treatment selection for chemical abusers: A diagnostic based model. *American Journal of Orthopsychiatry, 61,* 358–371.

Shaffer, J. W., Nurco, D. N., & Kinlock, T. W. (1984). A new classification of narcotic addicts based on type and extent of criminal activity. *Comprehensive Psychiatry, 25,* 315–328.

Shapiro, D. (1965). *Neurotic styles.* New York: Basic Books.

Shedler, J., & Block, J. (1990). Adolescent drug use and psychological health: A longitudinal inquiry. *American Psychologist, 45,* 612–630.

Siever, L. J., & Davis, K. L. (1991). A psychobiological perspective on the personality disorders. *American Journal of Psychiatry, 148,* 1647–1658.

Stabenau, J. R., & Hesselbrock, V. M. (1984). Psychopathology in alcoholics and their families and vulnerability to alcoholism: A review and new findings. In S. M. Mirin (Ed.), *Substance abuse and psychopathology* (pp. 132–168). Washington, DC: American Psychiatric Press.

Sutker, P. B., & Allain, A. N. (1988). Issues in personality conceptualizations of addictive behaviors. *Journal of Consulting and Clinical Psychology, 56,* 172–182.

Sutker, P. B., & Archer, R. P. (1984). Opiate abuse and dependence disorders. In H. E. Adams & P. B. Sutker (Eds.), *Comprehensive handbook of psychopathology* (pp. 585–621). New York: Plenum Press.

Sutker, P. B., Archer, R. P., & Kilpatrick, D. G. (1983). Sociopathy and antisocial behavior: Theory and treatment. In S. M. Turner, K. S. Calhoun, & H. E. Adams (Eds.), *Handbook of clinical behavior therapy* (pp. 665–712). New York: Wiley.

Sutker, P. B., & King, A. R. (1985). Antisocial personality disorder: Assessment and case formulation. In D. Turkat (Ed.), *Behavioral case formulation* (pp. 113–153). New York: Plenum Press.

Tomas, J. M., Vlahov, D., & Anthony, J. C. (1990). Association between intravenous drug use and early misbehavior. *Drug and Alcohol Dependence, 25,* 79–89.

Trulson, M. E., Heym, J., & Jacobs, B. L. (1981). Dissociations between the effects of hallucinogenic drugs on behavior and raphe unit activity in freely moving cats. *Brain Research, 215,* 275–293.

Uddo, M., Malow, R. M., & Sutker, P. B. (1993). Opioid and cocaine abuse and dependence disorders. In P. Sutker & H. Adams (Eds.), *Comprehensive handbook of psychopathology* (2nd ed., pp. 477–503). New York: Plenum Press.

Wallace, J. (1988). The relevance to clinical care of recent research in neurobiology. *Journal of Substance Abuse Treatment, 5,* 207–217.

Weddington, W. W., Brown, B. S., Haertzen, C. A., Cone, E. J., Dax, E. M., Herning, R. I., & Michaelson, B. S. (1990). Changes in mood, craving, and sleep during short-term abstinence reported by male cocaine addicts: A controlled residential study. *Archives of General Psychiatry, 47,* 861–868.

Weiss, R. D., Mirin, S. M., Michael, J. L., & Sollogub, A. C. (1986). Psychopathology in chronic cocaine abusers. *American Journal of Drug and Alcohol Abuse, 12,* 17–29.

Whitters, A. C., Cadoret, R. J., & McCalley-Whitters, M. K. (1987). Further evidence for heterogeneity in antisocial alcoholics. *Comprehensive Psychiatry, 28,* 513–519.

Wishnie, H. (1977). *The impulsive personality: Understanding people with destructive character disorders.* New York: Plenum Press.

Woody, G. E., McLellan, A. T., Luborsky, L., & O'Brien, C. P. (1985). Sociopathy and psychotherapy outcome. *Archives of General Psychiatry, 42,* 1081–1086.

Zaczek, R., Hurt, S., Culp, S., & De Souza, E. B. (1989). In K. Asghar & E. De Souza (Eds.), *Pharmacology and toxicology of amphetamine and related designer drugs* (NIDA Research Monograph 94, Department of Health and Human Services, pp. 223–239). Washington, DC: U.S. Government Printing Office.

Zhukov, V. N., Varkov, A. I., & Burov, Y. V. (1987). Effect of destruction of the brain serotonergic system on alcohol intake by rats in early stages of experimental alcoholism. *Biogenic Amines, 4,* 201–204.

12

PERFECTIONISM AND GOAL ORIENTATION IN IMPULSIVE AND SUICIDAL BEHAVIOR

PAUL L. HEWITT and GORDON L. FLETT

The importance of perfectionism as a relevant factor in clinical processes has recently begun to be established. Various research programs have demonstrated that components of perfectionism are germane to different types of maladjustment processes (e.g., Alden, Wallace, & Bieling, 1992; Frost, Marten, Lahart, & Rosenblate, 1990; Hewitt & Flett, 1990, 1991a, 1991b), and there have been suggestions that perfectionism is an important factor in the assessment of vulnerabilities (Hewitt & Flett, 1993). Strategies for the treatment of perfectionism have also been developed (Barrow & Moore, 1983). Because of the linkage between perfectionistic tendencies and individuals who have traditionally been viewed as impulsive (e.g., individuals with suicidal tendencies [Shaffer,

This research was supported by grant #410-91-1690 from the Social Sciences and Humanities Research Council of Canada. The data were collected while the first author was at the Brockville Psychiatric Hospital. We would like to thank Marjorie Cousins, Lois Ritchie, and Wendy Turnbull for their help in data collection.

247

1974] and individuals in the criminal justice system [Pacht, 1984]), we focus in this chapter on individual differences in perfectionism as a means of understanding some clinically relevant aspects of impulsive behavior.

One way to conceptualize impulsive actions is to examine impulsivity with respect to planful goal striving and goal setting. Impulsive individuals have been frequently described as lacking in specific goals (Svebak & Kerr, 1989) and unlikely to make plans (Barratt, 1985; S. B. G. Eysenck & Eysenck, 1977; Gerbing, Ahadi, & Patton, 1987). These people tend to live for the "here and the now" and to eschew the future. On the other hand, nonimpulsive individuals tend to be highly goal directed and to exhibit planful behavior (see Frese, Stewart, & Hannover, 1987) with detailed plans that focus on the attainment of personal future goals. Given that goal striving is regarded as nonimpulsive behavior, it follows that individuals who typically engage in the striving for self-related goals that are difficult to attain should also be characterized by low levels of trait impulsivity. Indeed, descriptions of perfectionists frequently allude to their methodical actions and the sometimes painstaking deliberations that occur before they take action (e.g., Missildine, 1963).

In this chapter, we examine the association between perfectionism and impulsivity, both at a theoretical and empirical level. Our orientation is based primarily on observations from the ego-control literature (see Block & Block, 1980) that focuses on standards and goal-setting behavior and levels of exerted control. Block and Block's (1980) longitudinal investigation of ego control in children suggests that high personal standard setting is associated with low impulsivity. For example, in their study of 4-year-olds, they reported correlations between a Q-sort method (teachers' ratings of the child's personality) and an experiment-based measure of ego impulse uncontrol. Children described as having high standards of performance for themselves were less likely to exhibit a lack of ego control in actual situations.

Block and Block (1980) contended that moderate levels of ego control were probably the most adaptive; maladaptive responses might result from too much or too little ego control. Similarly, our analysis of self-oriented perfectionism is based on the premise that a moderate level of goal striving and impulsivity is the most adaptive. Too high of a level of self-oriented perfectionism may serve as a diathesis that can lead to negative adjustment outcomes such as depression and suicide potential (Hewitt & Flett, 1993; Hewitt, Flett, & Turnbull-Donovan, 1992), whereas extremely low levels may lead to withdrawal and a decrease in intrinsic motivation (Hewitt & Flett, 1991b).

Any attempt to examine perfectionism and impulsivity must take into account salient differences between dimensions of perfectionism. Recent research in our laboratory and elsewhere has established that the perfectionism construct is multidimensional (Frost et al., 1990; Hewitt

& Flett, 1989, 1991b). A key distinction is made between self-oriented perfectionism (i.e., demanding that oneself be perfect) and two interpersonal dimensions referred to as socially prescribed perfectionism (i.e., the belief that significant others demand perfection) and other-oriented perfectionism (i.e., demanding that others be perfect). Whereas self-oriented perfectionism refers solely to personal standards that are under individual control, socially prescribed perfectionism is a social–cognitive variable that involves an external locus of control and a sense of hopelessness about the inability to please others by not meeting their expectations. Other-oriented perfectionism is a third dimension that is quite distinct in that the standards and expectations are directed outward rather than inward. This dimension is similar to authoritarianism and is associated with extrapunitiveness and other-directed blame (Hewitt & Flett, 1991a).

Clearly, the nature of the association between perfectionism and impulsivity should vary as a function of the dimension of perfectionism in question. We suggested earlier that self-oriented perfectionism is associated with an excessive degree of impulse control and extensive planning to meet one's goals. By contrast, the imposition of unrealistic standards of performance on the self by others (i.e., socially prescribed perfectionism) is associated significantly with an extreme lack of impulse control. The essence of our argument is that socially prescribed perfectionism and impulsivity should be positively associated because the lack of control inherent in socially prescribed perfectionism tends to undermine goal-directed activity and that impulsive acts are likely to be expressed as a form of reactance to the perceived imposition of unrealistic expectations on the self or to the realization of one's inability to meet others' expectations.

In this chapter, the link between perfectionism and impulsive behavior is examined from both theoretical and empirical perspectives. We report data from a variety of studies that indicate the presence of a direct link between dimensions of perfectionism and impulsivity, as well as an association between perfectionism and variables that reflect the presence or absence of impulsivity (e.g., goal-striving measures, planful problem-solving orientation, etc.). Finally, we examine applications of the association between perfectionism and impulsivity by presenting recent data on perfectionism and impulsive acts and wishes in the form of suicidal tendencies. We begin by outlining recent empirical work on the association between impulsivity and perfectionism.

DIMENSIONS OF PERFECTIONISM AND IMPULSIVITY

According to the conceptualization outlined earlier, self-oriented perfectionism is associated with intrinsic motivation and active striving to attain goals and master tasks. Given that people with impulsive tendencies are relatively less likely to be goal directed, self-oriented perfectionism should be associated with nonimpulsivity to the extent that individuals with extreme levels of self-oriented perfectionism should exhibit a high degree of planfulness and goal orientation with little indication of impulsiveness. This tendency should contrast sharply with the pattern expected with the social dimensions of perfectionism.

The link between other-oriented perfectionism and impulsivity is less clear, but a negative association with impulsivity might be expected because both other-oriented perfectionists and impulsive people have been described as hostile and punitive with respect to others, in part because of feelings of superiority (Hewitt & Flett, 1991b; Robbins, 1989). However, the major distinction with respect to impulsive tendencies should be between self-oriented versus socially prescribed perfectionism. Whereas self-oriented perfectionists actively work toward goals, socially prescribed perfectionists suffer motivational deficits and avoid rather than approach goals in a planful manner. Presumably, this avoidance behavior stems from the fear of failure that looms large when confronted by people who expect perfection (Flett, Blankstein, Hewitt, & Koledin, 1992; Flett, Hewitt, Blankstein, & Koledin, 1991).

PERFECTIONISM, CONSCIENTIOUSNESS, AND IMPULSIVITY

Given the only recent interest in empirical research on the perfectionism construct, it is not surprising that few published studies have examined perfectionism and impulsivity. However, an examination of indirect and direct evidence provides extensive support for our observations. Indirect support is provided by studies that have examined general levels of conscientiousness and impulsivity. For instance, Costa and McCrae (1988) administered the NEO Personality Inventory (Costa & McCrae, 1985) and the Personality Research Form (Jackson, 1984) to a sample of adults from the general population. One finding in this study was that conscientiousness was negatively correlated with impulsivity. Because other research in our laboratory has established that conscientiousness is associated with high levels of self-oriented perfectionism, it follows that self-oriented perfectionism may be associated with low levels of impulsivity.

In addition to the negative link between conscientiousness and impulsivity, Costa and McCrae (1988) also reported that impulsivity was correlated with greater neuroticism and greater depression. Both depression and neuroticism are variables that are associated with socially prescribed perfectionism (Hewitt & Flett, 1991a, 1991b; Hewitt, Flett, &

Blankstein, 1991). Thus, these data suggest indirectly the possibility of a positive link between socially prescribed perfectionism and impulsivity.

In a subsequent study, Costa and McCrae (1992) administered the NEO Personality Inventory and the Basic Personality Inventory (BPI; Jackson, 1989) to a sample of 117 adults. The BPI includes a measure of impulse expression. This measure was similarly found to be correlated negatively with conscientiousness and positively with neuroticism. Overall, then, the broad trait dimensions that most closely resemble self-oriented and socially prescribed perfectionism differ in the direction of their association with impulsivity.

PERFECTIONISM, PROBLEM SOLVING, AND IMPULSIVITY

Another link between perfectionism and impulsivity is suggested indirectly by research on problem solving. A common finding in the literature is that poor problem-solving ability is associated with impulsive tendencies. Initial work by Heppner, Hibel, Neal, Weinstein, and Rabinowitz (1982) on the development of the Problem Solving Inventory revealed that more effective problem solvers reported lower levels of impulsivity. Moreover, people who received problem-solving training were less likely to exhibit impulsive behavior (Dixon, Heppner, Petersen, & Ronning, 1979; Heppner, Baumgardner, Larson, & Petty, 1983). Therefore, problem-solving interventions may be the most effective with impulsive individuals (see Shure & Spivack, 1981).

Recently, we have begun to examine the association between dimensions of perfectionism and problem-solving orientation. The pattern of results is generally consistent with the view that socially prescribed perfectionism is associated with impulsivity, whereas self-oriented perfectionism is associated with nonimpulsivity. For instance, Flett, Hewitt, Blankstein, Solnik, and Van Brunschot (1992) had samples of college students complete the Multidimensional Perfectionism Scale (MPS; Hewitt & Flett, 1991b; Hewitt, Flett, & Turnbull, 1993) and the Social Problem Solving Inventory (D'Zurilla & Nezu, 1990). The SPSI is a multidimensional, self-report measure of problem-solving orientation and skills. The results showed that socially prescribed perfectionism was associated with a poorer problem-solving orientation in terms of cognitive, affective, and behavioral responses. By contrast, self-oriented perfectionism was associated with reports of a more positive problem-solving orientation. The negative reactions reported by the individuals with high levels of socially prescribed perfectionism reflect a diminished tendency to carefully weigh the alternatives to a problem in a goal-directed manner.

TABLE 1
First-Order Correlations Between Perfectionism and BPI Measures in Adolescents

	Perfectionism	
BPI measure	Social	Self
Hypochondriasis	.22**	−.08
Depression	.37*	−.29*
Denial	−.08	.05
Interpersonal Problems	.28*	−.26*
Alienation	.31*	−.37*
Persecutory Ideas	.31*	−.17**
Anxiety	.11	.06
Thinking Disorder	.32*	−.08
Impulse Expression	.21**	−.26*
Social Introversion	.31*	−.18**
Self-Depreciation	.39*	−.37*
Deviation	.39*	−.32*

Note. Correlations are two-tailed and are based on the responses of 107 subjects. BPI = Basic Personality Inventory.
*$p < .05$. ** $p < .01$.

DIRECT INVESTIGATIONS OF PERFECTIONISM AND IMPULSIVITY

In addition to indirect indications of a relationship between perfectionism and high or low levels of impulsivity, a variety of correlational studies in our laboratory have provided more direct evidence of the association between perfectionism and impulsivity. The findings tend to generalize across samples with various measures of impulsivity; perfectionism and impulsivity have been linked in adolescents, college students, and psychiatric patients.

For example, our research on impulsivity and perfectionism with adolescents has used the Child–Adolescent Perfectionism Scale (CAPS; Flett & Hewitt, 1990; Flett, Hewitt, Boucher, Davidson, & Munro, 1992). The CAPS is a 22-item measure that provides indexes of self-oriented perfectionism and socially prescribed perfectionism. Other-oriented perfectionism is not assessed by this scale. In one study reported by Flett, Hewitt, Boucher, et al. (1992), a sample of 107 adolescents completed the CAPS and the BPI. As noted previously, the BPI also measures other psychological problems such as hypochondriasis, depression, persecutory ideas, and social isolation. In this study, the proposed association between dimensions of perfectionism and impulsivity was particularly apparent when we examined the first-order correlations between the perfectionism dimensions and impulse expression after removing variance associated with the other perfectionism dimensions. The partial correlations are shown in Table 1, which shows small but significant correlations. It is most important to note that self-oriented perfectionism was correlated negatively with impulse expression, whereas socially prescribed perfectionism was correlated positively with impulse expression, confirming our expectations regarding self-oriented, socially prescribed perfectionism and impulsivity.

Existing research with college student samples has also tended to provide support for the hypothesized association between self-oriented perfectionism and low impulsivity. For example, Flett, Hewitt, Blankstein, and O'Brien (1991) examined perfectionism and learned resourcefulness. Subjects completed the MPS and Rosenbaum's (1980) Self-Control Scale. Although not generally regarded as a measure of impulsivity, results involving Rosenbaum's measure are relevant to our discussion because a factor analysis by Rude (1989) showed that the Self-Control Scale includes a Nonimpulsivity factor with items of direct relevance (e.g., "When I feel that I am too impulsive, I tell myself 'Stop and think before you do anything' "). The results of our study showed that self-oriented perfectionists did indeed report significantly higher levels of self-control, thus suggesting the presence of lower levels of impulsivity.

Another study examined the association between scores on the MPS and the Impulsiveness subscale of the Eysenck Personality Questionnaire (EPQ; H. J. Eysenck & Eysenck, 1975) in a sample of 107 college students. The EPQ Impulsivity scale is derived from items included in the Psychoticism scale (see Goh, King, & King, 1982; Howarth, 1986; Roger & Morris, 1991). The primary finding was that self-oriented perfectionism was negatively correlated with impulsivity in this sample. Previous analyses of the same data showed that socially prescribed perfectionism was associated primarily with neuroticism (see Hewitt et al., 1991).

Finally, we have assessed the extent to which the perfectionism dimensions are associated with the presence or absence of a planful goal orientation. A recent unpublished study with 294 college students examined the association between perfectionism and a measure of tenacious goal pursuit (see Brandtstadter & Renner, 1990). Not surprisingly, self-oriented perfectionists demonstrated a tendency toward low impulsivity by reporting a high level of tenacious goal pursuit. The association between self-oriented perfectionism and tenacious goal pursuit remained significant even after removing variance caused by the other MPS dimensions ($r = .43$, $p < .01$). Other partial correlations showed that other-oriented perfectionism was not correlated with tenacious goal pursuit but that socially prescribed was correlated negatively with tenacious goal pursuit ($r = -.32$, $p < .01$), once again suggesting a tendency for imposed standards to foster impulsivity.

Similar but not identical findings have been reported with psychiatric patients. In one study, we examined the responses of 84 psychiatric patients who had been administered the MPS and the Minnesota Multiphasic Personality Inventory (MMPI). This work is relevant in that the MMPI has an Impulsivity scale (Blackburn, 1971). These data revealed a nonsignificant association between self-oriented perfectionism and impulsivity. Analyses with the other MPS dimensions showed that both other-oriented perfectionism ($r = 27$, $p < .05$) and socially prescribed perfectionism ($r = .34$, $p < .01$) correlated significantly with impulsivity.

The findings obtained in the clinical sample are somewhat inconsistent with the findings obtained with other samples in that the association between self-oriented perfectionism and nonimpulsivity was not present. The most likely explanation for the lack of consistency across samples is that there are some important differences in the scale content of the various impulsivity measures used in these studies (see Gerbing et al., 1987). For example, the MMPI measure has no content that assesses the key areas of planfulness and goal orientation. Thus, it is not surprising that the expected negative association between self-oriented perfectionism and impulsivity was not obtained in the psychiatric sample. However, if the studies with the various populations are taken as a whole, it seems reasonable to conclude that self-oriented perfectionism tends to be associated with low impulsivity in the form of excessive goal striving, whereas socially prescribed perfectionism is associated with high impulsivity.

DIMENSIONS OF PERFECTIONISM AND SUICIDAL IMPULSES

In the previous sections, we have shown that perfectionism dimensions appear to be differentially associated with impulsive tendencies. The support for this association between perfectionism and impulsivity suggests that some perfectionistic individuals may engage in clinically relevant impulsive behaviors. Thus, although there has been some anecdotal evidence of the relationship between perfectionism and various impulsive acts (see Shaffer, 1974), our focus in this section is on recent research that we have conducted on perfectionism and suicidal impulses.

Suicide attempts have been described often as ultimate acts of impulsivity, and ample evidence indicates that suicide is often attempted and committed by individuals with impulsive temperaments (see Domino & Leenaars, 1989; Firth, Blouin, Natarajan, & Blouin, 1986; Plutchik & Van Praag, 1989; Plutchik, Van Praag, & Conte, 1989; Ramos-Brieva & Cordero-Villafafila, 1989; Tomlinson-Keasey, Warren, & Elliott, 1986). For instance, Withers and Kaplan (1987) conducted a chart review of adolescents who attempted suicide and found that impulsivity, depression, and anger were among the most common characteristics of these adolescents. Similarly, Firth et al. (1986) compared the dream content of a mixed sample of depressed and violent individuals who had or had not attempted suicide and found that impulsivity was one of the factors that best differentiated suicidal and nonsuicidal patients. Given these findings, it is not surprising that measures of impulsivity are often included on scales that are designed to assess vulnerability to suicide (e.g., Limbacher & Domino, 1985–1986).

Research in our laboratory has sought to examine the role of dispositional goal orientations in suicidal impulses by focusing on the role of the various perfectionism dimensions. There have been numerous reports that perfectionistic expectations, either stemming from the self or others, are implicated in the etiology and manifestation of suicidal behaviors (Baumeister, 1990; Orbach, Gross, & Glaubman, 1981; Stephens, 1987). One hypothesis guiding our work is that socially prescribed perfectionism should be associated directly and positively with suicidal impulses. This is generally consistent with the observation that the perceived imposition of perfectionistic expectations on the self facilitates feelings of alienation, helplessness, and lack of motivation (Hewitt & Flett, 1991b) and may provide an impetus for potentially impulsive acts. We would therefore expect that there is a direct association between socially prescribed perfectionism and suicidal tendencies.

Three recent studies have confirmed the role of perfectionism in suicidal tendencies. In fact, these studies have confirmed not only that perfectionism and suicidal impulses are related but that the link between perfectionism and suicidal wishes is robust enough to remain significant after removing variance caused by other well-known correlates of suicide and perfectionism.

In our initial study (Hewitt, Flett, & Turnbull-Donovan, 1992), a sample of 88 psychiatric patients completed the MPS, the MMPI, and the Beck Depression Inventory (BDI; Beck, Rush, Shaw, & Emery, 1979). The MMPI was scored to provide measures of suicidal impulses (Farberow & Devries, 1967). The data from this study confirmed the role of socially prescribed perfectionism in that this dimension was associated with suicidal impulses and depression. The other perfectionism dimensions were not correlated significantly with the suicide measure. In a secondary analysis, we conducted a hierarchical regression analysis that examined the link between perfectionism and suicide after including measures of depression and hopelessness in the model. Depression and hopelessness were included because they are two of the better psychological predictors of suicide (see Beck, Brown, Berchick, Stewart, & Steer, 1990), and it is important to determine whether perfectionism can account for unique variance after these factors have been controlled for. Our data showed that socially prescribed perfectionism did indeed account for a significant proportion of the remaining variance in suicide intent. The findings must be interpreted with caution because the hopelessness measure in this

TABLE 2
MPS Adjusted Means and Suicide for Subclinical and Clinical Samples

Perfectionism dimension	Suicide impulse		
	High	Medium	Low
Subclinical sample			
Self-oriented	71.72	—	61.95
Socially prescribed	57.58	—	51.45
Clinical sample			
Self-oriented	87.38	74.41	65.82
Socially prescribed	63.02	68.49	57.45

Note. The means reported for the Multidimensional Perfectionism Scale (MPS) dimensions are the means obtained from the discriminant function analysis after adjusting for prior entry of depression and hopelessness scores. The subclinical sample had a high and a low group, but no medium group, in terms of suicide impulse level.

study was a single-item measure, and the usefulness of the MMPI suicide measure has been questioned by some authors (Clopton, 1978; Waters, Sendbuehler, Kincel, Boodoosingh, & Marchenko, 1982). Nevertheless, the results of this study are relevant for our purposes in that they suggest that socially prescribed perfectionism is associated with impulsive urges that are highly maladaptive. Moreover, it appears that this association is not due simply to overlap between perfectionism and other constructs reflecting personal adjustment levels.

Subsequent research has sought to replicate these findings with a more acceptable measure of hopelessness and alternative measures of suicide ideation and intent. Additional studies have been conducted with a sample of 117 college students and 96 psychiatric patients (see Hewitt & Flett, 1992). Subjects in both studies were administered suitable measures of perfectionism, suicidal impulses, depression, and hopelessness. Because analyses of the suicide measures indicated that the data were not normally distributed, the subjects in both studies were divided into discrete groups. The distribution in the subclinical group was bimodal, so subjects were split into either the low- or high-suicide impulse group. The distribution in the clinical sample was trimodal, so subjects were classified into groups that corresponded to no suicidal impulses, moderate suicidal impulses, or extreme suicidal impulses. Discriminant function analyses with forced entry were then conducted. The question addressed by these analyses was the following: Would the perfectionism dimensions be significant in differentiating these groups after depression and hopelessness had been taken into account?

The results of these analyses are summarized in Table 2. Regarding the subclinical sample, the results with socially prescribed perfectionism were consistent with those reported by Hewitt, Flett, and Turnbull-Donovan (1992). That is, socially prescribed perfectionism was not only correlated with level of suicidal impulse, but this dimension was able to differentiate between the high and low groups after controlling for levels of depression and hopelessness (assessed respectively by the BDI and the Beck Hopelessness Scale). In contrast to the results reported by Hewitt et al. (1992), it was found that self-oriented perfectionism also differentiated between the two groups even after taking hopelessness and depression into account.

A similar pattern of findings was obtained with the clinical sample (see Table 2). Both self-oriented and socially prescribed perfectionism were correlated with suicidal impulses, and both measures differentiated the various groups after taking depression and hopelessness levels into account. Examination of the adjusted means in Table 2 shows that patients with moderate- and high-suicidal impulse levels, relative to those with a low level, reported higher scores on the MPS Socially Prescribed subscale. The most notable finding involved self-oriented perfectionism, in that the high-suicide-impulse group reported a substantially higher mean level of self-oriented perfectionism than did the moderate- and low-impulse groups.

Taken together, these results are consistent with Baumeister's (1990) escape-from-self model, which suggests that the impulsive act of suicide stems in part from personal and social standards that are too high, resulting in failure episodes. In terms of our analysis, these data confirm the role of both self-oriented and socially prescribed perfectionism in impulsive tendencies.

The findings with socially prescribed perfectionism are certainly consonant with our expectations regarding perfectionism and impulsive tendencies. They are also consistent with other research showing that individuals characterized by impulsive suicide attempts (i.e., patients with a diagnosis of borderline personality disorder) have substantially higher socially prescribed perfectionism scores than other patient groups (Hewitt et al., 1992). This provides more support that socially derived unrealistic expectations may be involved in suicidal impulses.

It is interesting that self-oriented perfectionism was a relevant factor in suicide. Although various researchers have suggested that excessive self-oriented perfectionism is also associated with suicidal tendencies (Baumeister, 1990), the findings reported here give rise to the following question: How can behavior, such as suicide attempts, stem from a perfectionism dimension such as self-oriented perfectionism that is associated commonly with low impulsivity rather than high impulsivity? One possible explanation is that not all potential suicide attempts are impulsive. Some attempts are planned with a great deal of forethought and with various motives (see Berman & Jobes, 1991). It may be that the suicide attempts made by the highly self-oriented perfectionists are more planned and potentially more effective, whereas suicide attempts made by highly socially prescribed perfectionists tend to be more impulsive and less thought out and planned. Although this is speculative, it might suggest that suicide vulnerability may differ as a function of the type of perfectionism that is elevated, with socially prescribed perfectionists being less planful and organized. Certainly, future research should be dedicated to determining whether impulsive suicides as opposed to planned suicides differ in terms of perfectionism.

Alternatively, the conscientious, self-oriented perfectionist may shift from nonimpulsive to impulsive tendencies at some point, perhaps as a function of environmental contingencies or feedback. Research on perfectionism and depression from a diathesis–stress perspective provides some insight into this potential shift from extreme nonimpulsivity (at least in the form of a high-goal and planning orientation) to impulsivity in terms of suicidal tendencies. Research in this area examines the possibility that perfectionism is a diathesis or vulnerability factor that requires the experience of life stress or failure in order for the perfectionist to experience depression. In various studies, we have administered the MPS, a measure of life stress, and a measure of depressive symptomatology to subjects (Flett, Hewitt, Blankstein, & Mosher, 1991; Hewitt & Flett, 1993). Consistent support for the diathesis–stress approach has been obtained with the self-oriented perfectionism dimension in both clinical and subclinical samples. Regression analyses with depression as the outcome measure have consistently shown that the interaction of self-oriented perfectionism and life stress accounts for unique variance, over and above the variance caused by the two main effect terms (Flett, Hewitt, Blankstein, & Mosher, 1991; Hewitt & Flett, 1993). Evidence for the diathesis–stress model is particularly strong when the focus is the interaction of self-oriented perfectionism and the presence of personal stressors in the achievement domain (Hewitt & Flett, 1993). On the basis of this evidence, we have concluded that the experience of failure or life stress disrupts the planfulness, control, and achievement striving that is central to self-oriented perfectionism and nonimpulsive, purposive goal striving (Flett, Hewitt, Blankstein, & Mosher, 1991).

CONCLUSION

Overall, the data reported earlier provide important insight into the nature of impulsivity. For example, some of the acting-out and rebellious behavior that characterizes impulsivity may stem in part from perceived exposure to unfair and inappropriate social expectations and the resultant failure. Exposure to controlling behavior in the form of socially prescribed perfectionism appears to undermine goal striving and conscientious attempts to act in a nonimpulsive fashion.

The findings with self-oriented perfectionism were particularly revealing and highlight the need to examine impulsivity from a self-regulation or a diathesis–stress perspective. The results of various studies combine to suggest that stable individual differences in excessive goal striving are associated with low impulsivity in the form of excessive planning and tenacious goal pursuit. At the same time, individuals with extremely low levels of impulsivity may be particularly susceptible to adjustment problems when their attempts to attain goals are disrupted by life events and personal failures. Clearly, research needs to directly examine how individuals with low levels of impulsivity respond to situations involving a lack of control or situations that call for a more spontaneous and impulsive orientation.

Given the preliminary nature of this work, it is not surprising that many other research issues remain to be examined. One potentially important line of research involves the distinction between dysfunctional versus functional impulsivity (Dickman, 1990; Heaven, 1991). Our analysis has focused on dysfunctional impulsivity, with little consideration of the positive aspects of impulsivity. On a related note, the impulsivity construct is complex and has many different components, including motor and cognitive components (see Barratt, 1985). Research on the role of extreme standards in cognitive styles may be particularly informative, especially from a developmental perspective (Siegler, 1988). Similarly, a focus on perfectionism may shed some light on the nature of anxious impulsivity, given the fact that anxiety and fear of failure are often exhibited by perfectionists (Hewitt & Flett, 1991b). Most important, research needs to examine dimensions of perfectionism with respect to impulsive behavior in controlled situations in order to obtain a much broader understanding of the role of personal and social goals in impulsivity.

In conclusion, we have shown that impulsive tendencies tend to be associated differentially with aspects of goal-related behavior in the form of perfectionism dimensions. Moreover, we argue that perfectionism may be an important variable in impulsive characteristics such as suicide vulnerability. This suggests that perfectionism dimensions are important in initial assessments of vulnerability and that social issues with respect to expectancies certainly may be important factors in dealing with impulsive individuals.

REFERENCES

Alden, L., Wallace, S., & Bieling, P. (1992). Perfectionism and goal-setting in dysphoric and socially anxious individuals. In P. L. Hewitt (Chair), *Per-*

fectionism and psychopathology. Symposium conducted at the annual convention of the Canadian Psychological Association, Quebec City, Quebec, Canada.

Barratt, E. S. (1985). Impulsive subtracts: Arousal and information processing. In J. T. Spence & C. E. Izard (Eds.), *Motivation, emotion, and personality* (pp. 137–146). Amsterdam: Elsevier.

Barrow, J. C., & Moore, C. A. (1983). Group interventions with perfectionistic thinking. *Personnel and Guidance Journal, 61,* 612–615.

Baumeister, R. (1990). Suicide as escape from self. *Psychological Review, 97,* 90–113.

Beck, A. T., Brown, G., Berchick, R. J., Stewart, B. L., & Steer, R. A. (1990). Relationship between hopelessness and ultimate suicide: A replication with psychiatric outpatients. *American Journal of Psychiatry, 147,* 190–195.

Beck, A. T., Rush, A. J., Shaw, B. F., & Emery, G. (1979). *Cognitive therapy of depression: A treatment manual.* New York: Guilford Press.

Berman, A. L., & Jobes, D. A. (1991). *Adolescent suicide: Assessment and intervention.* Washington, DC: American Psychological Association.

Blackburn, R. (1971). MMPI dimensions of sociability and impulse control. *Journal of Consulting and Clinical Psychology, 37,* 166.

Block, J. H., & Block, J. (1980). The role of ego-control and ego-resiliency in the organization of behavior. In W. A. Collins (Ed.), *The Minnesota Symposium on Child Psychology* (Vol. 13, pp. 39–101). Hillsdale, NJ: Erlbaum.

Brandtstadter, J., & Renner, G. (1990). Tenacious goal pursuit and flexible goal adjustment: Explication and age-related analysis of assimilative and accommodative strategies of coping. *Psychology and Aging, 5,* 58–67.

Clopton, J. R. (1978). A note on the MMPI as a suicide predictor. *Journal of Consulting and Clinical Psychology, 46,* 335–336.

Costa, P. T., Jr., & McCrae, R. R. (1985). *The NEO Personality Inventory manual.* Odessa, FL: Psychological Assessment Resources.

Costa, P. T., Jr., & McCrae, R. R. (1988). From catalog to classification: Murray's needs and the five-factor model. *Journal of Personality and Social Psychology, 55,* 258–265.

Costa, P. T., Jr., & McCrae, R. R. (1992). Normal personality assessment in clinical practice: The NEO Personality Inventory. *Psychological Assessment: A Journal of Consulting and Clinical Psychology, 4,* 5–13.

Dickman, S. J. (1990). Functional and dysfunctional impulsivity: Personality and cognitive correlates. *Journal of Personality and Social Psychology, 58,* 95–102.

Dixon, D. N., Heppner, P. P., Petersen, C. H., & Ronning, R. R. (1979). Problem-solving workshop training. *Journal of Counseling Psychology, 26,* 133–139.

Domino, G., & Leenaars, A. A. (1989). Attitudes toward suicide: A comparison of Canadian and U.S. college students. *Suicide and Life-Threatening Behavior, 19,* 160–172.

D'Zurilla, T. J., & Nezu, A. M. (1990). Development and preliminary evaluation of the Social Problem-Solving Inventory (SPSI). *Psychological Assessment: A Journal of Consulting and Clinical Psychology, 2,* 156–163.

Eysenck, H. J., & Eysenck, S. B. G. (1975). *Manual for the Eysenck Personality Questionnaire (Junior and Adult).* San Diego, CA: EDITS.

Eysenck, S. B. G., & Eysenck, H. J. (1977). The place of impulsiveness in a dimensional system of personality. *British Journal of Social and Clinical Psychology, 16,* 57–68.

Farberow, N. L., & Devries, A. G. (1967). An item differentiation analysis of MMPIs of suicidal neuropsychiatric hospital patients. *Psychological Reports, 20,* 607–617.

Firth, S. T., Blouin, J., Natarajan, C., & Blouin, A. (1986). A comparison of the manifest content in dreams of suicidal, depressed and violent patients. *Canadian Journal of Psychiatry, 31,* 48–53.

Flett, G. L., Blankstein, K. R., Hewitt, P. L., & Koledin, S. (1992). Components of perfectionism and procrastination in college students. *Social Behavior and Personality, 20,* 85–94.

Flett, G. L., & Hewitt, P. L. (1990). The Child–Adolescent Perfectionism Scale: Development and association with measures of adjustment. In R. Frost (Chair), *Perfectionism: Meaning, measurement, and relation to psychopathology.* Symposium conducted at the meeting of the Association for the Advancement of Behavior Therapy, San Francisco.

Flett, G. L., Hewitt, P. L., Blankstein, K. R., & Koledin, S. (1991). Dimensions of perfectionism and irrational thinking. *Journal of Rational–Emotive and Cognitive–Behavior Therapy, 9,* 185–201.

Flett, G. L., Hewitt, P. L., Blankstein, K. R., & Mosher, S. W. (1991). Perfectionism, life events, and depression: Testing a diathesis-stress model [Abstract]. *Canadian Psychology, 32,* 311.

Flett, G. L., Hewitt, P. L., Blankstein, K. R., & O'Brien, S. (1991). Perfectionism and learned resourcefulness in depression and self-esteem. *Personality and Individual Differences, 12,* 61–68.

Flett, G. L., Hewitt, P. L., Blankstein, K. R., Solnik, M., & Van Brunschot, M. (1992). *Perfectionism and appraisals of social problem-solving ability.* Manuscript submitted for publication.

Flett, G. L., Hewitt, P. L., Boucher, D. J., Davidson, L. A., & Munro, Y. (1992). *The Child and Adolescent Perfectionism Scale: Development, validation, and association with adjustment.* Manuscript submitted for publication.

Frese, M., Stewart, J., & Hannover, B. (1987). Goal orientation and planfulness: Action styles as personality concepts. *Journal of Personality and Social Psychology, 52,* 1182–1194.

Frost, R. O., Marten, P. A., Lahart, C., & Rosenblate, R. (1990). The dimensions of perfectionism. *Cognitive Therapy and Research, 14,* 449–468.

Gerbing, D. W., Ahadi, S. A., & Patton, J. H. (1987). Toward a conceptualization of impulsivity: Components across the behavioral and self-report domains. *Multivariate Behavioral Research, 22,* 357–379.

Goh, D. S., King, D. W., & King, L. A. (1982). Psychometric evaluation of the Eysenck Personality Questionnaire. *Educational and Psychological Measurement, 42,* 297–309.

Heaven, P. C. L. (1991). Personality correlates of functional and dysfunctional impulsiveness. *Personality and Individual Differences, 12,* 1213–1217.

Heppner, P. P., Baumgardner, A. H., Larson, L. M., & Petty, R. E. (1983, August). *Problem-solving training for college students with problem-solving deficits.* Paper presented at the 90th Annual Convention of the American Psychological Association, Anaheim, CA.

Heppner, P. P., Hibel, J. H., Neal, G. W., Weinstein, C. L., & Rabinowitz, F. E. (1982). Personal problem-solving: A descriptive study of individual differences. *Journal of Counseling Psychology, 29,* 580–590.

Hewitt, P. L., & Flett, G. L. (1989). The Multidimensional Perfectionism Scale: Development and validation [Abstract]. *Canadian Psychology, 30,* 339.

Hewitt, P. L., & Flett, G. L. (1990). Dimensions of perfectionism and depression: A multidimensional analysis. *Journal of Social Behavior and Personality, 5,* 423–438.

Hewitt, P. L., & Flett, G. L. (1991a). Dimensions of perfectionism in unipolar depression. *Journal of Abnormal Psychology, 100,* 98–101.

Hewitt, P. L., & Flett, G. L. (1991b). Perfectionism in the self and social contexts: Conceptualization, assessment, and association with psychopathology. *Journal of Personality and Social Psychology, 60,* 456–470.

Hewitt, P. L., & Flett, G. L. (1992). Dimensions of perfectionism and suicide intent in psychiatric patients [Abstract]. *Canadian Psychology, 33,* 510.

Hewitt, P. L., & Flett, G. L. (1993). Dimensions of perfectionism, daily stress, and depression: A test of the specific vulnerability hypothesis. *Journal of Abnormal Psychology, 102,* 58–65.

Hewitt, P. L., Flett, G. L., & Blankstein, K. R. (1991). Perfectionism and neuroticism in psychiatric patients and college students. *Personality and Individual Differences, 12,* 273–279.

Hewitt, P. L., Flett, G. L., & Turnbull, W. (1993). *Levels of perfectionism in borderline personality disorder.* Manuscript submitted for publication.

Hewitt, P. L., Flett, G. L., & Turnbull-Donovan, W. (1992). Perfectionism and suicide potential. *British Journal of Clinical Psychology, 31,* 181–190.

Howarth, E. (1986). What does Eysenck's Psychoticism scale really measure? *British Journal of Psychology, 77,* 223–227.

Jackson, D. N. (1984). *Personality Research Form manual* (3rd ed.). Port Huron, MI: Research Psychologists Press.

Jackson, D. N. (1989). *Basic Personality Inventory manual.* Port Huron, MI: Sigma Assessment Systems.

Limbacher, M., & Domino, G. (1985–1986). Attitudes toward suicide among attempters, contemplators, and nonattempters. *Omega Journal of Death and Dying, 16,* 325–334.

Missildine, W. H. (1963). *Your inner child of the past.* New York: Pocket Books.

Orbach, I., Gross, Y., & Glaubman, H. (1981). Some common characteristics of latency-age suicidal children: A tentative model based on case study analyses. *Suicide and Life-Threatening Behavior, 11*, 180–190.

Pacht, A. R. (1984). Reflections on perfection. *American Psychologist, 39*, 386–390.

Plutchik, R., & Van Praag, H. (1989). The measurement of suicidality, aggressivity, and impulsivity. *Progress in Neuropsychopharmacology and Biological Psychiatry, 13*(Suppl.), 23–34.

Plutchik, R., Van Praag, H., & Conte, H. R. (1989). Correlates of suicide and violence risk: III. A two-stage model of countervailing forces. *Psychiatry Research, 28*, 215–225.

Ramos-Brieva, J. A., & Cordero-Villafafila, A. (1989). Aggressiveness or low control of impulsiveness? Two hypotheses about suicide in depressives. *Medical Science Research, 17*, 229–230.

Robbins, S. B. (1989). Validity of the Superiority and Goal Instability Scales as measures of deficits in the self. *Journal of Personality Assessment, 53*, 122–132.

Roger, D., & Morris, J. (1991). The internal structure of the EPQ scales. *Personality and Individual Differences, 12*, 759–764.

Rosenbaum, M. (1980). A schedule for assessing self-control behaviors: Preliminary findings. *Behavior Therapy, 11*, 109–121.

Rude, S. S. (1989). Dimensions of self-control in a sample of depressed women. *Cognitive Therapy and Research, 13*, 363–376.

Shaffer, D. (1974). Suicide in childhood and early adolescence. *Journal of Child Psychology and Psychiatry, 15*, 275–291.

Shure, M. B., & Spivack, G. (1981). The problem-solving approach to adjustment: A competency-building model of primary prevention. *Prevention in Human Services, 1*, 87–103.

Siegler, R. S. (1988). Individual differences in strategy choices: Good students, not-so-good students, and perfectionists. *Child Development, 59*, 833–851.

Stephens, B. J. (1987). Cheap thrills and humble pie: The adolescence of female suicide attempters. *Suicide and Life-Threatening Behavior, 17*, 107–118.

Svebak, S., & Kerr, J. H. (1989). The role of impulsivity in preference for sports. *Personality and Individual Differences, 10*, 51–58.

Tomlinson-Keasey, C., Warren, L. W., & Elliott, J. E. (1986). Suicide among gifted women: A prospective study. *Journal of Abnormal Psychology, 95*, 123–130.

Waters, B. G. H., Sendbuehler, J. M., Kincel, R. L., Boodoosingh, L. A., & Marchenko, I. (1982). The use of the MMPI for the differentiation of suicidal and nonsuicidal depressions. *Canadian Journal of Psychiatry, 27*, 663–667.

Withers, L. E., & Kaplan, D. W. (1987). Adolescents who attempt suicide: A retrospective clinical chart review of hospitalized patients. *Professional Psychology: Research and Practice, 18*, 391–393.

13

PROCRASTINATION AND IMPULSIVENESS: TWO SIDES OF A COIN?

JOSEPH R. FERRARI

Procrastination may be defined as strategically delaying the beginning or completion of tasks. It has been estimated that as many as 70% of college students procrastinate on academic tasks (Ellis & Knaus, 1977). Furthermore, one in four adults identify themselves as chronic procrastinators (McCown, Johnson, Carise, 1991). There are occasions in which postponing tasks facilitates performance (e.g., when prioritizing assignments or waiting to obtain all relevant information before completing a task or making a decision). Nonetheless, there are numerous situations in which chronic procrastination is inappropriate because it may inhibit optimal task performance and ultimately be self-defeating.

Chronic procrastination has been associated with a number of affective, behavioral, and cognitive characteristics. For example, procras-

I thank Tracey Clark, Pam Webber, and Kim Brown for collecting, coding, and suggesting data processes and Bill McCown and Jennifer Posa for data analysis. I also thank my colleagues in psychology for donating valuable class time for the collection of students' responses.

265

tination has been related to low self-confidence and decreased self-esteem, as well as to states of high anxiety, neurosis, diffuse identity, forgetfulness, self-presentation concerns, perfectionism, disorganization, and noncompetitiveness (Beswick, Rothblum, & Mann, 1988; Effert & Ferrari, 1989; Ferrari, 1991c, 1991d, 1992a; Lay, 1986, 1987, 1988; McCown, Johnson, & Petzel, 1989). When compared with nonprocrastinators, procrastinators have also reported significantly greater levels of public self-consciousness, social anxiety, and self-handicapping tendencies, but they are not significantly different in verbal or abstract intelligence (Ferrari, 1991a).

Research also indicates that procrastinators spend less preparation time on tasks that were likely to succeed and more time on projects likely to fail (Lay, 1990); they also tend to underestimate the time required to complete tasks (McCown, Petzel, & Rupert, 1987). Procrastinators also promise themselves they will not postpone future tasks (Lay, Edwards, Parker, & Endler, 1989). Additionally, procrastinators attempt behavioral self-handicapping (Ferrari, 1991b), engage in impression management (Ferrari, 1991c), actively avoid receiving self-relevant information (Ferrari, 1991d), and behave in a perfectionistic manner for ingratiation purposes (Ferrari, 1992a). Finally, procrastinators recommend severe reprimands to peers who might demonstrate poor task performance (Ferrari, 1992b). In general, these studies demonstrate that chronic procrastinators tend to have substantially more maladaptive behavior patterns and personality tendencies than nonprocrastinators.

A frequently cited theoretical model regarding procrastination was based on the observations and analog studies of chronic procrastinators by Burka and Yuen (1983). These clinical psychologists stated that procrastinators aim to protect a "vulnerable self-esteem." Procrastinators view their self-worth solely on the basis of task ability, and ability is determined by one's performance on completed tasks. By delaying task completion, the procrastinator's perceived (or actual) inability regarding task completion is never tested. Consequently, the person may maintain an illusion of ability, and procrastination serves to foster this sense.

There is support for this theory. For example, Ferrari (1991b) examined whether procrastinators and nonprocrastinators would choose the presence of an environmental obstacle (bogus distracting, debilitating noise) while completing a task perceived to assess cognitive thinking ability. Procrastinators were found to self-handicap their task performance to protect social and self-esteem, yet only under certain conditions. Self-handicapping by procrastinators occurred when performance was kept private and the task was highly self-diagnostic (self-esteem protection) and when the performance was public to others and the task was nondiagnostic of ability (social esteem enhancement by avoiding a "silly" task). In another study, Ferrari (1991d) asked procrastinators and nonprocrastinators to create a task that would assess their own cognitive ability. Task items were described as easy or difficult and either nondiagnostic or diagnostic of cognitive ability. Participants were also told they would or would not receive performance feedback. Results indicated that in the feedback conditions, procrastinators chose most items from the easy-nondiagnostic set of items, whereas nonprocrastinators predominantly chose easy-diagnostic items. Apparently, procrastinators prefer to remain unaware of their cognitive abilities, presumably as a way to protect or shield their vulnerable self-esteem.

In summary, research conducted by Ferrari and others indicates that chronic procrastination is neither adaptive nor effective. Correlational data suggest that procrastinatory behavior is related to low self-worth and high self-presentation, self-awareness, and self-defeating tendencies. Experimental data suggest that procrastinators, compared with nonprocrastinators, actively seek to protect their self-esteem and enhance their social esteem. Taken together, these recent empirical results suggest that chronic procrastination may be more than an inability to manage one's time; it may be an inefficient life-style.

IS PROCRASTINATION THE OPPOSITE OF IMPULSIVENESS?

Impulsiveness may be defined as the tendency to spend less preparation time than most people of equal ability before taking action (Dickman, 1990). Substantial literature exists on individual differences in impulsivity (as this book demonstrates), and impulsiveness has been closely associated with several personality traits including sociability, venturesomeness, high activity level, and proneness to boredom (S.B. G. Eysenck & Eysenck, 1963, 1977; 1978; Gerbing, Ahadi, & Patton, 1987). Recent work on the relationship between impulsivity and cognitive processing variables suggests that the consequences of impulsivity are not always negative. For example, Dickman (1985) found that when an experimental task was highly difficult, impulsive responding produced relatively few errors. Similarly, Dickman and Meyer (1988) reported that high- compared with low-impulsive subjects were actually more accurate under conditions of limited time availability for decision making.

Dickman (1990) investigated the circumstances under which a general tendency to respond quickly and inaccurately would exist (i.e., to be impulsive) and demonstrated such tendencies to be either a source of difficulty or benefit. In his study, a self-report scale was developed to discriminate between functional and dysfunctional impulsivity. *Functional impulsivity* occurs when a person responds quickly and inaccurately, yet this style of responding is optimal. *Dysfunctional impulsivity* occurs when a person responds in a rapid, inaccurate manner to situations, with detrimental results. Both traits appear to involve the tendency to deliberate less than most people of equal ability prior to taking action.

Functional impulsivity has positive consequences and is a source of pride, but dysfunctional impulsivity leads to difficulty. Dickman (1990) compared individuals who scored high and low on Functional and Dysfunctional Impulsivity subscales in several studies. In one study, scores on the Functional Impulsivity scale were correlated positively with venturesomeness, risk taking, enthusiasm, and optimism; scores on the Dysfunctional Impulsivity scale were correlated negatively with a concern about keeping materials methodically organized and a desire to make decisions on the basis of complete knowledge of results. In another study, impulsive responding led to errors in a speed–accuracy task, yet high-functional impulsive subjects obtained more correct answers than did slower low-functional impulsive subjects. High- and low-dysfunctional impulsive subjects did not demonstrate any relationship to speed–accuracy trade-offs associated with task performance.

What, then, is the relationship between impulsiveness and procrastination? On the surface, it would appear that these personality variables are opposites. Impulsivity seems to be the tendency to spend relatively *little* time before responding; procrastination, by contrast, may be considered the tendency to spend relatively too *much* time before responding. Procrastinators may seek to increase accuracy over speed. Functionally, but not dysfunctionally, impulsive individuals seek to increase speed and to maintain an accurate level of performance.

Some evidence suggests that procrastinators misjudge the amount of time required for tasks (Lay, 1988; McCown et al., 1987). These errors in judgment may involve either perceptual or memory failures. Procrastinators, as compared with nonprocrastinators, may delay tasks simply because they fail to assess all relevant bits of information. Broadbent, Cooper, Fitzgerald, and Parkes (1982) adopted the term *cognitive failure* to refer to lapses in perception, memory, or, put simply, "absent-mindedness." In fact, Effert and Ferrari (1989) found that indecision, or "decisional procrastination," was related to self-reported cognitive failures (e.g., frequent memory loss and forgetfulness) among a sample of college students. Procrastinators, then, may take longer to complete tasks (i.e., slow speed) because they fail to weigh or consider all cognitive information; therefore, their "speed" may be decreased *and* their performance "accuracy" may be poor. It may be that procrastination relates to dysfunctional impulsiveness (in which error rates, or task failures, occur often) but that it is unrelated to functional impulsiveness (wherein rapid response rates do not result in performance decrements).

Furthermore, motives for procrastination may vary. As noted earlier, decisional procrastination (indecision) may be attributable to cognitive inabilities. Therefore, dysfunctional, but not functional, impulsivity may be related to decisional procrastination. However, some individuals delay tasks in order to avoid assessments of their performance (Ferrari, 1991b, 1991d). Evaluation apprehension motivates these people to chronic procrastination. To the extent that these motives influence procrastination tendencies, one would expect dysfunctional impulsivity to be related to avoidant procrastination. Functional impulsivity should be unrelated to either cognitive (i.e., decisional) or behavioral (e.g., avoidant) procrastination tendencies. Finally, individuals who report frequent tendencies toward both decisional and avoidance procrastination may report higher rates of dysfunctional impulsiveness than do individuals who report infrequent procrastinatory tendencies. In the remainder of this chapter, I focus on the results of research undertaken to investigate the relationship between impulsivity and decisional and avoidant procrastinations.

THE PROCRASTINATION–IMPULSIVENESS RELATIONSHIP: SOME EMPIRICAL DATA

A set of procrastination and impulsiveness scales were administered to a sample of 136 college students (114 women and 22 men) enrolled in introductory psychology at a small, private, rural, residential institution. Students were either in their first (76.5%) or second (23.5%) year; they had a mean age of 18.2 years ($SD = 0.5$). None of these respondents had any prior experience as research participants. Respondents were first asked to sign, date, and return a consent form that indicated that a set of standard personality scales would be completed anonymously. Extra course credit was awarded to respondents who completed all self-report inventories (including other psychometric scales not discussed here) within a 65-minute class session. Respondents completed each inventory in the same order. I now discuss the psychometric properties of the inventories, listed in the order of completion.

McCown and Johnson's (1989) Adult Inventory of Procrastination

This 15-item, 5-point scale (1 = *very untrue*, 5 = *very true*) assesses avoidant procrastination tendencies across a variety of everyday tasks. Scores on items such as "I pay my bills on time" and "I don't get things done on time" range from 15 to 75. Normative data with undergraduates ($N = 431$; $M = 56.2$, $SD = 14.1$; McCown & Johnson, 1989) indicated a Cronbach alpha of .79 and a retest reliability of .71. With the present sample, the mean score was 37.9 ($SD = 10.0$). McCown and Johnson (1989) reported several validity studies on this scale, including that high scorers delayed returning materials and filing income taxes. In addition, Ferrari (1992c) found that high scores were related to delays in returning completed scales and to avoidance of cognitive information and challenges.

Mann's (1982) Decisional Procrastination Scale

This scale is embedded among five other scales measuring coping patterns in dealing with decisional conflicts and yields a 31-item total inventory. Scores on the five-item Procrastination scale range from 5 to 25 in which respondents indicate on 5-point Likert scales (1 = *not true*, 5 = *true*) their tendency to put off decisions by doing other tasks. Procrastination scale items include "I delay making decisions until it is too late" and "I put off making decisions." Normative data that were based on a sample of undergraduates (N = 212; M = 12.9, SD = 2.2; L. Mann, personal communication, January 5, 1988) indicated that the scale had a Cronbach alpha of .80 and a retest reliability of .69. With the present sample, the mean score was 13.7 (SD = 3.9). Scale scores were predictive of academic and everyday procrastination rates and related to locus of control, social anxiety, low self-esteem, and forgetfulness (Beswick et al., 1988; Effert & Ferrari, 1989; Ferrari, 1989).

Dickman's (1990) Functional–Dysfunctional Impulsivity Inventory

The Functional Impulsiveness subscale is an 11-item, 5-point Likert scale (1 = *low*, 5 = *high*) designed to assess high-speed performance with little decrement in response quality. Items include "I would enjoy working at a job that required me to make a lot of split-second decisions" and "People admire me because I can think quickly"; scores range from 11 to 55. The Dysfunctional Impulsiveness subscale contains 12 items rated on a 5-point Likert scale (1 = *low*, 5 = *high*) developed to assess high-speed–high-error performances. Sample items include "I often say and do things without considering the consequences"; scores range from 12 to 60. Normative data on the subscales were not provided, although the Cronbach alphas with undergraduate samples suggest that both subscales have acceptable internal consistency (Sample 1 n = 477, Functional subscale = .83, Dysfunctional subscale = .86, interscale = .22; Sample 2 n = 188, Functional subscale = .74, Dysfunctional subscale = .85, interscale = .23; Dickman, 1990). With the present sample, the mean scores for the Functional Impulsiveness and the Dysfunctional Impulsiveness subscales were 32.1 (SD = 4.0) and 35.6 (SD = 6.7), respectively.

Dickman (1990) reported that functional impulsiveness was closely associated with enthusiasm, adventuresomeness, extraversion, and activity, whereas dysfunctional impulsiveness was closely associated with disorderliness, conscientiousness, and a lack of concern for hard facts when making decisions. High-functional impulsive subjects also had a shorter time between scanning geometric figures and higher accuracy scores than did low-functional or high- and low-dysfunctional impulsive subjects. Heaven (1991), using Australian high-school-age students, found that Dysfunctional Impulsiveness scores were negatively related to self-esteem and positively related to psychoticism and other traditional measures of impulsiveness. Functional Impulsiveness scores were related positively to self-esteem and traditional measures of impulsiveness for women but not men.

The Broadbent et al. (1982) Cognitive Failures Questionnaire (CFQ)

The CFQ is a 25-item, 5-point Likert scale (1 = *never*, 5 = *very often*) that assesses the frequency with which a person makes mistakes because of lapses of memory, forgetfulness, or absent-mindedness. Scores range from 25 to 125, and sample items include "Do you fail to notice signposts on the road?" and "Do you find you forget appointments?" Normative data that were based on samples of undergraduates (Sample 1 $n = 124$, $M = 79.2$, $SD = 13.2$; Sample 2 $n = 221$, $M = 62.5$, $SD = 14.5$; Broadbent et al., 1982) indicated coefficient alphas between .79 and .89 and retest reliabilities of .80–.82. With the present sample, the mean score was 63.3 ($SD = 15.2$). Broadbent et al. reported that cognitive failure scores were related to external locus of control, state and trait anxiety, defensiveness, forgetfulness, difficulty in making up one's mind, depression, and obsessive thoughts.

No differences between men and women were found between scores on either procrastination measure in the present study. This finding is consistent with previous research on decisional procrastination (Effert & Ferrari, 1989) and other types of procrastination (see Ferrari, 1991c, 1991d, 1992b). Consequently, I do not present any additional comparisons between men and women; I also collapsed all further analyses across gender.

Table 1 shows the intercorrelations between scale scores. Decisional and avoidant procrastinators were related to functional impulsivity, although the coefficients were small. Moreover, as expected, both types of procrastination were significantly related to dysfunctional impulsiveness. Again, the correlation coefficients were low in magnitude, suggesting that much of the variance between the concepts remains to be explained. Nonetheless, with decisional procrastination and avoidant procrastination, the correlation coefficients for dysfunctional impulsiveness were larger than the coefficients for functional impulsiveness. In addition, decisional and avoidant procrastination were significantly related, yet func-

TABLE 1
Simple Correlation Coefficients Between
Self-Reported Personality Measures

Self-reported measures	Adult inventory of procrastination	Decisional procrastination	Functional impulsivity	Dysfunctional impulsivity
Decisional procrastination	.446**			
Functional impulsivity	.233*	.253*		
Dysfunctional impulsivity	.410**	.466**	.189	
Cognitive failures	.176	.441**	.233**	.409**

Note. N = 136 participants.
* $p < .05$. ** $p < .005$.

tional and dysfunctional impulsiveness were relatively independent concepts. Decisional procrastination and cognitive failure were also positively related. Cognitive failure was unrelated to avoidant procrastination but was related to functional and dysfunctional impulsivity.

Comparing Procrastinators and Nonprocrastinators on Scale Scores

The mean score across all participants on the Adult Inventory of Procrastination was 38.1 ($SD = 10.0$) and 13.7 ($SD = 3.9$) on the Decisional Procrastination Scale. Respondents were classified as frequent procrastinators if their score was 1 SD above the mean on both procrastination measures. This procedure produced 17 respondents categorized as "procrastinators." Respondents were classified as infrequent procrastinators if their score was 1 SD below the mean on both procrastination measures, producing 21 individuals categorized as "nonprocrastinators." With these 38 individuals, avoidant procrastination and decisional procrastination were highly related ($r = .91, p < .001$). Table 2 shows the mean score on each personality measure for procrastinators and nonprocrastinators.

Of course, selecting extreme scores on both procrastination measures produced samples that were significantly different on avoidant procrastination, $t(42) = 4.56, p < .001$, and decisional procrastination, $t(42) = 5.23, p < .001$. Chronic procrastinators, compared with nonprocrastinators, reported significantly more dysfunctional impulsiveness, $t(42) = 2.1, p < .05$, and cognitive failure, $t(42) = 2.2, p < .05$. There

TABLE 2
Mean Scores for Procrastinators and Nonprocrastinators on Each of the Major Personality Variables

Self-reported variables	Procrastinators (n = 17)		Nonprocrastinators (n = 21)	
	Score	SD	Score	SD
Adult inventory of procrastination	51.86	3.93	25.57	1.90
Decisional procrastination	19.14	1.21	6.29	1.38
Functional impulsivity	28.57	3.78	30.14	3.63
Dysfunctional impulsivity	37.43	6.32	31.00	4.16
Cognitive failure	80.00	8.54	56.71	9.83

was no significant difference between procrastinators and nonprocrastinators on self-reported functional impulsiveness.

PROCRASTINATION AND DYSFUNCTIONAL IMPULSIVENESS

On the basis of this study, it appears that frequent procrastination and at least one form of impulsiveness (dysfunctional) are related—not opposite—concepts. Chronic procrastinators, compared with nonprocrastinators, also reported a higher tendency for dysfunctional impulsiveness but not for functional impulsiveness. These results suggest an interesting (but hypothetical) sequence of events when procrastinators approach a task. Consider the following scenario: Some procrastinators (i.e., individuals who are indecisive and who avoid completing tasks because they do not want challenging and perhaps diagnostic information), by definition, wait to complete a target task. As the task deadline approaches, however, these individuals "speed up" and work faster to complete the task (i.e., they act more impulsively). They work rapidly just before the task deadline, which, in turn, increases their error rate, and they thus experience dysfunctional impulsiveness.

Furthermore, decisional procrastination was related to cognitive failure, and chronic procrastinators claimed high rates of poor processing ability (e.g., failure to attend to all necessary information). Effert and Ferrari (1989) also found that decisional procrastination was related to cognitive failure, speed, and impatience and inversely related to competitiveness. Together with the study presented in this chapter, these results suggest that procrastinators who act hastily as a task deadline approaches increase the probability of errors, resulting in poor performance (i.e., of experiencing dysfunctional impulsiveness). Indeed, these "procrastinators turned (dysfunctional) impulsives" might have failed to process fully all of the necessary information required for successful performance. Consequently, poor task performance occurs. Of course, this scenario is speculative and requires further experimental evaluation. Nevertheless, results from this study suggest that procrastination and impulsiveness are *not* opposite personality concepts. Indeed, in some respects they are related constructs and seem to be examples of self-defeating, self-destructive tendencies.

Finally, the study presented in this chapter indicates that decisional and avoidant procrastination types were related. This finding supports research involving these and other types of procrastination (e.g., Beswick et al., 1988; Ferrari, 1989; Lay, 1988). Additionally, the fact that functional and dysfunctional impulsiveness were not significantly related is consistent with Dickman (1990) and Heaven (1991).

CONCLUSION

Results of the study presented in this chapter suggest that procrastination and impulsiveness are not simply two sides of a coin (i.e., just opposites), nor are they orthogonal. Instead, frequent procrastination (e.g., indecisiveness and pronounced tendencies to avoid threatening situations) is related to dysfunctional impulsiveness (high speed–high error) but not to functional impulsiveness (high speed–low error). To the extent that these procrastinators possess deficits in cognitive processing abilities, tendencies to speed up and work faster to complete a task by deadline will likely result in poor performance because of the subsequent lack of sufficient time and ability to perform efficiently. Chronic procrastinators, in turn, may attribute their task failures to the lack of time and not to their lack of ability (because they tried, although at the last minute). Because procrastination is such a pervasive and disconcerting behavior, further experimental research with these variables is warranted to clarify causal relationships and to explore just how procrastinators respond under various circumstances.

REFERENCES

Beswick, G., Rothblum, E. D., & Mann, L. (1988). Psychological antecedents to student procrastination. *Australian Psychologist, 23,* 207–217.

Broadbent, D. E., Cooper, P. F., Fitzgerald, P., & Parkes, K. R. (1982). The Cognitive Failures Questionnaire (CFQ) and its correlates. *British Journal of Clinical Psychology, 21,* 1–16.

Burka, J. B., & Yuen, L. M. (1983). *Procrastination: Why you do it and what to do about it.* Reading, PA: Addison-Wesley.

Dickman, S. J. (1985). Impulsivity and perceptional individual differences in the processing of the local and global dimensions of stimuli. *Journal of Personality and Social Psychology, 48,* 133–149.

Dickman, S. J. (1990). Functional and dysfunctional impulsivity: Personality and cognitive correlates. *Journal of Personality and Social Psychology, 58,* 95–102.

Dickman, S. J., & Meyer, D. E. (1988). Impulsivity and speed–accuracy trade-offs in information processing. *Journal of Personality and Social Psychology, 54,* 274–290.

Effert, B., & Ferrari, J. R. (1989). Decisional procrastination: Examining personality correlates. *Journal of Social Behavior and Personality, 4,* 151–156.

Ellis, A., & Knaus, W. J. (1977). *Overcoming procrastination.* New York: Signet Books.

Eysenck, S. B. G., & Eysenck, H. J. (1963). On the dual nature of extraversion. *British Journal of Social and Clinical Psychology, 2,* 46–55.

Eysenck, S. B. G., & Eysenck, H. J. (1977). The place of impulsiveness in a dimensional system of personality description. *British Journal of Social and Clinical Psychology, 16,* 57–68.

Eysenck, S. B. G., & Eysenck, H. J. (1978). Impulsiveness and venturesomeness: Their position in a dimensional system of personality description. *Psychological Reports, 43,* 1247–1253.

Ferrari, J. R. (1989). *Self-handicapping by procrastinators: Effects of performance privacy and task importance.* Unpublished doctoral dissertation, Graduate School of Arts and Sciences, Adelphi University, Garden City, NY.

Ferrari, J. R. (1991a). Compulsive procrastination: Some self-reported characteristics. *Psychological Reports, 68,* 455–458.

Ferrari, J. R. (1991b). Self-handicapping by procrastinators: Protecting self-esteem, social-esteem, or both? *Journal of Research in Personality, 25,* 245–261.

Ferrari, J. R. (1991c). A preference for a favorable public impression by procrastinators: Selecting among cognitive and social tasks. *Personality and Individual Differences, 12,* 1233–1237.

Ferrari, J. R. (1991d). Procrastination and project creation: Choosing easy, non-diagnostic items to avoid self-relevant information. *Journal of Social Behavior and Personality, 6,* 619–628.

Ferrari, J. R. (1992a). Procrastination and perfect behavior: An exploratory factor analysis of self-presentation, self-awareness, and self-handicapping components. *Journal of Research in Personality, 26,* 75–84.

Ferrari, J. R. (1992b). Procrastination in the workplace: Attributions for failure among individuals with similar behavioral tendencies. *Personality and Individual Differences, 13,* 315–319.

Ferrari, J. R. (1992c). Psychometric validation of two procrastination inventories for adults: Arousal and avoidance measures. *Journal of Psychopathology and Behavioral Assessment, 14,* 97–110.

Gerbing, D., Ahadi, S., & Patton, J. (1987). Toward a conceptualization of impulsivity: Components across the behavioral and self-report domains. *Multivariate Behavioral Research, 22,* 1–22.

Heaven, P. C. L. (1991). Personality correlates of functional and dysfunctional impulsiveness. *Personality and Individual Differences, 12,* 1213–1217.

Lay, C. H. (1986). At last, my research in procrastination. *Journal of Research in Personality, 20,* 479–495.

Lay, C. H. (1987). A modile profile analysis of procrastinators: A search for types. *Personality and Individual Differences, 8,* 705–714.

Lay, C. H. (1988). The relationship of procrastination and optimism to judgements of time to complete an essay and anticipation of setbacks. *Journal of Social Behavior and Personality, 3,* 201–214.

Lay, C. H. (1990). Working to schedule on personal projects: An assessment of person–project characteristics and trait procrastination. *Journal of Social Behavior and Personality, 5,* 91–104.

Lay, C. H., Edwards, J. M., Parker, J. D. A., & Endler, N. S. (1989). An assessment of appraisal, anxiety, coping, and procrastination during an examination period. *European Journal of Personality, 3,* 195–208.

Mann, L. (1982). *Decision Making Questionnaire.* Unpublished scale, Flinders University of South Australia, Bedford Park.

McCown, W., & Johnson, J. (1989). *Validation of an adult inventory of procrastination.* Unpublished manuscript, Department of Psychology, Villanova University, Villanova, PA.

McCown, W., Johnson, J., & Carise, D. (1991). Trait procrastination in self-described adult children of excessive drinkers: An exploratory study. *Journal of Social Behavior and Personality, 6,* 149–153.

McCown, W., Johnson, J., & Petzel, T. (1989). Procrastination, a principal component analysis. *Personality and Individual Differences, 10,* 197–202.

McCown, W., Petzel, T., & Rupert, P. (1987). An experimental study of some hypothesized behaviors and personality variables of college student procrastination. *Personality and Individual Differences, 8,* 781–786.

III

TREATMENT OF IMPULSIVITY

14

IMPULSIVITY IN CHILDREN AND ADOLESCENTS: MEASUREMENT, CAUSES, AND TREATMENT

AILEEN D. FINK and WILLIAM G. McCOWN

Even a cursory review of the behavioral science literature indicates that there is little agreement on an operational definition of impulsivity, especially regarding this behavior in children and adolescents. According to Milich and Kramer (1984), impulsivity has been defined as "poor judgment," "weak restraints," and "inability to delay gratification." Those authors noted that researchers working in the area of impulsivity have typically adopted a definition suited to the area under study or employed a measure or procedure used in prior research, largely for the sake of convenience. Other definitions seem to be more related to familiarity with the dependent measures rather than to theoretical reasons. Given the variety of definitions and measures of impulsivity in children and adolescents, it is surprising that there has been any consistency of findings across laboratories.

In psychology, as in most sciences, operational definitions of dependent variables are an essential step before research can proceed (Cook

279

& Campbell, 1979). There are a number of ways in which researchers and clinicians have attempted to assess impulsivity operationally. These include direct observations, self-reports, and paper-and-pencil measures. Milich and Kramer (1984) pointed out a number of similarities among various measures of impulsivity that are worth noting: Many tests of child or adolescent impulsivity incorporate a measure of rate of response, and number of errors is also often considered important. However, a pessimistic review of the literature regarding tests of impulsiveness has been furnished by Oas (1985a), and it remains controversial to what extent different methods of assessment of impulsiveness actually measure the same construct.

CLASSES OF OPERATIONAL DEFINITIONS

There are at least nine different classes of operational definitions that have been involved in the construct of impulsiveness and its measurement. The recognition of differences between these types of dependent measures is important for both clinicians and researchers. Subtle changes in operational definitions of impulsiveness may substantially alter both the denotative and the connotative meaning of the research findings. We now review these classes.

Gross Motoric Measures

It is probably most typical to find a lay definition of impulsivity that focuses on the failure to inhibit motor responses (e.g., "Johnny has problems getting out of his seat all the time"). This is especially true for observers of preschool and early primary-age children. This type of definition has a particular appeal to researchers interested in children because it can be easily operationally defined and measured, generally with a high degree of interrater reliability (e.g., number of times the child's buttocks is not in contact with the seat). Another advantage of gross motoric measures is that they can be collected fairly inexpensively and are amenable to computer-assisted data collection, thus circumventing human error and systematic bias.

Disadvantages of gross motoric measures lie in their potential lack of validity and generalizability. Perhaps the major problem involves construct validity. Motoric movement may not correlate with other indexes of the construct. This is because motor restlessness may have little to do with the absence of thought or of unplanned behavior (Johnson & McCown, in press). Furthermore, reliability may be low because of a number of factors, including time of day, time since last meal, the child's level of alertness, or even his or her present mood state.

Inhibition of Common or Situational Motor Responses

A second class of operational definitions of impulsiveness emphasizes the capacity for inhibition of common motor responses elicited explicitly in the test-taking situation. On the basis of this motor response definition, tests have been developed to assess an individual's ability to inhibit specific motor responses, such as the Draw-A-Line-Slowly test (Maccoby, Dowley, Hagan, & Degerman, 1965) and the Walk-A-Line-Slowly test. Although used less frequently with children, the Stroop Color and Word test or one of its variants also requires inhibition of a specific response. Another frequently used type of task demands that the child inhibit a response to a simple go signal when a simple no-go signal is also presented (Schachar & Logan, 1990).

An obvious advantage of this class of tests is their potential for ecological validity. However, the use of this class of dependent measures has many disadvantages similar to those of gross motoric measures. Although reliability may be more adequate, construct validity remains problematic. According to Milich and Kramer (1984), there appears to be some research support for the Draw-A-Line-Slowly test and less support for the Walk-A-Line-Slowly test, although most studies failed to control for level of intellectual ability, which may influence these performance measures. Hence, the results of these studies are often difficult to interpret, and of little value to the practitioner.

Reinforcement Schedules

Researchers have also combined operant schedules of reinforcement to measure children's capacities to inhibit responses. For example, the ability of a child to refrain from approaching some desirable object (Mischel, 1958; Sears, Rau, & Alpert, 1965) has been used as a measure of impulsiveness. In operant terms, this is an estimate of the child's ability to escape stimulus control of a powerful reinforcer. A disadvantage is that stimuli may vary idiosyncratically in their reinforcement value and response to one stimuli or class of stimuli may therefore not generalize to other stimuli, settings, or occasions.

Gordon (1979) proposed the use of the differential rate of low responding procedure (DRL) as an assessment tool in the identification of impulsivity in children with attention-deficit hyperactivity disorder (ADHD). Essentially, the procedure involves setting up a schedule whereby the subject is reinforced only if he or she inhibits responding between designated intervals. For example, the child is instructed to push a button on a console and then to wait for awhile before pushing the button again. If the child waits for the specified period (e.g., 6 seconds), then he or she is provided with a reward.

Gordon (1979) found that children who were diagnosed as hyperactive were significantly less efficient at performing the task than were nonhyperactive children, regardless of age or intelligence level. The results also suggested that the level of performance on the DRL task was related to the kind of mediating behaviors a child engaged in. For example, children who used physical strategies (e.g., jumping jacks) performed more poorly than those who used cognitive strategies (e.g., counting out loud). Although these results are important, the importance of this method should also be emphasized. The use of DRL and other schedules of reinforcement is popular in neurosciences, and the use of this class of variables may be of assistance for people who wish to provide transitional models from animals to children.

Artificial Motoric Tests

Other measures involve having the child or adolescent complete a paper-and-pencil or computerized test that is based on inhibition of motor responsiveness for a task with a less natural counterpart in everyday behavior. Tests of these types include the Matching Familiar Figures Test (MFFT; Kagan, 1966; Kagan, Rossman, Day, Albert, & Phillips, 1964) and the Porteus Maze Test (Porteus, 1955, 1968). The MFFT is probably the most widely used measure of impulsivity, and Milich and Kramer (1984) reported that hundreds of articles have been published in which the MFFT was used as either the primary or sole measure of impulsivity. Because the MFFT and Porteus Maze Test are used in many research studies and are commonly used by clinicians, we briefly review these instruments.

The MFFT was based on Kagan's (1966) notion of cognitive tempo and reflection–impulsivity. According to Messer (1976), "reflection–impulsivity" describes the tendency to reflect adequately in a specific condition, namely, where there are a variety of possible alternatives and the appropriate response is not clear. Individuals described as "reflective" are those who are relatively slow and accurate in selecting a response alternative, whereas "impulsive" individuals are relatively fast and inaccurate. The MFFT is a 12-item matching-to-sample task in which a subject is asked to select one of six alternative pictures that matches a standard picture. The subject continues to select until the identical match is found. The latency to first response as well as the number of errors are scored; individuals whose score is above the median for number of errors, below the median on latency, or both are generally considered to be impulsive.

Although the MFFT was widely adopted as a measure of impulsivity, the reliability and validity of the instrument were challenged by a number of researchers, including Block, Block, and Harrington (1974) and Block, Gjerde, and Block (1986). The classification of subjects into one of four categories (fast accurate, slow accurate, fast inaccurate, and slow inaccurate) has been criticized as being unreliable by Becker, Bender, and Morrison (1978), who found that only 47% of the subjects were classified into the same category at a 1-year follow-up test. Furthermore, the fact that two of the classifications (fast accurate and slow inaccurate) are generally ignored in the interpretation of the test results may indicate a weakness in the validity of the instrument. Regardless, the MFFT remains extremely popular and has been used as a dependent measure in more than 800 published or unpublished studies.

The Porteus Maze Test (Porteus, 1968), originally developed as a measure of intelligence (Porteus, 1955), was adopted as a measure of impulsivity with the addition of the Q score (qualitative score). Rather than focus on the number of mazes solved, the Q score examines the individual's approach to completing the maze. The Q score is computed on the basis of the occurrence of behaviors, including lifting the pencil during drawing (the pretest instructions prohibit this) and crossing a boundary line while completing a maze. Research examining the Porteus Maze Test as a measure of impulsiveness has produced mixed results. Although some studies suggested that the test could accurately discriminate delinquents from nondelinquents (Purcell, 1956), these studies did not always control for level of intelligence. According to Milich and Kramer (1984), the evidence for this measure as a valid determinant of impulsiveness is inferential, and care is needed if this score is used for clinical purposes.

Direct Observation

Direct observation has become common among clinicians and researchers, perhaps as part of the "antitrait" behavioral zeitgeist or perhaps as an indication of the extent to which physicians and others who are customarily without extensive psychometric training are now contributing to the behavioral science literature. A number of research studies conducted in the area of impulsivity have used direct observation of behaviors. The notable difference among studies using direct observation in this manner concerns the relationship of the observer to the subject. Some studies have used therapists as observers (e.g., Kendall & Wilcox, 1980), whereas others have used psychiatric nurses (Oas, 1983), teachers, or parents (e.g., Klinteberg, Magnusson, & Schalling, 1989). In addition, studies vary with regard to the degree of training provided to raters prior to observation. Some studies have included a specific training protocol prior to observation, and others have used raters who had not received any additional training as part of the study.

The setting under which observation is conducted has also varied across studies. For example, Hinshaw, Henker, and Whalen (1984) used naturalistic observation of a group of hyperactive boys on a playground to assess the effect of reinforcement plus self-evaluation versus reinforcement alone and Ritalin versus placebo on social behavior. Kendall and Wilcox (1980) included therapist observation (therapy setting) of improvement in impulse control as a dependent measure in a study that examined the effect of cognitive–behavioral treatment for impulsivity. Breen (1989) used a laboratory playroom as the setting to observe and compare the behavior of normal children and those with attention deficits. An advantage of this type of research is that it has ecological validity, although external validity and reliability may be more problematic.

Psychometric Instruments Rated by Others Who Are Familiar With the Child or Adolescent

A common means used to assess impulsivity has been paper-and-pencil ratings by individuals other than the child or adolescent. A number of instruments have evolved that allow the clinician to rate impulsiveness or proximal behaviors of impulsiveness (e.g., the Gordon Diagnostic System; Gordon, 1983). There are numerous forms and checklists designed to measure impulsive behavior in children. These include the Self-Control Rating Scale (Kendall & Wilcox, 1980), the Conners Teacher's Questionnaire (Conners, 1969), the Child Behavior Checklist (Achenbach & Edelbrock, 1983), and the School Situations Questionnaire (Barkly, 1981). The Connors rating form seems especially popular in schools because it allows teachers to compare behavior with that of other children. These scales typically request that the rater indicate the frequency (ranging from number of occurrences within an hour to a week) with which the child exhibits restless, excitable, inattentive, and impulsive behaviors.

One of the most widely researched scales is the Behavior Problem Checklist (Quay & Peterson, 1983). The revised version of this scale consists of 89 items rated on a 3-point severity index. Six primary dimensions are identified, including motor excessiveness, conduct problems, and socialized aggressiveness. Reliability and validity appear to be adequate for both research and clinical use.

The Child Behavior Checklist (Achenbach & Edelbrock, 1983) is a 118-item measure also rated on a 3-point scale. Although factor analyses of the findings result in age-dependent solutions, in general, a two-factor solution emerges that represents the consistent dimensions of child behavior problems of externalizing (i.e., impulsive acting out) and internalizing (withdrawal and anxiety). The first factor most clearly corresponds to impulsivity. Again, reliability and validity appear adequate for both clinical and research use.

The Personality Inventory for Children (PIC; Wirt, Lachar, Klinedinst, & Seat, 1984) is an example of a measure that incorporates behavioral assessment by parents with a concern with psychometrics that is more common in true self-report instruments. The PIC contains 600 true–false items answered by a knowledgeable adult, usually the mother. It includes 3 validity scales and 12 clinical scales. One scale includes the measure of hyperactivity. This instrument is also used routinely in research and clinical treatment.

Self-Report

Complete or true self-report is not as popular in children as in adults because it is assumed that children are less accurate at assessing their own behaviors. Perhaps the most common measure of self-report of impulsiveness in children is S. B. G. Eysenck and Eysenck's (1980) Impulsivity Questionnaire. (This test is described in chapter 8 by S. B. G. Eysenck, this volume.) The Minnesota Multiphasic Personality Inventory is also used with adolescents to determine impulsiveness. Other popular instruments include the Impulse Control Categorization Instrument (Matsushima, 1964), a 24-sentence instrument that presents situations and then asks children to rate, on a continuum, their choice between impulsive–aggressive behavior and behavior requiring impulse control to address the situation. In general, children's self-reports are used for research purposes only, whereas adolescents' self-reports are more acceptable for clinical purposes.

Projective Measures

The use of projective measures as indexes of impulsiveness in children has declined. It is relatively rare to see projective tests being used as a true criterion measure. The Rorschach Inkblot Test, in particular, has become less of a measure of change than an experimental tool. This is true despite some evidence that the Rorschach is useful for determining impulsiveness in children and adolescents (Exner & Weiner, 1982). McCown and Barkhausen (1993) have designed a Rasch-scaled measure of impulsiveness that is based on the Thematic Apperception Test. However, this scoring system remains experimental and has not been extensively validated with children or adolescents.

Although the use of projective drawings is popular, their use with children has not, in our opinion, received sufficient support to justify their popularity for assessing impulsiveness. For example, it is common for clinicians to interpret a hastily drawn Bender-Gestalt Test as an index of impulsiveness. However, findings on the use of the Human Figures Drawing Test and other projective drawings as a method of assessing impulsiveness and other factors remain inconclusive despite extensive literature (Adler, 1970). When reliable indexes exist (e.g., Robins, Blatt,

& Ford, 1991), they may measure other constructs, such as maturity or adequacy of object relations and not impulsiveness per se. The clinical use of projective methods should probably be relegated to screening instruments until better and replicated evidence emerges regarding their psychometric integrity.

Physiological Indexes

Researchers occasionally use psychophysiological measures as a method of assessing impulsiveness. These include the latency of response to external audio or visual stimuli, or the evoked potential. Kagan, Lapidus, and Moore (1979) have found a correlation between electromyographic (EMG) results and impulsiveness. (A further discussion of the history of physiological measures of impulsiveness is presented in chapter 3 by Barratt and chapter 2 by Daruna and Barnes, this volume.)

Need for Heteromethods

Performance in the tasks used to measure impulsivity in children tends to be correlated with one's cognitive level (Olson, Bates, & Bayles, 1990; Silverman & Ragusa, 1990). However, correlations between the various methods of measuring impulsivity are generally low (Gaddis & Martin, 1989; Olson, 1989), suggesting that impulsivity is multidimensional and only partially captured by any single method of assessment. This was illustrated in a study by Olson (1989). Multiple measures of impulsivity were assessed in 79 preschoolers (aged 48–68 months). Measures included the Peabody Picture Vocabulary Test, tests of motor inhibition and delay of gratification, and teacher ratings. Different measures of impulsivity were largely independent, clustering according to shared method variance. One exception to this was that teacher ratings of social cooperativeness contributed significantly to the factor indexing good delay abilities in a performance task. Follow-up analyses 1 year later revealed a methodologically homogeneous delay-of-gratification dimension and a second cross-measure dimension defined by scores on tests of cognitive and motor impulsiveness. Findings suggest that there may be several different types of impulsiveness in early childhood, with different implications for social development.

Similarly, Gaddis and Martin (1989) examined the relations among various measures of impulsivity appropriate for use with preschool children: the Kansas Reflection-Impulsivity Scale for Preschoolers, the Mazes subtest of the Wechsler Preschool and Primary Scale of Intelligence, the Goodman Lock Box, and the Kaufman Assessment Battery for Children. Subjects' teachers also completed a preschool rating scale. The two teacher rating scales were highly correlated, but few significant relationships were obtained among the performance measures or between teacher ratings and performance measures. We argue that with children, it is important for multiple dependent measures to be used for both research and clinical purposes (i.e., a multimethod approach). For clinical purposes, it is inappropriate for assessment to be made on the basis of one person's (e.g., a parent's) accounts regardless of his or her reliability. Finally, because impulsive children may perform well in novel situations (e.g., the first interview in a clinician's office), clinical data—including structured interviews—must be weighed carefully with results from other domains.

IMPULSIVITY AND CHILDHOOD DYSFUNCTION

The current psychiatric diagnostic system, the revised third edition of the *Diagnostic and Statistical Manual of Mental Disorders* (DSM–III–R; American Psychiatric Association, 1987), lists poor impulse control as a symptom of many disorders usually first evident in infancy, childhood, or adolescence (i.e., mental retardation, ADHD) and implicates its existence in others (i.e., conduct disorder, oppositional defiant disorder). However, there is no separate diagnostic category for impulse control disorder per se as there is in the adult diagnostic section. Although the discontinuity between childhood and adult disorders promoted by the DSM–III–R may be somewhat puzzling, it may also reflect empirical findings. According to Zametkin and Borcherding (1989), the overlap of symptoms of a variety of impulse-related disorders of childhood remain problematic, and syndromes frequently coexist.

Given the heterogeneity of diagnosable syndromes associated with impulsiveness, it is not surprising that impulsiveness has been associated with a variety of childhood dysfunction. The ensuing discussion is simply a sample of some of the areas in which impulsiveness in child and adolescent behavior has been implicated.

Antisocial Behavior

Impulsivity has often been linked to delinquent behavior in children and adolescents. For example, Vinogradov, Dishotsky, Doty, and Tinklenberg (1988) conducted a prospective study of 63 male adolescents (aged 13–20 years; mean age = 16.6 years) who admitted to committing rape. From the files of the adolescents, a broad composite was developed to describe the "typical" adolescent rape episode. Of particular relevance was the finding that 79% of the adolescents reported that the rape was not premeditated (suggesting impulsiveness). Twenty-seven percent of the adolescents reported that they were engaged in committing another crime (e.g., burglary) prior to the rape. It should be noted that in more than half of the rape episodes, the adolescent reported that he was under the influence of one or more psychoactive drugs, making it difficult to separate the adolescent's predisposition to impulsive behavior from the tendency of drugs to alter behavioral inhibition.

In general, the literature supports a relation between impulsivity and adolescent criminality (but see Oas, 1985b). However, the exact relation between impulsivity and antisocial behavior is currently unclear (Rigby, 1986). There is certainly no one-to-one correlation between impulsive behaviors and antisocial behaviors. Furthermore, the relation may vary across the life span. For example, Loeber (1990) discussed the role of impulsiveness and various risk factors for antisocial behavior throughout the life span. Clearly, both social and biological factors are implicated in a complex interaction (Stoff, Friedman, Pollock, & Vitiello, 1989). Current research suggests that impulsiveness may have both positive and negative features (see Dickman, chapter 9, this volume), but the type of impulsivity that correlates with antisocial behavior and why that is the case have not been addressed. Regardless, impulsiveness remains associated with antisocial behavior (Rigby, Mak, & Slee, 1989), and some of the treatments that are pertinent to the latter are also indicated in the former (Lochman, White, & Wayland, 1991).

ADHD (discussed in the next section) has been found to increase the risk for delinquency when combined with conduct problems, especially aggressiveness (Frick, Strauss, Lahey, & Christ, 1993). ADHD may also increase the risk of substance abuse in late adolescence (Gittelman, Mannuzza, Shenker, & Bonegura, 1985).

Attention-Deficit Hyperactivity Disorder

Perhaps more confusion and controversy surrounds this diagnostic category than any other childhood disorder. In the past, several labels have been applied to an apparently common syndrome involving high activity level, difficulty in concentrating, restlessness, verbal and behavioral impulsiveness, and occasionally antisocial behavior as well. These labels have included hyperactivity, hyperkinetic syndrome, minimal brain disorder, and minimal brain dysfunction. However, the condition of hyperactivity is nonspecific and may be related to a number of causes, including endocrine disorders (i.e., hyperthyroidism), toxic conditions (i.e., lead poisoning), central nervous system impairment (i.e, acute encephalitis), and even anxiety.

Consequently, some authors have been critical of this diagnostic category, and the situation has not been made less complicated by the fact that the American Psychiatric Association has changed its definitional and subtypology criteria regarding the syndrome. However, most current researchers believe that ADHD exists as a separate and diagnosable syndrome apart from the aforementioned etiologies. Its causation, like many complex and poorly understood psychiatric syndromes, is probably a multidetermined skein of genetics, organic factors, and psychosocial variables (Frick et al., 1993).

Often, the diagnosis of ADHD is made on the basis of medication trials (Johnson & McCown, in press). Psychostimulants (discussed shortly) are often, but not universally, effective in reducing symptomatology. If the child or adolescent responds to a trial of psychostimulants, a positive diagnosis is confirmed. If these classes of drugs fail in therapeutic action, the child or adolescent's impulsiveness is seen as "functional" or more "psychological" in nature. As careful psychological assessments become rationed along with other aspects of health care, useless or even harmful trials of medication will probably become more commonplace as proxy diagnostic "challenges" as well as treatments. This is a trend that health-care providers and the general public should view with alarm.

Learning Disabilities

Impulsivity has also frequently been described as a component of the behavior of learning-disabled children (e.g., Gilger, Eliason, & Richman, 1989). According to Ross (1980), many learning-disabled children are prone to "blurt out" the first answer that comes to mind; as with other students, this first answer is more often wrong than right. Because of this, these children tend to be labeled "impulsive" and are said to have difficulty with impulse control. Clinically, it is very common for teachers and other professionals to confuse impulsiveness with the diagnosis of learning disability. Indeed, children who are impulsive but score in the adequate-to-superior range of intellectual and academic functioning may be labeled "learning disabled" primarily on the global negative clinical impression that their impulsivity creates.

Although there is some evidence of a link between learning disabilities and cognitive impulsivity, such a connection is clouded by past methodological inconsistencies (Walker, 1985). It is also likely that the nebulous definition of "learning disability" is more likely in children who behave impulsively, because these children are more likely to be labeled pejoratively. Furthermore, because impulsiveness may influence academic competence through reduction in attention span, it is difficult to determine whether impulsive behaviors are a core component of learning disabilities, an antecedent or a consequence. Regardless, clinicians should not assume that learning-disabled children are necessarily impulsive or the reverse. Despite some overlap, careful and separate diagnostic workups are required for each syndrome (Johnson & McCown, in press).

Alcohol and Other Drug (AOD) Usage

Substance abuse problems, especially in adolescence, are one of the most pressing public health concerns (Johnson & McCown, 1992). Impulsive rebelliousness and alienation from dominant society values predict AOD usage in adolescence (Jessor & Jessor, 1977; Kandel, 1982). Zuckerman (1987) suggested that the relation between impulsiveness and substance use is mediated by a common biological mechanism responsible for both drug abuse and sensation seeking. However, it is also possible that impulsive children lose friends and social supports, gradually affiliating with a deviant subculture because of a decreasing lack of satisfying options. Neither mechanism is necessarily mutually exclusive, and current research seems to suggest that both biological and social factors contribute to substance abuse.

Unfortunately, most comprehensive research regarding predictors of AOD (as well as other childhood and adolescent psychopathologies) have made poor use of individual-differences constructs. Research containing comprehensive measures of social support, peer interaction, and other variables that are important from a social learning perspective often fails to consider the importance of psychometrically and biometrically assessed variables. The reverse is also true. Therefore, the relative importance of biological and social factors—as well as the potential protection of risk factors in biologically vulnerable youths—remains obscure. One solution is to have researchers who represent or are familiar with a broader range of biopsychosocial constructs.

Posttraumatic Stress Disorder (PTSD)

The *DMS–III–R* lists impulsive behaviors as a symptom of PTSD. However, few studies have examined children's impulsiveness following acute or chronic traumatic stress. This is true despite the fact that PTSD has been well documented in children (Pynoos & Nader, 1989). One study (Adams & Adams, 1984) noted an increase in juvenile arrests in the acute postdisaster period of Mount Saint Helens, although those authors did not directly measure impulsivity. Areas that have received study are childhood sexual abuse and physical abuse, which have been linked to aggression and impulsiveness in children. This is discussed in a later section.

Depressive and Anxiety Disorders

Impulsive behavior is common in dysthymic children (Leon, Kendall, & Garber, 1980). The popular psychodynamic concept of "acting out" seems to reflect this behavior in children and adolescents who are depressed or anxious. Stark, Rouse, and Livingston (1991) reported data suggesting that anger, frustration intolerance, and psychomotor agitation are quite common in children who are depressed. In fact, anger was reported in 33% of the patients tested. Clinicians should carefully rule out these two conditions when impulsiveness is noted in children.

Personality Disorders

Borderline, hysteric, and narcissistic disorders are associated with increased impulsiveness. There is also mounting evidence that despite the *DSM–III–R*'s restrictions, these disorders are evident in late childhood and early adolescence. Borderline personality disorders have, in particular, been described in children by various authors (Petti & Law, 1982; van der Kolk, 1987), and this constellation of impulsive traits may be stable from childhood on. Longitudinal data are needed regarding children at risk to determine whether early identification and treatment can interrupt this pernicious personality disorder.

Developmental Disabilities and Mental Retardation

Impulsiveness is often assumed to be a characteristic of individuals with developmental disabilities or mental retardation (Messer, 1976), although this is not necessarily true (Borys & Spitz, 1978). Mechanisms regarding the manner in which low intellectual functioning influences impulsiveness have not yet been elucidated, although hypotheses abound. Regardless, at least a group of people with developmental disabilities appear to behave impulsively and could benefit from treatment aimed at reducing these global tendencies (Whitman, Scherzinger, & Sommer, 1991).

Summary

Impulsivity in children and adolescents appears to be optimally conceptualized as a symptom rather than a syndrome. Impulsivity is generally associated with certain disorders, such as conduct disorder and ADHD. However, it may accompany a variety of clinical syndromes. As such, the following review of etiology, assessment, and treatment of impulsivity encompasses studies of children with a variety of diagnoses.

ETIOLOGY

A number of possible explanations for the occurrence of impulsive behavior in children and adolescents have been advanced. These include neurological, biochemical, personality, and cognitive processing theories. Furthermore, increasing concern is being directed at childhood impulsivity acquired through prenatal substance abuse, a growing social factor in North American and European societies.

Neurological Etiologies

For some children or adolescents, impulsive behavior may be attributable to a single biological cause, as in the case of head trauma or other central nervous system dysfunction. At least a subset of children who are identified as impulsive by their parents and teachers show a pattern of neuropsychological and neurological performance suggestive of evidence of "soft" neurological problems (Vitiello, Stoff, Atkins, & Mahoney, 1990). However, for most impulsive children (e.g., children with ADHD), the etiology of their behavior is difficult to pinpoint and may be attributed to more than one cause (e.g., both biochemical and environmental). The difficulty in identifying the etiology of impulsivity in a child or adolescent points to the importance of a thorough history taking before deciding on a course of treatment.

According to Lezak (1983), damage to the executive functions of the brain include functions that govern independent and purposive behavior that can adversely affect impulse control. Specifically, focal damage to the orbital frontal or temporal areas of the brain have been implicated in motor restlessness, excitability, and poor impulse control. Damage to the brain can result from head injury (as in a car accident), lesion, or degenerative disease (as in Pick's disease and Huntington's disease). With regard to children and adolescents, although it is possible that head trauma or lesion may lead to impulsive behavior, the majority of children with impulse control difficulties do not have obvious brain damage or demonstrable neurological deficits. Whether this is attributable to an insensitivity of the instruments or actual etiological factors remains to be seen. Regardless, it is probable that as neuropsychological and neuropsychiatric assessment becomes more refined, etiologies of many more disorders of impulsiveness will be identified (Johnson & McCown, in press).

Neurohormonal and Neurotransmittal Etiologies

Issues regarding the link between biological mechanisms and impulsiveness are highly complex (see Zuckerman, chapter 5, this volume, for a further discussion). The search for biological factors relating to childhood impulsivity is complicated by the fact that many neurohormones do not cross the blood–brain barrier, and metabolite measures that can be measured (i.e., "proxy measures") may or may not exist (see

Daruna & Barnes, chapter 2, this volume, for a brief discussion of problems in measuring brain monoamine oxidase levels). Activity of subjects, diet, time of day, and other possible threats to internal validity may also affect metabolites or other dependent measures. It is therefore not surprising that the literature is rife with contradictory findings.

For example, some researchers have attempted to identify biochemical foundations of impulsivity in children and adolescents and have focused on the hypothalamic–pituitary–adrenal axis. One promising area has been the reported association between lower levels of urinary free cortisol (UFC) and aggression and impulsive behavior (Tennes & Kreye, 1985). However, as is typical, other evidence has been equivocal. Kruesi, Schmidt, Donnelly, Hibbs, and Hamburger (1989) compared the UFC levels of children with attention deficit disorder, conduct disorder, or both with those of a matched control group. Although significant differences were found between the patient and control groups on some measures, no difference was found between UFC for the two groups. The authors concluded that their study did not provide evidence to support the belief that UFC levels can be useful in differentiating disturbed (with aggressive, impulsive behavior) from nondisturbed children and adolescents.

Other researchers have attempted to understand the etiology of impulsivity associated with ADHD by studying the mechanisms of action of medications shown to be effective for the disorder. To date, studies in this domain have also produced conflicting results; no single neurotransmitter has been found that can account for ADHD. However, according to Zametkin and Borcherding (1989), dopamine and norepinephrine systems are "clearly involved." This is not surprising given the fact that these two neurotransmitters are involved in a variety of behavioral systems. It remains for future researchers to specify where and how these and other neurotransmitters interact to produce impulsive behaviors in children and adolescents.

Trait and Temperament Explanations

Trait and temperament explanations do not necessarily differ from other biological explanations. They may simply represent a different level of abstraction (Brody, 1988). However, they postulate that at least some of the variance of the behavior in question is related to individual-differences factors that can be readily, reliably, and validly measured. In this sense, traits are said to possess explanatory power. Historically, traits were usually reserved for the measurement of behaviors that were normally distributed, whereas temperament corresponded more with the notion of fixed "types" popularized by the ancient Greeks. More recently, temperament has been used to indicate more biologically related attributes of behavior, although some authors seem to use the terms interchangeably (Brody, 1988). Regardless, this class of explanation usually implies that much of what is labeled "abnormal" is simply a deviation of normal behavior from the statistical mean. In this manner, trait or temperament theories may be seen to possess at least some explanatory power.

One example of a useful theory of temperaments was posited by Buss, Plomin, and their associates. Buss and Plomin (1975) suggested that individuals are born with a set of temperaments (i.e., innate tendencies that underlie personality traits) that include activity, emotionality, sociability, and impulsivity. These temperaments can be described on a continuum whereby an individual may possess anywhere from a high to a low degree of that temperament. In the case of impulsivity, the extremes of the continuum can be described as "impulsive" (high degree of temperament) and "deliberate" (low degree of temperament). Buss and Plomin suggested that individual differences in impulsivity relate to genetic predisposition.

Other well-known theorists who have addressed the relationship between personality and impulsivity are Hans and Sybil B. G. Eysenck, Gray, and Zuckerman. H. J. Eysenck (see chapter 4, this volume) postulated a three-factor theory of personality—neuroticism, extraversion, and psychoticism—in which personality traits are related to biological and genetic influences. Gray (1982) rotated the Eysenckian axis to make impulsiveness a primary dimension of personality. There has been considerable controversy in the literature regarding which model is the most useful. Heaven (1989) noted that there is some research evidence to support the idea that impulsiveness is related to Zuckerman's notion of sensation seeking (see chapter 5, this volume, for a further discussion). S. B. G. Eysenck and her associates (H. J. Eysenck & Eysenk, 1985; S. B. G. Eysenck, Easting, & Pearson, 1984; S. B. G. Eysenck & Eysenck, 1980) have explored the factor structure of impulsiveness in children and its relationship to empathy and desire for novelty.

Developmental Aspects

A developmental model of impulsivity is presented by Daruna and Barnes (chapter 2, this volume). Longitudinal data have been collected by Olson et al. (1990) examining early mother–child interaction as a predictor of children's later self-control capabilities. Seventy-nine children were assessed at 6 months, 13 months, 24 months, and 6 years of age. Responsive and stimulating parent–toddler interactions in the 2nd year predicted later measures of cognitive nonimpulsivity and ability to delay gratification. Security of mother–infant attachment predicted the same outcomes only for boys. Child cognitive competence in the 2nd year also consistently predicted children's later impulse control capabilities. These findings support a multidimensional and developmental conceptualization of the early antecedents of childhood impulsivity.

Kagan et al. (1979) conducted a follow-up investigation of 35 male and 33 female 10-year-olds who had been assessed originally at 4, 8, 13, and 27 months of age. The research did not reveal a strong relation among infant variables such as attentiveness, vocal excitability, irritability, activity, and reflection–impulsivity (the Haptic Matching Task), IQ (the Wechsler Intelligence Scale for Children–Revised), and reading ability (a modified version of Spache's Diagnostic Reading Scales) at age 10. A suggestive relation was found between assimilative smiling during infancy and a reflective attitude on the MFFT, and also between a slow tempo of play during infancy and longer response time on a specially constructed Embedded Figures Test at age 10. Although attentiveness during infancy predicted IQ and reading ability at age 10, both infancy and childhood variables were positively correlated with social class. This suggests that experiences associated with the social class of the parents, rather than particular infant qualities, were the more important predictors of cognitive variables at age 10. Children with high EMG levels from the flexor forearm tended to be slightly more impulsive at age 10 and less attentive at 27 months, even when the effect of social class was removed.

Cognitive Processes

Many researchers have focused on the relationship between impulsivity and cognitive processing deficits. According to Kendall and Wilcox (1980), research evidence has suggested that impulsive children demonstrate difficulties with problem solving, verbal mediation, and information seeking. Furthermore, a number of research studies, drawing on the observation that impulsive children often evidence difficulty in peer relationships, have focused on the relationship between impulsivity and social decision making. For example, Dodge and Newman (1981) examined the speed of responding of aggressive and nonaggressive boys who played a "detective game." Subjects heard a story about a boy who might have committed a crime and were then given the opportunity to listen to up to five testimonials to gather evidence.

Results indicated that aggressive boys listened to 30% fewer testimonials then did the nonaggressive boys. A comparison of the fast- and slow-responding aggressive boys indicated that the fast-responding aggressive boys were more likely to choose a guilty judgment even when the evidence suggested innocence. The authors interpreted their results to suggest that fast-responding aggressive boys may overattribute hostile behaviors to peers in unwarranted circumstances. Thus, impulsive children and adolescents may be at increased risk for making incorrect judgments about the behavior of peers, which, in turn, impairs their social functioning.

Some practical implications of these findings have been explored by McCown, Johnson, and Austin (1986). Those authors have shown that delinquent youths are more likely to be inaccurate regarding facial expression recognition in that they attribute negative meaning to ambiguous expressions. Whether this leads to more incidences of impulsive violence following misinterpretation of social cues, as the authors implied, is not certain and remains a question for further research.

Psychodynamic Theories

In classic psychoanalysis, impulsive behavior represents a lack of superego development (Freud, 1938). The hydraulic model suggests that the unrestrained id surfaces with impunity. Because the superego develops slowly and continues development through latency age, impulsiveness is a normal state for children (J. Brown, 1940).

Excessive impulsiveness in childhood or adolescence represents a failure of superego constraints (Cameron & Rychlak, 1985). More contemporary developments in analytic theory deemphasize drive and emphasize the role of early experience in relating to the world instead (i.e., object relations).

Prenatal Substance Abuse

Knowledge regarding the effects of prenatal exposure to cocaine, opiates, tobacco, alcohol, and marijuana on the neurodevelopment and behavioral outcome of infants and young children is suggestive of a potential role these drugs play in causing childhood impulsivity (Hutchings, 1989). There is little doubt that the use of AOD by pregnant women has a variety of negative physical and behavioral sequelae (Oppenheimer, 1991), including an apparent link between children who are more irritable and impulsive. However, despite lurid accounts in the popular press, it is presently unknown which, if any, deficits extend beyond early childhood (Fried, 1992). It is also very difficult to separate the effects of prenatal substance abuse from deficient child-rearing practices; lack of adequate prenatal care; poverty; a lack of maternal bonding; and, in the case of minority-group members, societal racism.

There does appear to be a link between genetic factors associated with substance abuse and impulsivity, especially in men (Cloninger, 1987). These factors include a neuropsychological impairment that may preexist or be exacerbated by early alcohol use in male offspring (Shuckit, 1985).

Social Factors

There is much literature that suggests that delay of gratification is partially acquired through social learning. Indeed, much of the current research regarding impulsiveness is attributable to the initial research efforts of Mischel (1958). It is well-known that specific behaviors associated with impulsiveness may also be socially learned (Bandura, 1973). In general, however, many social learning theorists have denied the importance of consistency of personality across situations (Brody, 1988). Despite evidence that specific types of antisocial behavior may be acquired in youth through modeling (e.g., sexual abuse [Conte, 1986]; domestic violence [Johnson, 1989]), there has been little research in recent years regarding the role of modeling and generalized trait impulsiveness.

However, social learning theorists have fruitfully suggested that human impulsiveness can result from a lowered expectancy (subjective probability) of reinforcement (Bandura, 1973). This has prompted a number of laboratory experimenters to attempt to model impulsiveness. For example, Navarick (1987) reported an experiment in which the effects of probability and delay were assessed by asking 90 college students to make repeated choices between reinforcement schedules in which the reinforcers were slides of entertainment figures. An immediate 5-second reinforcer was consistently chosen over an immediate 40-second reinforcer when the probability of receiving the large reinforcer had previously been low, thus implying that impulsiveness can occur without time-based discounting. A choice between a certain, small reinforcer and an uncertain, large reinforcer varied according to which reinforcer was immediate and which was delayed.

Regarding broader and more macrosocial factors, there is a consistent body of research linking poverty and family stress with impulsive behaviors such as substance abuse and unpremeditated crime. Indeed, this has been well-known for almost 100 years (Hill, 1949). Individuals from multiproblem families facing poverty, discrimination, poor living conditions, and low likelihood for social mobility are also likely to behave impulsively (Kaplan, 1986), perhaps because such behavior is reinforced in such environments. It is also presumed that the construct of "undersocialization" accounts for impulsiveness in lower socioeconomic family members (Parsons & Bales, 1955) and that this is mediated through cognitive processes (Spivack & Shure, 1974). However, some authors believe that impulsive behavior is a response to the extraordinary family stress encountered in disempowered families (McCown & Johnson, 1993). The role of social factors in impulsiveness is discussed further by L'Abate (chapter 6, this volume).

TREATMENT FOR CHILDHOOD AND ADOLESCENT IMPULSIVENESS

Earlier reviews of the literature regarding the treatment for impulsiveness summarize a number of useful intervention strategies (Margolis,

Brannigan, & Poston, 1977; Readence & Bean, 1978). These include the following: (a) reinforcement techniques to induce short-term changes in an impulsive cognitive style; (b) reflective modeling known to significantly modify impulsive behavior (this is especially true for peer modeling); (c) instruction in scanning strategies that alter impulsiveness on problem-solving tasks or in those that involve visual discrimination; (d) training in discrimination of distinctive features, which aids impulsive learners; (e) self-instruction and covert and overt rehearsal of strategies; and (f) deliberately increasing concern over being correct.

An early and somewhat prototypic study that examined the influence of training was conducted by Egeland (1974). The results of this study suggested that children who received some type of training in how to restrain their behavior evidenced both a significant increase in latency to response and a decrease in error rate on the MFFT relative to a control group of children who had received no training. Perhaps more interesting was the finding that children who received training in the use of specific scanning strategies (e.g., search for similarities and differences) had maintained their improved performance, whereas the group that received training in a general strategy (i.e., "take your time") did not maintain improvement. Five months following training, the group that received specific strategies improved significantly on a reading comprehension test, but the other two groups did not show improvement.

During the past 15 years, there has been an explosion in research regarding cognitive–behavioral intervention for treating impulsiveness in children (see Kendall & Braswell, 1993, for a review). For example, Kendall and Wilcox (1980) compared the efficacy of two forms of cognitive–behavioral intervention and an attention-placebo control for a sample of children identified as "noncontrolled." One treatment group received concrete self-instructional training (i.e., directions applied only to the task at hand, such as the child learning to say to himself or herself, "Look at the picture"), whereas the other received conceptual self-instructional training (i.e., strategies applied to a wide range of situations, such as the child learning problem-solving skills). Dependent measures included the MFFT, the Porteus Maze Test, self-report via ICCI, and a number of rating scales (including the Conners Teachers Rating Scale); assessment was conducted at pretreatment, posttreatment, and follow-up one year later.

The results indicated that there was no significant difference among the three groups on either the self-report, MFFT, or Porteus Maze scores (all three groups showed improvement in MFFT and Porteus Maze scores). However, there was a significant difference in teacher ratings; the conceptual group evidenced improvement across assessment relative to the other two groups. The authors concluded that their study may provide evidence that metacognitive training (conceptual) may improve the generalization of the training.

Another study supporting the effectiveness of cognitive–behavioral interventions was conducted by Hinshaw et al. (1984). Those authors examined the effect of methylphenidate and placebo, and cognitive–behavioral self-evaluation and extrinsic reinforcement, for a sample of boys diagnosed with ADHD. The primary dependent measure was direct observation of playground behavior, which was coded by raters as either appropriate, negative, or nonsocial.

The results suggested that methylphenidate was associated with significantly more positive behaviors when compared with placebo and that self-evaluation training was associated with more positive behaviors than extrinsic reinforcement. However, it is noteworthy that the difference between medication and placebo only held for younger children. When the four combinations of treatment were rank ordered according to frequency of positive social behavior, medication plus self-evaluation was ranked the highest.

Reid and Borkowski (1987) assessed the effectiveness of attribution training and self-control training in a sample of hyperactive children. Subjects were divided into three groups: The control group received strategy training but no self-control or attribution training; the self-control group received training in self-regulation; and the self-control plus attribution group received the same training in self-regulation plus training in making attributions. Dependent measures included the MFFT, the Conners Teachers Rating Scale, the Self-Control Rating Scale, the Specific Attribution Questionnaire, and the General Personal Causality Questionnaire.

The results of the study suggested the self-control plus attribution training corresponded to improvement in several of the dependent measures, including a decrease in impulsivity. This improvement persisted at a 10-month follow-up assessment. Reid and Borkowski (1987) suggested that the addition of attributional retraining to the more traditional cognitive–behavioral treatment (e.g., training in self-control) may enhance the efficacy of treatment.

On the basis of findings such as these, there seems little doubt that when the conceptualization of impulsivity is related to conduct disorder or cognitive processing deficits, cognitive–behavioral strategies are clearly the treatment of choice (Kendall, 1991). Generally, these interventions focus on teaching strategies to help children restrain the tendency to respond without reflecting on the situation. For example, the Think Aloud program (Camp & Bash, 1981) was developed for use by teachers and psychologists to improve children's cognitive problem–solving abilities. The Think Aloud program was based partly on the concepts of self-instruction (Meichenbaum & Goodman, 1971), cognitive problem solving (Spivack & Shure, 1974), and modeling (Bandura, 1969).

This review of research studies examining the effectiveness of treatment for impulsivity highlights a few important points. First, few, if any, studies have been conducted exclusively on impulsivity per se. Rather, many treatment outcome studies include many dependent measures, with a few of these assessing impulsivity. Second, the diversity in dependent measures used in the various studies makes it difficult to draw conclusions across studies. Third, although many studies purport to examine the effectiveness of "cognitive–behavioral" strategies, each study appears to use a different cognitive–behavioral intervention (e.g., self-evaluation, self-control, self-instructional training). Again, the difference in the type of cognitive–behavioral strategy limits the generalizability of the findings to the general category of cognitive–behavioral interventions.

Medication

If the impulsivity is associated with an ADHD, pharmacological treatment is common. According to Zametkin and Borcherding (1989), "the efficacy and safety of a variety of medications in reducing inattention, motor restlessness, impulsivity, aggression and socially inappropriate behavior is dramatic and powerful" (p. 447). Medications commonly prescribed to decrease impulsivity include methylphenidate, dextroamphetamine, and pemoline. To date, only a handful of studies have specifically examined the efficacy of these medications in the treatment of impulsivity per se.

One such study conducted by R. T. Brown and Sexson (1987) sought to examine the effect of treatment with methylphenidate for a sample of Black male adolescents. The authors used a crossover design in which each subject received (at separate intervals) a placebo and three methylphenidate doses (0.15 mg/kg, 0.30 mg/kg, and 0.50 mg/kg). Dependent measures used to assess impulsivity included the MFFT, a behavioral rating, and the Gordon Diagnostic System (the other dependent measures, such as weight, are not reviewed here). A comparison across treatment conditions suggested that the 0.30- and 0.50-mg/kg doses were associated with a decrease in impulsivity relative to the placebo and lower dose of methylphenidate. In addition, the data suggested a linear trend between dosage and improvement on dependent measures.

Psychodynamic Therapy

Psychodynamic therapies remain immensely popular for childhood impulsve control disorders. This is true despite the evidence that meta-analysis suggests that psychodynamic therapies are the least effective treatment for children compared with behavioral and cognitive–behavioral treatment (Casey & Berman, 1985). For example, Kendall, Reber, McLeer, Epps, and Ronan (1990) compared the efficacy of cognitive–behavioral training with supportive–psychodynamic therapy for the treatment of conduct-disordered children using a crossover design. Among the dependent measures used were the Conners Teacher's Questionnaire and the Self-Control Rating Scale to assess impulsivity. The results suggested, overall, that cognitive–behavioral intervention was associated with a significant decrease in impulsivity (as assessed by the Connors Teacher's Questionnaire and the Self-Control Rating Scale) when compared with the supportive therapy. However, the authors noted that on some dependent measures, both treatments produced significant improvement, whereas on other measures, there was no improvement associated with either treatment.

Treatment of Trauma-Based Impulsiveness

There is a growing literature suggesting that incest or severe physical abuse can result in impulsive behaviors. Aggressive and even assaultive behavior has been noted in toddlers (George & Main, 1979) and in abused school-age children (Livingston, 1987). Harney (1992) noted that children who are physically abused often engage in disruptive behavior in school and behave antisocially. Harney further suggested that impaired impulse control and cognitive and developmental impairments are primary symptoms of child abuse.

Extreme childhood trauma may cause multi-impulsive or dissociative behavior (Herman, Perry, & van der Kolk, 1989). Adult patients with such histories usually receive the diagnosis of borderline, antisocial, or even multiple personality disorders, and their pathology is usually recognized at a much earlier age than the *DSM–III–R* allows the diagnosis to be given. Such children usually receive psychodynamic or psychoanalytic treatment, and behavior modification and cognitive–behavioral interventions are less commonly used with these patients.

The reasons for this are not clear. There has been a dearth of studies using cognitive–behavioral studies with sexually abused and severely traumatized children. Our own and our colleagues' clinical experiences suggest somewhat less success using cognitive–behavioral techniques with sexually abused children, with an effect size about half of that associated with impulsiveness related to other etiologies. This may indicate the intractableness of trauma-related disorders or may suggest the need for psychodynamic approaches. However, more data are needed before this issue can be resolved. Given the ubiquitousness of trauma-related disorders, especially in increasingly violent urban areas, comparison of cognitive–behavioral and dynamic therapies for severely traumatized children would be beneficial for practicing clinicians.

CONCLUSION

Currently, it appears that impulsivity is best viewed as a construct describing behavior associated with a number of diagnostic categories rather than a psychiatric condition in and of itself. This suggests that merely labeling a child or adolescent "impulsive" does little to advance the understanding of that individual's primary problem. A review of the literature suggests that there are many unanswered questions regarding impulsivity in children and adolescents. How does one define the construct of impulsivity? What is the most appropriate way to assess impulsivity? Why do some children and adolescents evidence impulsive behaviors? Although some of the current research begins to address these questions, more well-designed research studies are needed. We hope that future studies that address these issues will be conducted and provide this much-needed information to those who work with (or live with) impulsive children and adolescents.

REFERENCES

Achenbach, T., & Edelbrock, C. (1983). *Manual for the Child Behavior Checklist and Revised Child Behavior Profile*. Burlington, VT: University Associates in Psychiatry.

Adams, P., & Adams, G. (1984). Mt. St. Helens' ashfall. *American Psychologist, 39*, 252–260.

Adler, P. T. (1970). Evaluation of the figure drawing technique: Reliability, factorial structure, and diagnostic usefulness. *Journal of Consulting and Clinical Psychology, 35*, 52–57.

American Psychiatric Association. (1987). *Diagnostic and statistical manual of mental disorders* (3rd ed., rev.). Washington DC: Author.

Bandura, A. (1969). *Principles of behavior modification*. New York: Holt, Rinehart & Winston.

Bandura, A. (1973). *Aggression: A social learning analysis*. Englewood Cliffs, NJ: Prentice Hall.

Barkly, R. A. (1981). *Hyperactive children: A handbook for diagnosis and treatment*. New York: Guilford Press.

Becker, L., Bender, N., & Morrison, G. (1978). Measuring impulsivity-reflection: A critical review. *Journal of Learning Disabilities, 11*, 626–632.

Block, J., Block, J. H., & Harrington, D. M. (1974). Some misgivings about the Matching Familiar Figures Test as a measure of reflection–impulsivity. *Developmental Psychology, 10*, 611–632.

Block, J., Gjerde, P. F., & Block, J. H. (1986). More misgivings about the Matching Familiar Figures Test as a measure of reflection–impulsivity: Absence of construct validity in preadolescence. *Developmental Psychology, 22*, 820–831.

Borys, S. V., & Spitz, H. H. (1978). Reflection–impulsivity in retarded adolescents and nonretarded children of equal MA. *American Journal of Mental Deficiency, 82*, 601–604.

Breen, M. J. (1989). Cognitive and behavioral differences in ADHD boys and girls. *Journal of Child Psychology and Psychiatry, 30,* 711–716.

Brody, N. (1988). *Personality: In search of individuality.* San Diego, CA: Academic Press.

Brown, J. (1940). *The psychodynamics of abnormal behavior.* New York: McGraw-Hill.

Brown, R. T., & Sexson, S. B. (1987). A controlled trial of methylphenidate in black adolescents. *Clinical Pediatrics, 27,* 74–81.

Buss, A., & Plomin, R. (1975). *A temperament theory of personality development.* New York: Wiley–Interscience.

Cameron, N., & Rychlak, J. (1985). *Personality development and psychopathology: A dynamic approach* (2nd ed.). Boston: Houghton Mifflin.

Camp, B. W., & Bash, M. S. (1981). *Think Aloud.* Champaign, IL: Research Press.

Casey, R., & Berman, J. (1985). The outcome of psychotherapy with children. *Psychological Bulltein, 98,* 388–400.

Cloninger, C. (1987). Neurogenetic adaptive mechanisms in alcoholism. *Science, 236,* 410–416.

Conners, C. (1969). A teacher rating scale for use in drug studies with children. *American Journal of Psychiatry, 126,* 884–888.

Conte, J. (1986). Sexual abuse and the family: A critical analysis. *Journal of Psychotherapy and the Family, 2,* 113–126.

Cook, T., & Campbell, D. (1979). *Quasi–experimentation: Design and analysis issues for field settings.* Boston: Houghton Mifflin.

Dodge, K. A., & Newman, J. P. (1981). Biased decision-making processes in aggressive boys. *Journal of Abnormal Psychology, 90,* 375–379.

Egeland B. (1974). Training impulsive children in the use of more efficient scanning techniques. *Child Development, 45,* 165–171.

Eysenck, H. J., & Eysenck M. (1985). *Personality and individual differences: A natural science approach.* New York: Plenum Press.

Eysenck, S. B. G., Easting, G., & Pearson, P. R. (1984). Age norms for impulsiveness, venturesomeness and empathy in children. *Personality and Individual Differences, 5,* 315–321.

Eysenck, S. B. G., & Eysenck, H. J. (1980). Impulsiveness and venturesomeness in children. *Personality and Individual Differences, 1,* 21–29.

Exner, J. E., & Weiner, I. B. (1982). *The Rorschach: A comprehensive system: Vol. 3. Assessment of children and adolescents.* New York: Wiley

Freud, S. (1938). *The basic writings of Sigmund Freud.* New York: Modern Library.

Frick, P., Strauss, C., Lahey, B., & Christ, M. A. (1993). Behavioral disorders of children. In P. Sutker & H. Adams (Eds.), *Comprehensive handbook of psychopathology* (pp. 765–789). New York: Plenum Press.

Fried, P. (1992, July). *The effects of prenatal drug exposure on infants and children.* Paper presented at the First Link One Office of Substance Abuse Prevention Conferences, Washington, DC.

Gaddis, L., & Martin, R. (1989). Relationship among measures of impulsivity for preschoolers. *Journal of Psychoeducational Assessment, 7,* 284–295.

George, C., & Main, M. (1979). Social interactions of young abused children: Approach, avoidance and aggression. *Child Development, 50,* 306–318.

Gilger, J., Eliason, M., & Richman, C. (1989). A comparison of cognitive and behavioral patterns in learning–disabled children: Subtype and sex differences. *Developmental Neuropsychology, 5,* 227–243.

Gittelman, R., Mannuzza, R., Shenker, R., & Bonegura, N. (1985). Hyperactive boys almost grown up: I. Psychiatric status. *Archives of General Psychiatry, 42,* 937–947.

Gordon, M. (1979). The assessment of impulsivity and mediating behaviors in hyperactive and non–hyperactive boys. *Journal of Abnormal Child Psychology, 7,* 317–326.

Gordon, M. (1983). *The Gordon Diagnostic System.* Boulder, CO: Clinical Diagnostic Systems.

Gray, J. A. (1982). *The neuropsychology of anxiety: An inquiry into the functions of the septo–hippocampal system.* Oxford, England: Oxford University Press.

Harney, P. A. (1992). The role of incest in developmental theory and treatment of women diagnosed with personality disorder. *Women and Therapy, 12,* 39–57.

Heaven, P. C. (1989). Venturesomeness and impulsiveness: Their relation to orientation to authority among adolescents. *Personality and Individual Differences, 10,* 1205–1208.

Herman, J., Perry, J., & van der Kolk, B. (1989). Childhood trauma in borderline personality disorder. *American Journal of Psychiatry, 146,* 490–495.

Hill, R. (1949). *Families under stress.* New York: Harper & Row.

Hinshaw, S. P., Henker, B., & Whalen, C. K. (1984). Cognitive–behavioral and pharmacologic interventions for hyperactive boys: Comparative and combined effects. *Journal of Consulting and Clinical Psychology, 52,* 739–749.

Hutchings, D. (Ed.). (1989). *Prenatal abuse of licit and illicit drugs.* New York: New York Academy of Sciences.

Jessor, R., & Jessor, S. L. (1977). *Problem behavior and psychosocial development: A longitudinal study of youth.* San Diego, CA: Academic Press.

Johnson, J. (1989). *The relation of observation of parental violence in family of origin, support network variables, social support and alcohol abuse in male spouse abusers.* Unpublished doctoral dissertation, Loyola University, Chicago.

Johnson, J., & McCown, W. (1992). An overview of substance abuse. In P. Sutker & H. Adams (Eds.), *Comprehensive handbook of psychopathology* (2nd ed., pp. 437–450). New York: Plenum Press.

Johnson, J., & McCown, W. (in press). *The family treatment of neurobehavioral disorders.* Binghamton, NY: Haworth Press.

Kagan, J. (1966). Reflection–impulsivity: The generality and dynamics of conceptual tempo. *Journal of Abnormal Psychology, 71,* 17–24.

Kagan, J., Lapidus, D., & Moore, M. (1979). Infant antecedents of cognitive functioning: A longitudinal study. *Annual Progress in Child Psychiatry and Child Development*, 46–77.

Kagan, J., Rossman, B., Day, D., Albert, J., & Phillips, W. (1964). Information processing in the child: Significance of analytic and reflective attitudes. *Psychological Monographs*, 78(Whole No. 578).

Kandel, D. (1982). Epidemioloigcal and psychosocial perspectives on adolescent drug use. *Journal of American Academic Clinical Psychiatry*, 21, 328–347.

Kaplan, L. (1986). *Working with multiproblem families*. Lexington, MA: Lexington Books.

Kendall, P. (1991). Guiding theory and therapy for children and adolescents. In P. Kendall (Ed.), *Child and adolescent therapy: Cognitive behavioral procedures* (pp. 3–22). New York: Guilford Press.

Kendall, P., & Braswell, L. (1993). *Cognitive–behavioral therapy for impulsive children* (2nd ed.). New York: Guilford Press.

Kendall, P., Reber, M., McLeer, S., Epps, J., & Ronan, K. R. (1990). Cognitive-behavioral treatment of conduct disordered children. *Cognitive Therapy and Research*, 14, 279–297.

Kendall, P., & Wilcox, L. (1980). Cognitive–behavioral treatment for impulsivity: Concrete versus conceptual training in non-self-controlled problem children. *Journal of Consulting and Clinical Psychology*, 48, 80–91.

Klinteberg, B., Magnusson, D., & Schalling, D. (1989). Hyperactive behavior in childhood and adult impulsivity: A longitudinal study of male subjects. *Personality and Individual Differences*, 10, 43–50.

Kruesi, M., Schmidt, M., Donnelly, M., Hibbs, E., & Hamburger, S. (1989). Urinary free cortisol output and disruptive behavior in children. *Journal of the American Academy of Child and Adolescent Psychiatry*, 28, 441–443.

Leon, G., Kendall, P., & Garber, J. (1980). Depression in children: Parent, teacher, and child perspectives. *Journal of Abnormal Child Psychology*, 8, 221–235.

Lezak, M. (1983). *Neuropsychological assessment* (2nd ed.). New York: Oxford University Press.

Livingston, R. (1987). Sexually and physically abused children. *Journal of the American Academy of Child and Adolescent Psychiatry*, 26, 413–415.

Lochman, J., White, K., & Wayland, K. (1991). Cognitive–behavioral assessment and treatment with aggressive children. In P. Kendall (Ed.), *Child and adolescent therapy: Cognitive–behavioral procedures* (pp. 25–65). New York: Guilford Press.

Loeber, R. (1990). Development and risk factors of juvenile antisocial behavior and delinquency. *Clinical Psychology Review*, 10, 1–41.

Maccoby, E. E., Dowley, E. M., Hagan, J. W., & Degerman, R. (1965). Activity level and intellectual functioning in normal preschool children. *Child Development*, 44, 274–279.

Margolis, H., Brannigan, G., & Poston, M. (1977). Modification of impulsivity: Implications for teaching. *Elementary School Journal, 77,* 231–237.

Matsushima, J. (1964). An instrument for classifying impulse control among boys. *Journal of Consulting Psychology, 28,* 87–90.

McCown, W., & Barkhausen, D. (1993). *A Rasch scaled measure of impulsiveness suitable for use with Thematic Apperception tests.* Unpublished manuscript.

McCown, W., & Johnson, J. (1993). *The treatment resistant family: A consultation/crisis intervention treatment model.* Binghamton, NY: Haworth Press.

McCown, W., Johnson, J., & Austin, S. (1986). Inability of delinquents to decode facial expressions of emotion. *Journal of Social Behavior and Personality, 1,* 91–97.

Meichenbaum, D., & Goodman, J. (1971). Training impulsive children to talk to themselves: A means of developing self-control. *Journal of Abnormal Psychology, 77,* 115–126.

Messer, S. (1976). Reflection–impulsivity: A review. *Psychological Bulletin, 83,* 1026–1052.

Milich, R., & Kramer, J. (1984). Reflections on impulsivity: An empirical investigation of impulsivity as a construct. *Advances in Learning and Behavioral Disabilities, 3,* 57–94.

Mischel, W. (1958). Preference for delayed reinforcement: An experimental study of a cultural observation. *Journal of Abnormal and Social Psychology, 66,* 57–61.

Navarick, D. (1987). Reinforcement probability and delay as determinants of human impulsiveness. *Psychological Record, 37,* 219–225.

Oas, P. (1983). Impulsive behavior and assessment of impulsivity with hospitalized adolescents. *Psychological Reports, 53,* 764–766.

Oas, P. (1985a). The psychological assessment of impulsivity: A review. *Journal of Psychoeducational Assessment, 3,* 141–156.

Oas, P. (1985b). Impulsivity and delinquent behavior among incarcerated adolescents. *Journal of Clinical Psychology, 4,* 22–24.

Olson, S. (1989). Assessment of impulsivity in preschoolers: Cross–measure convergences, longitudinal stability, and relevance to social competence. *Journal of Clinical Child Psychology, 18,* 176–183.

Olson, S., Bates, J., & Bayles, K. (1990). Early antecedents of childhood impulsivity: The role of parent–child interaction, cognitive competence, and temperament. *Journal of Abnormal Child Psychology, 18,* 317–334.

Oppenheimer, E. (1991). Alcohol and drug misuse among women: An overview. *British Journal of Psychiatry, 158*(Suppl. 10), 36–44.

Parsons, T., & Bales, R. (1955). *Family, socialization and interaction process.* New York: Free Press.

Petti, A., & Law, W. (1982). Borderline psychotic behavior in hospitalized children: Approaches to assessment and treatment. *Journal of the American Academy of Child Psychiatry, 21,* 197–202.

Porteus, S. (1955). *The maze test: Recent advances.* Palo Alto, CA: Pacific Books.

Porteus, S. D. (1968). New applications of the Porteus Maze Tests. *Perceptual and Motor Skills, 26,* 787–798.

Purcell, K. (1956). A note on Porteus Maze and Wechsler-Bellevue scores as related to antisocial behavior. *Journal of Consulting Psychology, 20,* 361–364.

Pynoos, R., & Nader, K. (1989). Prevention of psychiatric morbidity in children after disaster. In D. Shaffer, I. Phillips, & N. Enzer (Eds.), *Prevention of mental disorders, alcohol and other drug use in children and adolescents* (pp. 225–272). Rockville, MD: Department of Health and Human Services.

Quay, H., & Peterson, D. (1983). *Interim manual for the Revised Behavioral Problem Checklist.* Unpublished manuscript, University of Miami, Miami, FL.

Readence, J., & Bean, T. (1978). Modification of impulsive cognitive style: A survey of the literature. *Psychological Reports, 43,* 327–337.

Reid, M., & Borkowski, J. (1987). Causal attributions of hyperactive children: Implications for teaching strategies and self-control. *Journal of Educational Psychology, 79,* 296–307.

Rigby, K. (1986). Orientation to authority: Attitudes and behavior. *Australian Journal of Psychology, 38,* 153–160.

Rigby, K., Mak, A., & Slee, P. (1989). Impulsiveness, orientation to institutional authority, and gender as factors in self-reported delinquency among Australian adolescents. *Personality and Individual Differences, 10,* 689–692.

Robins, C., Blatt, S., & Ford, R. (1991). Changes in human figure drawings during intensive treatment. *Journal of Personality Assessment, 57,* 477–497.

Ross, A. (1980). *Psychological disorders of children: A behavioral approach to theory, research and therapy* (2nd ed.). New York: McGraw-Hill.

Schachar, R., & Logan, G. D. (1990). Impulsivity and inhibitory control in normal development and childhood psychopathology. *Developmental Psychology, 26,* 710–720.

Sears, R., Rau, L., & Alpert, R. (1965). *Identification and child rearing.* Stanford, CA: Stanford University Press.

Shuckit, M. (1985). Behavioral effects of alcohol in sons of alcoholism. In M. Galanter (Ed.), *Recent developments in alcoholism* (Vol. 3, pp. 11–19). New York: Plenum Press.

Silverman, I. W., & Ragusa, D. M. (1990). Child and maternal correlates of impulse control in 24-month-old children. *Genetic, Social, and General Psychology Monographs, 116,* 435–473.

Spivack, G., & Shure, M. (1974). *Social adjustment of young children.* San Francisco: Jossey-Bass.

Stark, K., Rouse, L., & Livingston, R. (1991). Treatment of depression during childhood and adolescence: Cognitive–behavioral procedures for the individual and family. In P. Kendall (Ed.), *Child and adolescent therapy: Cognitive–behavioral procedures* (pp. 165–208). New York: Guilford Press.

Stoff, D., Friedman, E., Pollock, L., & Vitiello, B. (1989). Elevated platelet MAO is related to impulsivity in disruptive behavior disorders. *Journal of the American Academy of Child and Adolescent Psychiatry, 28,* 754–760.

Tennes, K., & Kreye, M. (1985). Children's adrenocortical response to classroom activities and tests in elementary school. *Psychosomatic Medicine, 47,* 451–460.

van der Kolk, B. (1987). *Psychological trauma.* Washington, DC: American Psychiatric Press.

Vinogradov, S., Dishotsky, N. I., Doty, A. K., & Tinklenberg, J. R. (1988). Patterns of behavior in adolescent rape. *American Journal of Orthopsychiatry, 58,* 179–187.

Vitiello, B., Stoff, D., Atkins, M., & Mahoney, A. (1990). Soft neurological signs and impulsivity in children. *Journal of Developmental and Behavioral Pediatrics, 11,* 112–115.

Walker, N. W. (1985). Impulsivity in learning disabled children: Past research findings and methodological inconsistencies. *Learning Disability Quarterly, 8,* 85–94.

Whitman, T., Scherzinger, M., & Sommer, K. (1991). Cognitive instruction and mental retardation. In P. Kendall (Ed.), *Child and adolescent therapy: Cognitive–behavioral procedures* (pp. 276–315). New York: Guilford Press.

Wirt, R. D., Lachar, D., Klinedinst, J. K., & Seat, P. D. (1984). *Multidimensional description of child personality: A manual for the Personality Inventory for Children, Revised 1984.* Los Angeles: Western Psychological Services.

Zametkin, A. J., & Borcherding, B. G. (1989). The neuropharmacology of attention-deficit hyperactivity disorder. *Annual Review of Medicine, 40,* 447–451.

Zuckerman, M. (1987). Biological connection between sensation seeking and drug abuse. In J. Engel & L. Oreland (Eds.), *Brain reward systems and abuse* (pp. 165–176). New York: Raven Press.

15

IMPULSIVITY IN ADULT NEUROBEHAVIORAL DISORDERS

JUDITH A. HOLMES, JUDITH L. JOHNSON, and ANN L. ROEDEL

IMPULSIVITY AND FAILURE OF REGULATION

Disorders of impulse control are frequently associated with adult psychiatric diagnoses (e.g., borderline personality disorder, kleptomania, explosive disorder, pyromania) as well as disorders with a typical onset in childhood (e.g., Tourette's syndrome, tic disorders, attention-deficit hyperactivity disorder). Behavioral and personality changes are also commonly found in adult neurological diseases. In fact, an inability to restrain one's impulses and control behavior may be the initial symptom or primary complaint noted by concerned family members or by neurologists when examining a patient for potential central nervous system (CNS) dysfunction. In this chapter we focus on the etiology and clinical presentation of impulsiveness associated with adult neurobehavioral disorders.

Pathological impulsivity can be conceptualized as a failure to regulate, monitor, or control behavior and emotional expression. Because of an inability to exercise normal controls, behavior becomes disruptive and may consist of either excesses or deficits. Some authors have suggested that in some cases, these changes may actually represent preexisting tendencies that were simply not given public expression (Lishman, 1987). Therefore, premorbid personality traits may be accentuated, and emotions, behaviors, and basic drives of sexuality and aggressiveness may emerge (Bond, 1984). In other instances, personality changes may be striking, with the appearance of inappropriate behavior that is completely uncharacteristic of the individual prior to the onset of the disorder.

The effects of impulsivity in neurobehavioral disorders may be seen across areas of functioning that include cognition, behavior, and emotional expression. The impact on cognition leads to a change in functioning that may be characterized by a failure to consider, review, or appreciate the nature of one's actions. Poor judgment is common, as are limited insight, diminished self-monitoring capability, and a lack of concern about the impact of one's actions on others.

Behavioral changes consist of inattention, distractibility, increased psychomotor activity, and a wide range of inappropriate actions and comments. Indeed, the deterioration of social behavior and diminished awareness of the needs and feelings of others may be the first noticeable symptom of neurological disease and the most distressing to family and friends. The individual becomes indifferent about the consequences of his or her behavior and exhibits a decreased ability to appreciate the impact of that behavior on others. In interpersonal interactions, there may be a lack of normal, adult restraint and tact and a weakening of ethical controls. In addition, the individual may become sexually preoccupied, which may manifest itself through lewd comments, inappropriate sexual advances, promiscuity, or perverse behavior.

Emotional expression is labile because of an inability to modulate or control the expression of feelings. Thoughts and emotions ranging from irritability to jocularity to euphoria may readily surge to the surface and be spontaneously expressed. Aggressive acts are also common, and even minor provocation may result in sudden outbursts of violence. All of these changes within the spheres of cognition, behavior, and emotion are usually compounded by the individual's lack of awareness of any difficulty (anosognosia) and subsequent inability to acknowledge deficits or changes.

NEUROPSYCHOLOGICAL CAUSES OF BEHAVIORAL DYSREGULATION

Disordered behavioral control can result from pathology at any level of the CNS (Woodcock, 1986). Hence, impulsivity is a frequent com-

TABLE 1
Neurobehavioral Disorders Presenting With Impulsivity as a Primary Symptom or Major Feature

Etiology	Diagnosis
Degenerative diseases	Frontal lobe dementias, Huntington's chorea, and Alzheimer's disease
Infectious diseases	General paresis and AIDS
Trauma	Closed-head injury and penetrating wounds
Vascular	Particularly frontal lesions
Temporal neoplasms	Particularly frontal and lobe tumors
Metabolic	Wilson's disease and hepatic disease
Temporal neurosurgery	Particularly frontal and lobes
Organic personality syndromes	Pseudopsychopathic personality disorder and episodic dyscontrol syndrome

Note. AIDS = acquired immunodeficiency syndrome.

ponent in a wide range of neurological disorders, including degenerative diseases, trauma, neoplasms, metabolic disorders, infectious disease, vascular disorders, and neurosurgery. Additionally, a host of other disorders with varying etiologies may be present with impulsivity as an associated feature, although our discussion is limited to only those major areas listed in Table 1.

Inasmuch as the pathological effects of cerebral insult on personality are not well-known, specific anatomical–behavioral correlations are often speculative. Despite this, there are several well-documented areas of cerebral involvement frequently associated with changes in behavior and personality. In particular, frontal lobe lesions may produce personality changes with a fairly well-recognized pattern of behavior, frequently with impulsivity as a typical feature (Lishman, 1987).

Lesions in the diencephalon and brain stem may also produce symptoms similar to those observed in frontal lobe lesions, including behavioral disinhibition (Lishman, 1987). A lack of social concern and sudden outbursts of violence are characteristic in hypothalamic lesions. In addition, lesions involving the thalamus, midbrain, and pons have been implicated in personality changes that include behavioral and emotional dyscontrol (Elliott, 1986). For example, Strub and Black (1981) discussed an atypical presentation of an Alzheimer's type of dementia (DAT) with hyperactivity and uninhibited, impulsive behavior. At autopsy, their patients were found to have considerable atrophy of the locus coeruleus that was likely caused by a noradrenergic deficit. Thus, personality changes and an increase in impulsive tendencies can be associated with varying areas and different processes within the brain.

Frontal Lobe Syndromes and Impulsivity

Although behavioral and affective changes may occur with cerebral insult to a number of areas, damage to the frontal lobes is particularly implicated in personality changes (Lishman, 1987). Conditions producing damage to this region have been known to lead to dramatic changes in personality and behavior that may be the most prominent aspect of the patient's presentation (Cummings, 1985).

There is a general consensus that damage to this area leads to readily identifiable changes in emotions, behavior, and cognition (Lipowski, 1978). The "frontal syndrome" is a constellation of symptoms used to refer to a heterogeneous group of disorders producing such personality and behavioral changes. Typical features include impulsivity, disinhibition, emotional lability, irresponsibility, lack of concern and tact, poor judgment, and occasional marked aggressiveness. Although damage to one lobe may produce an alteration in personality, bilateral lesions result in the most severe changes, particularly those involving the orbital and frontal areas of the brain (Lishman, 1987). Damage to this area is thought to interfere with the frontal monitoring system by disrupting the connections between the limbic system, thalamus, and frontal convexity (Cummings, 1985). The result is a disinhibited behavioral syndrome, with impulsivity as the outstanding feature.

In frontal lobe syndromes, the individual exhibits a "coarsening of the personality" (Cummings, 1985), and impulses are acted on without consideration of the consequences. In a study of 20 patients with frontal lobe dysfunction admitted to a psychiatric facility, McAllister and Price (1987) noted that disinhibited behavior coupled with affective lability were the primary symptoms leading to admission. Indeed, because of the predominant symptoms in personality and behavior, the term *pseudopsychopathic* (Blumer & Benson, 1975) was coined to describe the changes associated with frontal lobe damage.

Numerous disorders and diseases may result in damage to the frontal lobes, including neoplasms, degenerative diseases, cerebral vascular accidents, and traumatic damage from neurosurgery, penetrating wounds, or closed-head injuries.

Trauma

Trauma is a common cause of frontal lobe dysfunction, particularly open-head injuries produced by high-velocity missile wounds. Such injuries may result in relatively localized damage, leaving the individual with primarily behavioral and personality deficits while other functioning re-

312   *HOLMES, ROEDEL, AND JOHNSON*

mains intact. The earliest and perhaps best known case of frontal lobe dysfunction resulting from a missile wound is that of Phineas Gage, a 25-year-old railroad worker described by Harlow in 1848. Mr. Gage suffered a penetrating bifrontal injury when an iron rod was accidentally dynamited through his skull. Although his physical recovery was good and his cognitive functioning remained grossly intact, major personality and behavioral changes ensued. Specifically, he became impulsive, emotionally labile, insightless, and inconsiderate (Harlow, 1868). Although disturbing to those who knew him to be a formerly pleasant and well-controlled individual, the patient himself remained unconcerned regarding these changes.

Jarvie (1954) found disinhibition to be the central feature of the mental status changes in a study of 6 patients with frontal head wounds. He argued that disinhibition could be conceptualized as a primary release phenomenon that unleashes changes in behavior, affect, and cognition. He observed that despite relatively spared intellect, these patients' main difficulty was the loss of ability to control abrupt shifts in mood, basic drives, and behavior. He further observed lack of insight into these changes and difficulties.

Closed-Head Injury

Closed-head injury, which is frequently caused by motor vehicle accidents, is a common cause of frontal lobe dysfunction. In particular, damage to the orbital aspect of the frontal lobes frequently results in personality and behavioral changes, with disinhibition as the outstanding feature. These patients are often impulsive, lack tact and concern, have reduced control, and may engage in antisocial acts. Sexual preoccupation is common, and conversation may be punctuated with inappropriate sexual comments. The following case history illustrates the effects of a closed-head injury on impulsive behavior.

> *Case 1.* A 20-year-old man involved in a motor vehicle accident sustained a closed-head injury to the frontal–orbital region. During the course of rehabilitation, his behavior was characterized by marked impulsivity that was often aggressive. His impulsiveness and lack of control resulted in immature behavior, constant childish demands, and episodes of unexpected belligerance that quickly dissipated once he was removed from the situation. On formal testing, he obtained a Wechsler Adult Intelligence Scale–Revised (WAIS–R; Wechsler, 1981) Full Scale score of 71, a Verbal scale score of 75, and a Performance scale score of 67; he also had a severe impairment in memory, as indicated by a Wechsler Memory Scale–Revised (WMS–R; Wechsler, 1987) score of 52. Although his cognitive and memory functioning showed gradual improvement over time, personality and behavioral changes persisted. Indeed, these posed the greatest barrier

to his recovery and were the most difficult symptoms for his family to deal with.

Degenerative Diseases and Impulsivity

Dementia of the Frontal Lobe Type

Certain degenerative diseases, particularly those with extensive frontal lobe involvement, commonly present with behavioral and personality changes. Indeed, in some disorders, the earliest symptoms may be a change in social behavior. For example, Neary, Snowden, Northern, and Goulding (1988) described 7 patients with a non-Alzheimer's dementia who primarily presented with personality and behavioral changes. This dementia of the frontal lobe type (DFT), which the authors indicated may represent a form of Pick's disease, is characterized by a breakdown in social behavior and personality that consists of impulsivity, disinhibition, and social misconduct. As noted, these changes may be the most prominent symptoms, with cognitive and memory functioning remaining relatively intact. Often, the course of changes may be gradual and progress over several years. The following case illustrates this process.

> *Case 2.* The patient was a 73-year-old man with a 5-year history of gradual personality and behavioral change. His wife described him as displaying significant changes in his personality consisting primarily of poor judgment and socially inappropriate behavior. This included a tendency to offer sexually and racially charged comments and doing things such as removing his clothes in public. On another occasion, he arrived home with a large quantity of alcohol and a group of youths he had just met, having invited them over for a party. He also displayed episodes of provocative and combative behavior and had thrown a glass of water in a waiter's face in response to a minor annoyance in a restaurant. Insight was lacking, and he displayed little concern about his behavior or his wife's distress. On neuropsychological testing, his impulsivity and lack of review interfered with his ability to perform adequately despite relatively intact cognitive functioning. He tended to blurt out the first thing that came to mind with little or no reflective thinking. The results indicated a WAIS–R Full Scale score of 90 and an overall WMS–R Memory score of 85. Although both of these scores are below average, they are not in the moderate or severely deficient range; hence, his primary difficulty was related more to personality changes than to cognitive deficiencies.

In addition, Strub and Black (1981) described a group of patients with an atypical Alzheimer's disease presenting hyperactivity and uninhibited behavior. Similarly, Kumar, Schapiro, Haxby, Grady, and Friedland (1990) described a group of patients with a diagnosis of DAT who exhibited frontal lobe behavior, including impulsivity. The authors found that these individuals had selective decrements in positron emission tomography-measured glucose metabolism in the orbitofrontal, prefrontal, and anterior cingulate regions. This differed from other DAT patients without frontal lobe behavior who showed diffuse cerebral decrements in glucose metabolism.

Huntington's Chorea

Huntington's chorea is a degenerative disease frequently presenting with personality and behavioral changes. Although cognitive skills remain grossly intact early in the disease, social and emotional behaviors show considerable deterioration. Impulsivity, emotional instability, impaired social skills, and explosive behavior are all frequently observed in these patients, who are often described as moody and aggressive. Insight is often retained early in the disease, and the resultant despair coupled with poor impulse control leads to a high rate of suicide in this population (Strub & Black, 1981).

Infectious Diseases

Personality changes often occur as a complication of infection in the CNS. For example, it is well documented that in the tertiary stage of syphillis, neurological sequelae appear that eventuate in a dementia typically referred to as general paresis. Common behavioral manifestations are those related to frontal lobe damage, including impulsivity, poor judgment, and limited insight.

Acquired immunodeficiency syndrome (AIDS) can affect cerebral functioning and lead to a change in behavior and personality. There are well-documented neurological and neuropsychological complications that have been noted as a common aspect of AIDS since the early reports of this disease. Psychological symptoms may in fact be observed before other neurological changes emerge (Perry & Jacobsen, 1986); in some cases, psychological symptoms may be the most prominent feature. Although such symptoms were initially thought to be an emotional response to a fatal disease, research has since demonstrated that the human T-cell lymphotropic virus-type III is toxic to cells of the CNS (Gallo, Salahuddin, & Popovic, 1984). Therefore, neuropsychological changes may well be attributable to cerebral lesions caused by the virus. The American Academy of Neurology (1991) has recently suggested guidelines for the diagnosis of dementia in AIDS that include behavioral symptoms. These guidelines incorporate changes in personality, including inappropriate behavior, disinhibition, affective lability, and impaired judgment.

Although depression, apathy, and withdrawal are all common psychological reactions to AIDS, some patients have been reported to experience maniclike episodes (Perry & Jacobsen, 1986). The following case illustrates this latter and somewhat atypical presentation.

Case 3. C. K. was a 36-year-old man who was infected with the human immunodeficiency virus for approximately 10 years. During that time, he experienced relatively few medical complications from the virus. He was referred for a neuropsychological evaluation because of his family's and friends' concern about his inappropriate behavior. The patient had 3 years of college education and managed his own business until 2 years before testing. During that time, he experienced a gradual progressive decline in cognitive functioning and memory in addition to impairment in motor skills. Most striking, however, was his behavioral change. Friends became concerned when he began to forage through neighborhood garbage cans and would bring home items that he found. His physician reported that during an office visit, the patient touched him in an overly familiar and inappropriate fashion. When seen for a neuropsychological evaluation, he presented as a jocular, euphoric individual who displayed both verbal and physical signs of dyscontrol. He remained mildly euphoric throughout and tended to joke, sing, make noises, and laugh unless involved in a task. At one point, he asked to use the restroom and began to unfasten his trousers while walking down the hall. He demonstrated no concern about his behavior. The results of neuropsychological testing revealed a WAIS–R Full Scale score of 81, a Verbal scale score of 79, and a Performance scale score of 84. Regarding memory, his WMS–R General Memory scale score was 51, indicating severe impairment.

Metabolic Disorders

Wilson's disease is a rare metabolic disorder caused by an inborn error of copper metabolism that affects the liver and CNS functioning. It has been noted that psychiatric and behavioral changes are common features and that personality changes may occur early in the course of the disease. It is estimated that approximately 25% (Strub & Black, 1981) of Wilson's disease patients present initially with behavioral abnormalities. In most cases, these changes consist of uninhibited social behavior, impulsivity, and emotional lability. These patients tend to be immature and emotionally labile, which may lead to outbursts of rage, destructive acts, or antisocial conduct (Lishman, 1987).

Hepatic encephalopathy, also known as portal–systemic encephalopathy, may present with striking personality changes, including impulsivity. Neuropsychological symptoms may dominate the clinical picture (Lishman, 1987; Summerskill, Davidson, Sherlock, & Steiner, 1956), and such changes may be evident long before the disease has been diagnosed. Although the exact symptomatology of the psychological presentation may vary, common elements include uninhibited behavior and exaggerated emotional expression, including irritability and joviality. Lishman (1987) described a characteristic personality style exhibited by those with hepatic disease that is similar to a frontal lobe disorder; the features include jocularity, tactlessness, poor social judgment, and impulsivity.

Miscellaneous

Vascular lesions, surgical resections, and neoplasms all may lead to personality changes, including impulsivity, particularly if the frontal lobes are involved. It has been estimated that between 60%–90% of patients with frontal lobe tumors have psychiatric symptoms (Direkze, Bayliss, & Cutting, 1971). Tumors involving the orbitofrontal cortex are associated with impulsivity, disinhibition, and heightened irritability.

Neurosurgical resection of the frontal lobes may also produce prominent alterations in behavior and personality, with disordered impulse control as a major feature. These symptoms may follow either unilateral or bilateral lobectomy.

In the case of vascular disorders, disruptions of cerebral anterior vessels in particular may result in impulsivity and other behaviors resembling a frontal lobe syndrome (Bigler, 1988). Along these lines, Steinman and Bigler (1986) described 7 patients who had experienced anterior communicating artery aneurysms; they maintained generally intact intellectual functioning but had marked behavioral and personality changes, including impulsivity. The following case illustrates this process.

> *Case 4.* The patient was a 49-year-old man who underwent a left frontal lobectomy to remove a subarrachnoid hemorrhage. A marked deterioration in behavior was noted following surgery, including both verbal and physical aggression. Although a tendency toward aggressive behavior was noted in his premorbid personality, such inclinations were now greatly exaggerated. His affective expression was labile and embellished, and low frustration tolerance often led him to shout obscenities in response to minor annoyances. At other times he became physically assaultive, and during the neuropsychological evaluation he struck the examiner when frustrated with a challenging task. On formal neuropsychological examination, he obtained a WAIS–R Full Scale score of 70. Memory was severely impaired with a profound retrograde amnesia and new learning deficits. Although he clearly had considerable cognitive deficits, much of his impairment was related to severe impulsivity and distractibility, which interfered with his ability to attend adequately to tasks. Despite his cognitive limitations, it was the behavioral aspects of his presentation that necessitated a specialized neurobehavioral management program with residential placement.

Organic Personality Syndromes

Organic personality syndromes produce a pathological change in personality and behavior that has an organic basis. These disorders are distinguished from other neurological causes of behavioral dysfunction discussed thus far because of the absence of significant cognitive impairment. In the organic personality disorders, disruption of behavior and a change in personality are the primary features. The etiology may be caused by almost any disease or injury that affects brain functioning, particularly the frontal and temporal lobes. The most common causes are head injury, neoplasms, temporal seizure disorder, vascular diseases, and demyelinating diseases such as multiple sclerosis (Stoudemire, 1987).

One form of organic personality disorder leading to a disruption in impulse control has been referred to as the pseudopsychopathic personality disorder, described earlier. This disorder is characterized by emotional lability, impulsivity, socially inappropriate behavior, and hostility. These individuals display an indifference for the consequences of their behavior and a failure to appreciate the impact of their actions on others. The most common cause of this disorder is a closed-head injury.

The episodic dyscontrol syndrome is a subtype of organic personality disorder characterized by sudden, destructive outbursts of aggression that may last from 15 minutes to 2 hours. In addition to physical assaultiveness, sexual impulsivity may be a feature that can occasionally be violent (Strub & Black, 1981). Although it is believed that there is a range of etiologies that may produce this disorder, damage to the temporal limbic structures has been identified as one common cause (Strub & Black, 1981). Predisposing factors include a history of hyperactivity and early CNS damage.

CONCLUSION AND TREATMENT STRATEGIES

There are a vast number of neurological disorders with implications for impulse and behavior control. Frequently, personality and behavioral change may be the most prominent aspect of the individual's presentation and may even be the initial symptom. An understanding of the impact of neurological disease on personality and behavior is critical for both the accurate diagnosis and appropriate management of the impulsive and neurobehaviorally disordered individual. A comprehensive theory of impulsivity should ideally account for both organically based impulsivity and impulsivity as a personality characteristic, inasmuch as differing etiologies yield strikingly similar behavioral patterns. This is likely to be one of the many exciting theoretical and research areas in impulsivity in future years.

Treatment efforts naturally begin by attending to any underlying medical disorder or condition producing the personality and behavioral changes. Once this has been accomplished, or if this is not feasible, efforts should then be directed at assisting the individual in recovering as much psychological functioning and autonomy as possible. Pharmacological interventions aimed at symptom management have been found to be of benefit in some cases, particularly when symptoms involve agitation, psychomotoric restlessness, excessive anxiety, or sleeplessness. For behavioral disturbances, psychological interventions may be particularly useful.

Because of the individual's typical lack of insight and awareness, psychologically based interventions often include environmental management and behavioral strategies. However, for some less severely affected individuals, counseling may be used to help raise awareness to the negative impact of their behavior and to develop methods of self-regulation and control.

Specific behavioral procedures that are based on principles of operant conditioning theory have been successfully used to control and reduce impulsivity. Such strategies identify and then target the problematic symptom, the conditions under which it occurs, and the consequent events. The behavior in question is then modified by altering either precipitants or consequences. For example, treatment may begin initially by focusing on unacceptable behavior in a controlled situation. Reinforcement and reward are liberally used, which may include an immediate reward for inhibiting responses. The length of time between the behavior and reward is then gradually lengthened. In conjunction with such techniques, the therapist should also maximize the individual's cognitive capacity for self-control (e.g., self-statement modification) to promote a sense of self-efficacy regarding the ability to delay response or inhibit impulses. Once some amount of self-control is obtained in the training session, efforts are then made to assist the individual in transferring this learning to other situations. A second method of intervention—behavioral contracts—has also been successfully used in some situations. For example, the patient who makes constant demands may be taught to limit his or her request to specific times of the day. Requests made at nondesignated times are not reinforced and are then eventually extinguished. By using this method, some patients are able to learn through constant reinforcement and reminders that certain behaviors and requests will be honored only at specific times.

In addition to specific techniques designed to modify certain behaviors, family involvement is critical. Treatment personnel should provide family members with specific information regarding the patient's symptoms and behavior, along with advice on how to manage inappropriate behaviors. If a behavioral contract or other behavior methods are used, family members and caregivers need to understand the contingencies in order to respond appropriately and consistently to the targeted behavior. Beyond this, family members and caregivers are typically required to make life-style adjustments to accommodate changes in the patient's functioning. Such changes are often permanent, and it is frequently frustrating and stressful for those who must live and interact with the neurobehaviorally disordered individual. Family therapy and individual counseling are important components in assisting family members to understand the prognosis for the patient as well as the nature of his or her symptoms. Too often, those involved with the behaviorally disordered individual do not understand the changes in brain functioning that produce the symptoms and ascribe motivational factors to account for inappropriate behavior. This often proves destructive, as well-meaning friends or relatives may assume that the patient could change his or her behavior if he or she wanted to; this assumption can then lead to anger or punishment by the friends and family. Such responses are seldom helpful and only serve to increase the overall level of stress for all concerned. Counseling or support-group involvement may also be helpful in assisting caregivers to manage their own emotional reactions to what is often a stressful situation. Treatment efforts should thus be directed at both the family and the disordered individual.

REFERENCES

American Academy of Neurology: Practice handbook. (1991). Minneapolis, MN: Author.

Bigler, E. (1988). *Diagnostic clinical neuropsychology.* Austin: University of Texas Press.

Blumer, D., & Benson, D. F. (1975). Personality changes with frontal and temporal lobe lesions. In D. F. Benson & D. Blumer (Eds.), *Psychiatric aspects of neurological disease* (pp. 151–169). New York: Grune & Stratton.

Bond, M. (1984). The psychiatry of closed head injury. In N. Brooks (Ed.), *Closed head injury: Psychological, social, and family consequences* (pp. 150–178). New York: Oxford University Press.

Cummings, J. (1985). *Clinical neuropsychiatry.* New York: Grune & Stratton.

Direkze, M., Bayliss, S. G., & Cutting, J. C. (1971). Primary tumors of the frontal lobe. *British Journal of Clinical Practice, 25,* 207–213.

Elliott, F. A. (1986). The episodic dyscontrol syndrome and aggression. *Neurology Clinics, 2,* 113–125.

Gallo, R. C., Salahuddin, S. Z., & Popovic, M. (1984). Frequent detection and isolation of cytopathic retroviruses (HTLV-III) from patients with AIDS. *Science, 224,* 500–503.

Harlow, J. M. (1848). Passage of an iron rod through the head. *Boston MS Journal, 34,* 389–393.

Harlow, J. M. (1868). Recovery from the passage of an iron bar through the head. *Publication of the Massachusetts Medical Society, 2,* 327–347.

Jarvie, N. F. (1954). Frontal lobe wounds causing disinhibition: A study of six cases. *Journal of Neurology, Neurosurgery and Psychiatry, 17,* 14–32.

Kumar, A., Schapiro, M. B., Haxby, J. V., Grady, C. L., & Friedland, R. P. (1990). Cerebral metabolic and cognitive studies in dementia with frontal lobe behavioral features. *Journal of Psychiatric Research, 24,* 97–109.

Lipowski, Z. J. (1978). Organic brain syndromes: A reformulation. *Comprehensive Psychiatry, 19,* 309–321.

Lishman, W. A. (1987). *Organic psychiatry: The psychological consequences of cerebral disorder.* Oxford, England: Blackwell Scientific.

McAllister, T. W. & Price, R. P. (1987). Aspects of the behavior of psychiatric inpatients with frontal lobe damage: Some implications for diagnosis and treatment. *Comprehensive Psychiatry, 28,* 14–21.

Neary, D., Snowden, J. S., Northern, B., & Goulding, P. (1988). Dementia of frontal lobe type. *Journal of Neurology, Neurosurgery and Psychiatry, 51,* 353–361.

Perry, S., & Jacobsen, P. (1986). Neuropsychiatric manifestations of AIDS-spectrum disorders. *Hospital Community Psychiatry, 37,* 135–142.

Steinman, D. R., & Bigler, E. D. (1986). Neuropsychological sequelae of ruptured anterior communicating artery aneurysm. *International Journal of Clinical Neuropsychology, 8,* 135–140.

Stoudemire, G. A. (1987). Selected organic mental disorders. In R. E. Hales & S. C. Yudofsky (Eds.), *Textbook of neuropsychiatry* (pp. 125–140). Washington, DC: American Psychiatric Press.

Strub, R. L., & Black, F. W. (1981). *Neurobehavioral disorders: A clinical approach.* Philadelphia: F. A. Davis.

Summerskill, W. H. J., Davidson, E. A., Sherlock, S., & Steiner, R. E. (1956). The neuropsychiatric syndrome associated with hepatic cirrhosis and an extensive portal collateral circulation. *Quarterly Journal of Medicine, 25,* 245–266.

Wechsler, D. (1981). *WAIS-R manual.* New York: Psychological Corporation.

Wechsler, D. (1987). *WMS-R: Wechsler Memory Scale—Revised manual.* New York: Psychological Corporation.

Woodcock, J. H. (1986). A neuropsychiatric approach to impulse disorders. In R. M. Restak (Ed.), *The psychiatric clinics of North America* (pp. 341–352). Philadelphia: Saunders.

16

MANAGEMENT OF THE ADULT IMPULSIVE CLIENT: IDENTIFICATION, TIMING, AND METHODS OF TREATMENT

MICHAEL BÜTZ and SEAN AUSTIN

Impulsiveness implies unplanned behavior that usually results in undesirable consequences (Barratt & Patton, 1983). As numerous authors have noted (e.g., Daruna & Barnes, chapter 2, this volume; H. J. Eysenck, chapter 4, this volume; Zuckerman, 1991), impulsiveness has both biological and social–cognitive components that likely represent a complex interaction between genetic and environmental factors. Impulsive behavior is a feature in many clinical syndromes, as indicated in the revised third edition of the *Diagnostic and Statistical Manual of Mental Disorders* (*DSM–III–R*; American Psychiatric Association, 1987) and numerous other studies (e.g., McCown, 1993; Romney & Bynner, 1992; Rutter, 1987). Because impulsive behavior is a central feature of many psychiatric syndromes, issues regarding appropriate psychological assessment and treatment of impulsiveness are critical.

In this chapter we discuss clinical strategies for assessing and treating impulsive behavior in adults. Because impulsiveness is a trait that is presumably relatively stable throughout the life span (Zuckerman, chapter 5, this volume), we have chosen to speak of "managing" impulsiveness rather than successfully "curing" it. This is an important concept for therapists to grasp and one that is antithetical to the medical model training of many practitioners. Regardless of the purported causation a theorist ascribes to, substantial data suggest that impulsivity as a personality trait manifests itself early in life and is very difficult to change (see H. J. Eysenck, chapter 4, this volume). More specifically, we could not find any studies in the psychological literature that show that trait measures of impulsiveness in adults can be substantially altered through psychotherapy or other psychological procedures. Our strategy is not to cure impulsive behavior, because this is presently impossible, but instead to change the expression of impulsiveness so that impulsive individuals may learn more appropriate and less self-damaging behaviors.

In this chapter we draw primarily on our clinical experience. There are two polarities of thought regarding expertise derived from such experience. One extreme discounts clinical experience completely, believing it represents a "pseudoknowledge." The other extreme exalts clinical experience over and above findings from well-designed and relevant empirical studies because clinical experience is somehow seen as "truer" than the cold and often abstract findings of controlled research. Our position is somewhat in the middle of this controversy. We believe that clinical experience is never an optimal substitute for empirically derived and externally valid findings. However, in the absence of such studies, clinical experience is a useful heuristic and may be all the practitioner has to rely on when the scientific literature is sparse or nonexistent.

ASSESSMENT OF IMPULSIVITY IN ADULT CLIENTS

Our experiences as clinicians and supervisors of clinician trainees suggest that the first mistake therapists typically make when treating the impulsive client is in failing to assess the severity and consistency of impulsive behaviors. Impulsiveness may be confined to a narrow range of environmental cues (Mischel, 1983) or it may be a broad, encompassing factor prevalent across a variety of behavioral situations (H. Eysenck & Eysenck, 1985). The clinical importance of recognizing impulsive behaviors depends, in large part, on the setting in which the practitioner is functioning. In the absence of environmental controls, it is valuable for the therapist and client to learn to anticipate or predict impulsive behavior. In an inpatient setting, this is somewhat less of an issue inasmuch as behavior is largely controlled through environmental management. However, in such settings, longer term therapeutic goals may include a greater focus on managing impulsivity.

Assessing impulsiveness in the "typical" client is a challenging task. A wealth of social–cognitive literature indicates that failure to inhibit impulsive behaviors can occur in a discrete and isolated manner and that instances of impulsive behavior do not necessarily presage evidence of a global personality factor (Mischel, 1961, 1986). Furthermore, as Zuckerman (chapter 5, this volume) indicates, impulsiveness can be either a first- or second-order personality trait. In other words, it may be either a broad encompassing factor with several subordinate constructs or a more narrow and specific personality characteristic, such as "academic impulsiveness" or "sexual impulsiveness." The *DSM–III–R* and the upcoming *DSM–IV* assume an even more limited view because these publications specify that impulsive behaviors are discrete and diagnosable syndromes. This is the case despite the fact there appear to be common interrelated factors in a number of personality disorders and discrete behaviors (Romney & Bynner, 1992). Impulsiveness may also be associated with certain acute states, such as when an angry or inebriated person does things he or she would not even think of doing otherwise (Johnson & McCown, in press). Thus, impulsiveness may not be identified as a problem, even if intermittent impulsive behaviors constitute significant sources of difficulty.

Methods of assessing impulsiveness and theories regarding impulsive behaviors are discussed throughout this volume. Regardless of the theoretical orientation regarding causation (and it is certain the causes of impulsiveness are multifactorial), it is important to recognize from the client's history the degree to which impulsivity is present within the client's behavioral repertoire. The reader should note that we did not state that impulsivity is "in" the client. Impulsive behavior is obviously something that is not "in" anyone but instead is a behavioral expression of past learning and genetics interacting with environmental and perceived environmental contingencies. The reader should also note that we did not use the phrase *diagnose impulsiveness* because such a phrase involves an arbitrary binary distinction for what is a somewhat normally distributed and dimensional phenomenon (see Barratt, chapter 3, this volume). Diagnosing psychiatric disorders with impulsive features from the checklist provided by the *DSM–III–R* is relatively easy. On the other hand, assessing the clinical and global significance of impulsiveness, although more useful, is often quite challenging.

There are several reasons for this, some of which we discuss shortly. One that warrants initial attention and discussion here is the fact that many people are dissuaded from attempting a psychological assessment of impulsive individuals because they suspect that impulsive clients will not remain in treatment long enough for a thorough evaluation. Second, impulsive individuals are more likely to have problems with substance abuse, antisocial behavior including violence to self and others, and other difficulties that may interfere with the accurate and comprehensive collection of information (see Johnson, Malow, Corrigan, & West, chapter 11, this volume, for a further discussion of this issue). Put simply, they may be either consciously or unconsciously deceptive, a validity problem that plagues both clinical interviews and psychometric tests (Hartmann, Roper, & Bradford, 1979).

Nonetheless, adequate assessment of problems associated with impulsiveness, although difficult, is critical for comprehensive treatment planning. Merely focusing on impulsivity as a clinical problem in and of itself may reduce the time involved in identifying therapeutic goals. This, logically, reduces the time spent in psychotherapy, or at least directs it toward more meaningful productivity. Perhaps as important from a therapeutic standpoint is that prediction of impulsive behaviors by the therapist may also help to convince the client of the legitimacy of the therapist and therapeutic methods. The importance of this last function should not be underestimated because it may be a potential method of keeping clients with multiple problems in therapy. Along these lines, McCown and Johnson (1993), in their book on resistant and impulsive families, suggested that "inoculation" of families to the idea they are at high risk for termination makes them less likely to abruptly terminate treatment. Moreover, helping individuals or families recognize that they may have a sudden desire to terminate therapy can itself be therapeutic inasmuch as frustrations and disillusionment commonly occurring in the early stages of the therapy can be anticipated within a framework that fosters anticipatory control of impulsive behavior such as premature termination.

Identifying Situational Specificity of Impulsive Behaviors

Even when individuals possess a high degree of trait impulsiveness, there is often a preference for particular circumstances for expression of impulsive behavior. In other words, highly impulsive people are not always impulsive in every situation. As Zuckerman (chapter 5, this volume) indicates, even individuals who are extraordinarily high in the trait of sensation seeking can become conscientious and even meticulous in preparing for certain experiences or when exposed to specific environmental situations. For example, mountain climbers or parachutists who score very high on measures of sensation seeking or disinhibition are typically quite conscientious regarding preparation for and implementation of these thrill-inducing activities.

Clinical experience, as well as a wealth of social learning theory-based literature, suggests that trait impulsivity may be even more selective (i.e., narrower) than is typically realized by trait-oriented clinicians. Although there is generally a low-to-moderate correlation between specific indexes of impulsiveness in different domains (at work, at school, with the family, in relationships, etc.), an understanding of measurement and multivariate personality theory indicates that people are capable of behaving in a highly impulsive manner in one or two areas of their lives, with simultaneous adequate functioning in other areas. Generally, trait measures of impulsiveness will fail to detect the specific domain of impulsive behavior that is affecting the client. To attain this degree of specificity, the use of broad measures of impulsiveness as well as highly tailored clinical inquiry is required.

Insensitivity of Clinicians to Impulsive Behaviors

An error that clinicians frequently make is assuming that impulsiveness will be evident from information obtained solely during the clinical interview. Impulsivity, whether confined to one specific setting or broad in its occurrence, is often far from obvious. Drogin and Drogin (chapter 17, this volume) reiterate that this is often the case for even experienced clinicians. Moreover, specific problems associated with impulsive behaviors may be even less manifest, particularly during the early phases of the therapy process. A careful case history may help to delineate problems with impulsiveness, providing that the client is forthcoming and capable of self-monitoring and insight. However, even well-intended and seasoned therapists may often fail to grasp the seriousness of impulse disorders in a client's presenting clinical picture.

There are numerous reasons why impulsive people can be somewhat less than straightforward during the early stages of psychotherapy. Impulsive people with an overlay of antisocial tendencies may simply confabulate, and this phenomenon may be particularly evident in substance abuse disorders (Forrest, 1992). Because learning to be honest is an important aspect of therapy for any disorder (Wallen, 1992), discrepancies between the client's verbal report and self-report instruments can be a useful starting point for therapeutic intervention.

Another reason it may be difficult to clinically ascertain impulsivity problems is that impulsive people may either consciously minimize the seriousness of their problems or lack awareness regarding their severity. It may be that much of the apparent denial associated with impulse control disorders (e.g., alcoholism) is related more to the unique thought processes of impulsive individuals than to evident specific dynamics of the clinical syndrome. In other words, denial is not limited to the diseases of alcoholism or drug abuse but is related to a variety of impulsive behaviors.

Although little clinical data exist regarding this issue, we do have a wealth of useful clinical experience regarding the elicitation of sensitive therapeutic material from impulsive clients in treatment. Clinical interviews are usually not sufficient and clearly do not supplant the advantages of objective psychological testing. Clients are often defensive about impulsive behaviors, especially during the early stages of therapy. Clients who are depressed, introverted, preoccupied, or neurotic may confabulate or even suppress any history of impulsive behaviors. Finally, there is the well-known problem of clinical bias. A host of factors distort both clinical judgment and the history obtained from the client (Lanyon & Goodstein, 1982). With these problems in mind, we strongly recommend the use of instrument-based psychological assessment, a position long advocated in the psychological literature (e.g., Meehl, 1960).

We found only one study that examined the ability of clinicians to adequately predict trait impulsiveness in their clients. Austin and McCown (in press) had 50 clinicians rate one client each on a version of S. B. G. Eysenck, Pearson, Easting, and Allsopp's (1985) I7 Questionnaire. Clients were administered the questionnaire prior to the first session, whereas clinicians completed a version designed to be answered in the manner in which the client would respond. Clinicians completed the questionnaires on three occasions: the first, third, and fifth therapy sessions. The correlations between the client responses and the therapist responses were .07, .15, and .22, respectively, none of which approached significance. Even correcting for this quasi-attenuation, none of these therapist–client relationships were significant. As a comparison, Austin and McCown found that the correlation between a *friend's* assessment and a target person's was .61, a highly significant relationship.

Findings such as these have suggested to us that the brief effort required for screening for elevated levels of trait impulsiveness are well worth the small effort. Clinicians simply are not as good at identifying this trait as they think. Although we are somewhat hesitant to suggest formal psychometric screening for every outpatient client entering treatment, it is much better to err on the side of "overtesting" than the reverse. The following case study illustrates the history of a client whose impulsive problems were not adequately assessed in the initial stages of therapy. The therapist attempted to rely on clinical impressions alone and was unsuccessful in treatment.

> Katy, a 39-year-old attorney, entered psychotherapy because of "problems with relationships." She had difficulty being more specific, except to say that in the past she had a history of numerous unsatisfactory relationships with heterosexual partners that left her feeling "empty and cynical." She also stated that she was depressed and wondering "if maybe I'm simply asexual or a lesbian. I just don't seem to get along with men." Katy stated that she had been married once, "right after college," to a career military officer she met while she was a junior. Although the couple remained friendly, the marriage

was brief. "It lasted about 2 years. He had problems with me wanting to go to law school. His and my view of what a good wife was kind of conflicted."

A clinical interview performed by an experienced clinician elicited a paternal history of alcoholism. Regarding her own substance use history, Katy stated, "Well, in college everybody smoked pot, so I was one of the crowd . . . it was never a problem though." In law school "the thing to do was cocaine. Even some professors did it with us. . . . Everyone just thought it was harmless." Katy reported stopping cocaine "when people began to find out it was fairly dangerous." Regarding bulimia, "sometimes I overeat when I'm stressed or down. I put some weight on when I quit smoking a couple of years ago. But I don't purge or anything."

Katy's history of psychotherapy was seemingly noncontributory for formulating a treatment plan. "I tried the college counseling center once or twice, to find out what I should do with my life. . . . A couple of times I wanted to find a therapist but I had an HMO [health maintenance organization insurance policy] and there was a big waiting list. I went to therapy for a few weeks when I lived in _____ [her second job after law school] but had to quit because I didn't have the time."

Psychotherapy with Katy floundered for more than 6 months. Gradually she became more comfortable discussing her depression and "pain" and eventually made the link between her depressive thoughts and a history of having been verbally abused by both her father and her older brother. However, insight seemed to have little effect on her functioning, and she continued to involve herself in a series of disastrous relationships to "numb" herself out.

The therapist next suggested a trial dose of a tricyclic antidepressant, which did not help. The client was then given monoamine oxidase inhibitor, to no avail. Seven months into treatment, the therapist decided that psychological testing might be of use and sent the client to a psychologist specializing in assessment.

Although Katy's Minnesota Muliphasic Personality Inventory-2 (MMPI-2) profile was unusual, an analysis immediately illuminated some of Katy's difficulties. Katy was indeed an impulsive, but introverted, and depressed woman with extraordinarily high defenses. She sought to minimize her problems and generally was capable of maintaining appropriate control. She was reluctant to admit to any distress at all, which might suggest a tendency to minimize symptoms of behavior already presented, yet this was not particularly salient to the clinician involved. Also, analysis of subscales using the procedures recommended by Butcher (1991) indicated a probable history of substance abuse problems. As a result of clinical reliance on interviewing in the absence of objective psychological testing, these difficulties were not fully identified, and Katy essentially wasted a good deal of time, effort, and money in psychotherapy.

Rapport: A Problem in Assessment

Another problem regarding the assessment of impulsive behaviors is that clinical interviews assessing such behavior are frequently alienating to the client. Interviewing a client about drug usage, sexual behavior, gambling, excessive eating, antisocial activities, and other actions that are generally considered "vices" may seem like routine or normal questioning to the clinician who repeats them daily. To the client, however, such questions may be threatening, and even shocking, and may also damage the client–therapist rapport. This is especially true for minority clients, lower socioeconomic status groups, depressed individuals, or any person who is apt to feel that people in authority are judgmental of their behaviors. A key problem for the clinician is how to inquire regarding difficult, sensitive areas and still maintain rapport with the client.

The clinician often deals with this apprehension regarding loss of rapport or client alienation simply by avoiding these questions. This commonly occurs with trainees but is also common among people who may be culturally incongruent with the client in question. For example, male therapists may feel that rapport would be compromised if they asked their female clients about sexual behaviors. Majority culture members may feel that it is racist or stereotypic to question minority individuals about substance abuse problems (Butler, 1992). Similarly, people of both sexes may feel uncomfortable interviewing those with different sexual orientations.

Clinicians who inquire about impulsive behavior and feel uncomfortable doing so often develop a number of personal distancing strategies. A common one occurs when the therapist develops a "speech" discounting the importance of this information. It often begins along the lines of, "Well, I really don't think this applies to you, but I have to ask these questions of everyone." When this occurs, it is not surprising that the clinical interview produces a paucity of useful data. Often, such clinicians are unpleasantly surprised when 2 or 3 months later critical information regarding impulsive behaviors "suddenly" becomes salient to the therapist.

Psychometric Assessment of Impulsiveness: Avoiding the Problems of a Clinical Interview

Formal psychological testing will probably never supplant an interview by a trained clinician. Regardless, it avoids many of the problems encountered in a purely clinical assessment. There is extensive literature—which we do not summarize here—that suggests that objectively scored psychological tests have more validity and reliability than clinical interviews. Given the overwhelming lack of accuracy in assessing impulsiveness found by Austin and McCown (in press), the necessity of objective psychological testing becomes clear. When there is any history at all suggestive of problems with impulsiveness, it is our opinion that the clinician should use a psychometrically valid instrument for assessment. Furthermore, if the clinician finds herself or himself unable to talk about key areas with the client because of resistance or sensitivity, a more accurate portrait of the client's functioning will be obtained by obtaining objective test data.

But which psychological test should the clinician use? Fink and McCown (chapter 14, this volume) outline nine classes of instruments used to assess impulsiveness in children. These authors argue that clinicians need to use a multimethod assessment strategy in determining childhood impulsive behavior. These same classes of instruments, along with the clinical interview, are also used with adults. However, the most widely used instrument to assess adult impulsiveness is self-report, and there is less controversy in the literature regarding the adequacy of self-report instruments with adults than with children. In our experience, a self-report measure is an appropriate instrument to use to detect global impulsive behaviors. Evidence obtained from its use may then provide the basis for further inquiry regarding exactly which spheres of client functioning are impaired by impulsive behaviors. This information may be obtained from either more specialized questionnaires or more specific clinical interviewing.

Our preference for self-report remains the MMPI, especially in its revised edition (Butcher, 1991). The MMPI provides an enormous amount of relevant data, is inexpensive, and is relatively easy to administer. Compared with the information developed from either shorter omnibus inventories or longer structured interviews, the MMPI represents a reasonable compromise between inclusiveness and diagnostic accuracy. The use of "critical items," statements that are atypically endorsed by the general population, may also serve as an excellent beginning point for a clinical interview. Furthermore, the MMPI may be especially useful in suggesting treatment strategies for people with impulsive problems (Svanum & Ehrmann, 1992). Only when time, ability, or rapport is critically limited does it seem warranted to use a less inclusive psychometric scale.

Our emphasis on objective psychological testing does not discount the importance of a good clinical interview, especially one that uses information obtained from critical items. Although information gained from the clinical interview may often be incomplete, the client's presenting problem, history, and self-disclosure should not be discounted as potential useful starting points for forming hypotheses regarding client characteristics. Clients frequently acknowledge impulse-related problems but provide incomplete descriptions or display limited insight into the ramifications of their impulsive behavior. Such information should never be disregarded. Any type of information possessed by the clinician should be used in a hypothetical–deductive manner to furnish a behavioral and dynamic portrait of the client's interpersonal or adjustment problems. That is, when a client acknowledges difficulties with impulsive behavior, the next clinical challenge becomes a hypothetical–deductive exercise in determining *where* impulsiveness is problematic. Similarly, if the client's history suggests a pattern of impulsive behaviors, then these data can be used as the beginning of hypothesis testing regarding the client's behavior and dynamics.

A helpful strategy that we often deploy is to use MMPI data obtained after a first clinical session to form the basis of the next interview. During the second interview, the clinician, armed with data from the MMPI, is able to gently assist the client in identifying specific behaviors that are likely to be problematic for impulsive individuals. For example, MMPI data can help the clinician to determine whether the client has a high likelihood of having experienced problems with family, substance abuse, authority, work, or sexuality. Once objective psychological data have been collected, it is usually much easier for impulsive clients to reveal information about specific areas of their lives where their impulsiveness is problematic.

TREATMENT CONSIDERATIONS

We set out to write a chapter that would discuss many of the common problems clinicians often face when attempting therapy with impulsive adults. It is our opinion that the specific types of treatment administered (e.g., behavioral, cognitive–behavioral, psychodynamic, etc.) are less important than are pragmatic concerns. These include topics such as how to schedule the recalcitrant client, how to bill impulsive clients for services, and how to manage countertransferential feelings of resentment and anger that seem to be so often encountered in treating such clients. Unfortunately, many of these topics are rarely discussed in psychotherapy textbooks or graduate school courses, and it is almost as if competent therapists are not troubled by these pragmatic concerns. Therefore, in the remainder of this chapter we will focus on these common problems that are typically encountered and offer suggestions for the clinician in handling them.

Scheduling

Scheduling is one of the most important problems a therapist who treats impulsive clients faces. It is also one of the first problems encountered. The first indicator that an individual may behave impulsively often occurs when the prospective client calls and requests to be seen immediately. Sometimes, these individuals will cajole or plead with the therapist to be seen at once, despite the fact the therapist's schedule may be completely full. Motivation seems genuine and the clinician is often placed in the uncomfortable position regarding changing his or her schedule in order to facilitate an "emergency" first session. Too often, such therapists feel "burned" when the client fails to show for the first session. If therapy ever does begin, the therapist may enter the relationship with some sense of being manipulated or other countertransference issues.

Guidelines regarding whether an individual or family should be immediately seen are not as straightforward as most clinicians would like, and clinical judgment is important. Suicidal or homicidal clients obviously should be seen as soon as possible, as should people with a history of impulsive behavior that is irrevocable in its consequences (McCown & Johnson, 1993). However, many clinicians note that individuals who demand to be seen immediately are also often the least likely to show for the initial therapy session and hence may precipitate the most frustration for the therapist.

In general, therapists attempting to treat impulsive clients and families should prepare themselves for a substantial crisis or emergency component in their work (McCown & Johnson, 1993). Additionally, such clients often call inappropriately and need to be "educated" about the rules of therapy. For example, impulsive people may call late in the evening to explain why they missed a therapy session several days ago. Questions regarding the unconscious motivation of such behaviors are interesting. However, interpretation of unconscious motives is usually not as useful as straightforward behavioral interventions designed to reduce the frequency of failures to show for scheduled appointments.

We have also found it useful to take an aggressive stance when scheduling impulsive clients, calling them two or three times before each appointment to confirm their attendance. "No shows" should be followed up by direct and immediate telephone contact, with a respectful yet factual tone and content.

McCown (1993) found that impulsiveness was the single most important predictor of early treatment termination, far outweighing the effects of demographic variables or personality factors such as conscientiousness. Part of the challenge of working with multi-impulsive clients and families is tempered by the fact they frequently terminate treatment early, abruptly do not show up for sessions, and generally conduct themselves as chaotically in therapy as they do in other aspects of their lives. Nonpunitive but consistent structure may be the most productive intervention with such clients.

Regarding impulsively placed, between-session calls, our clinical experience suggests that it is often helpful to politely but realistically limit such calls. In all but cases of borderline personality disorder, it is usually possible to set up rules before there is a problem and refer to them in later sessions. Of course, this should be tailored to individual circumstances of the case. For example, rule setting should be done cautiously with borderline personalities who may be hypersensitive to rejection.

Another question regarding scheduling is how frequently impulsive clients should be seen. To our knowledge, there are no data that suggest that it is helpful to see outpatient clients more frequently than the standard practice of one time per week, unless they are in crisis (discussed shortly). We can rarely think of an indication for seeing these clients on a more frequent basis, unless the modality of treatment requires such timing, such as in psychoanalytic therapy. More frequent intervention may increase dependency and actually decrease the amount of forethought and work accomplished by the client between sessions.

Fees

A major source of countertransference issues with impulsive clients regards payment of fees. Although financial matters are difficult for most therapists to discuss (Herron & Welt, 1992), they may pose an even greater challenge when treating impulsive clients. Impulsive people may have financial difficulties associated with impulsive spending or less-than-satisfactory work histories. They may also place a low priority on responsible behavior, such as paying debts on time. There is nothing more exasperating for the therapist than working intensively with a difficult, impulsive client who subsequently fails to meet his or her financial obligations for services received. Missed payments can strain a tenuous therapeutic relationship very quickly and may also be an excuse for the client to justify abrupt termination from treatment. It is not uncommon for impulsive clients who have failed to pay fees to drop out of treatment because they are embarrassed about their debt or too anxious to discuss it with their therapist.

A therapeutic approach emphasizing a time-limited series of intermediate goals (i.e., four therapy sessions with evaluation of effectiveness throughout) may make treatment with these patients more practical than an immediate commitment to longer term therapy. It can be reiterated to the client that such an arrangement may help both the therapist and the client decide whether treatment is mutually "workable." Behavioral patterns that clearly interfere with future therapy, such as missing sessions or showing up intoxicated, can be pointed out to the client and may be appropriate grounds for determination that treatment between the two parties would not be productive. The suggestion to such clients can then be referral to another treatment provider or advice that they begin therapy only when they are more prepared to accept the limitations demanded of them. Furthermore, a useful stipulation can be that the client make responsible provisions for fee payment. If fees that the client can afford are not paid within a specific interval, the therapeutic contract can be renegotiated with the possibility of resumption when the client's financial behavior is more responsible.

A policy regarding fees should be consistent and articulated clearly in advance. This includes areas such as late cancellations, emergency sessions, between-session telephone calls, and extrasession case management. It is wise to discuss these policies directly and to give a signed copy of them to the client.

Finally, the therapist who works with impulsive clients must realize that he or she will necessarily "write off" a number of the accounts receivable from such clients. This is simply a fact of therapy and should not cause undue distress. One of the more gratifying moments a therapist can have is when an impulsive client, several years later after an apparently unsuccessful treatment, begins paying for these long-past services. These payments usually are accompanied by a brief note from the client stating how well he or she is doing and how helpful the therapy was in getting the client on the right track. When this happens, the therapist is often at a loss to imagine what he or she could have possibly done to effect the beginning of a change process that sometimes proves to be quite dramatic. It is best not to worry too much and to realize that people sometimes change for reasons that are not understood. The therapist should just be glad that he or she was credited for being part of the process.

Choices of Specific Treatments for Impulsive Clients

In the treatment of impulsive clients, we wish that more data regarding optimal therapeutic interventions were available for the practitioner's use. As we have stated repeatedly, although there is a wealth of clinical experience and lore, little empirical data exist regarding preferred methods for intervention into impulsivity. Again, we are forced primarily to rely on "knowledge"—perhaps some will say mere opinions—generated from our successes and our failures. We believe, however, that this experience may be useful for other practitioners and may be the basis for future empirical studies that validate our beliefs.

One point is hardly arguable: Impulsive people are more apt to be involved in psychiatric or psychological crises (McCown & Johnson, 1993). They are also more likely to be involved in risky or even dangerous behaviors, which sometimes are potentially life threatening. When a patient is in crisis, it is of the utmost importance that he or she receive appropriate services regardless of past history, payment of fees, scheduling problems, or other confounding problems. Clear triaging of problems to be treated should occur in instances of life-threatening behaviors, such as suicidality, high-risk sexual behavior, or excessive ingestion of alcohol or illicit substances. The most dangerous problems need to be addressed first. Once they are under control, other useful aspects of therapy may proceed, such as a formal assessment. The therapist who treats impulsive clients soon learns that it is necessary to have exceptionally good crisis intervention skills. It is also necessary to know when to apply these skills to prevent potentially harmful behavior from occurring.

It is, however, a mistake to believe that therapy with impulsive individuals should primarily consist of long-term management with brief periods of crisis intervention. It may also be helpful to include a treatment that can be administered relatively briefly that empowers the client to develop better self-control or tendencies toward deliberation. For example, a multi-impulsive individual with a phobia can benefit greatly from attention paid to this "minor" and readily treatable problem. This enhanced efficacy is often a sufficient impetus to remain in psychotherapy for work on "larger" issues, and small therapeutic victories with difficult clients may be highly rewarding for the therapist as well (Kottler, 1992).

We now briefly discuss therapeutic modalities that practitioners who treat impulsive clients frequently find helpful. The discussion is clearly not exhaustive and can probably be practically supplemented by most clinicians who work with this population.

Medication

Therapy with impulsive clients often includes medication, which is a useful option for hypomanic impulsive individuals or clients with attention-deficit hyperactivity disorder. However, to our knowledge, there have been no extensive controlled trials of medications such as lithium carbonate for impulsive people. In the experience of our group and our colleagues, referral to a psychiatrist for a trial of lithium management is frequently helpful, particularly when the client has a history of impulsive behavior that has a bizarre or completely reckless quality. Clearly, better and more extensive clinical trials are needed. In the meantime, it does little harm to refer to a cooperative physician consultant.

Some limited evidence exists that impulsive individuals, particularly those diagnosed with borderline personality disorder, may experience mild symptom alleviation with lithium carbonate (Links, Steiner, Boiago, & Irwin, 1990) or carbamazepine (Cowdry & Gardner, 1988). This is particularly true for symptoms of anger, heightened irritability, and suicidal ideation. However, borderline patients have a very high attrition rate when receiving psychopharmacological intervention. For example, in the aforementioned studies, less than 60% of the patients completed treatment. In our clinical experience, "whether to stay on meds" usually becomes the salient therapeutic issue for impulsive clients, thus detracting from other topics that could be more fruitfully addressed. One advantage of the present situation regarding medication administration (which psychologists cannot prescribe) is that it may reduce expressions of such concern in therapy because the clinician can simply advise the client to discuss these issues with his or her prescribing physician (Bütz, 1993).

Behavioral and Cognitive Techniques

Behavioral techniques are often effective for discrete impulsive disorders (O'Leary & Wilson, 1987). An excellent review of the variety of behavioral interventions is provided in Bellack (1990). For example, individuals who behave without thinking about their substance abuse or fetish objects may profit from desensitization toward cues associated with their impulsive behaviors regarding these reinforcers. Furthermore, and as discussed earlier, clinical accounts (but no controlled studies) suggest that success in one area may generalize to other areas of impulse reduction and greater client confidence (Wolpe, 1990). This is especially true when anxiety is a prominent facet of the impulsive client's symptomatology.

Cognitive techniques for reducing impulsive behaviors are well-known and documented (Craighead, Craighead, Kazdin, & Mahoney, 1993), although they have received greater empirical focus with children and adolescents than with adults (Dobson, 1988). We share the concern of Fink and McCown (chapter 14, this volume) regarding the possible inadequacy of cognitive techniques with trauma-based impulsive behaviors, although more extensive research is needed. However, the efficacy of cognitive techniques is a strong argument for their usefulness and integration into a variety of settings (Giles, 1993).

Recently, Linehan (1993a) has developed a structured cognitive–behavioral approach for the treatment of borderline personality disorders. An accompanying manual (Linehan, 1993b) provides specific skills training for impulsive people. This treatment methodology is supported by very good data and should be useful to the practitioner. A similar, briefer volume (Layden, Newman, Freeman, & Morse, 1993) is also illuminative and may be of greater interest to clinicians attempting to integrate cognitive behavioral and psychodynamic therapies. "Manual-based" interventions for specific impulsive problems may hold great promise in assisting therapists to maintain the necessary quality of care for clients who usually present with multiple and difficult problems.

Psychodynamic Techniques

Psychodynamic and psychoanalytic approaches are often viewed as the treatment of choice for impulsive individuals, regardless of specific diagnosis (Wolman, 1992). However, it should be noted that this assertion has been made in the absence of substantial research literature (Aronson, 1989). Clearly, there is a large body of "clinical evidence" suggesting that psychodynamic techniques are appropriate for people who are characterologically impulsive (Marcus, 1992). Given the tenets of psychoanalytic theory, such approaches make logical sense for impulsive individuals (Auld & Hyman, 1991), and, at the very least, the therapy hour can serve as a safe time period to help the client regain ego controls. Clinical experience suggests the usefulness of dynamic and neoanalytic approaches, especially when the client's impulsiveness appears to be related to traumatic events (Sinclair, 1993).

If the clinical treatment of borderline personalities can be taken as prototypic of the treatment of impulsiveness, there are at least three different conceptual transitions for treatment of impulsiveness that are rooted in psychodynamic thinking (Aronson, 1989). The first conceptual framework was the most popular until approximately 1970 and involved an emphasis on supportive psychotherapies. Therapists using this framework sought to "limit regression, strengthen defenses, and prevent the transference from becoming too intense" (Aronson, 1989, p. 513). During the 1970s, therapists shifted to a second framework involving more expressive and exploratory features with an additional emphasis on limit setting, tolerance of hostility, and therapeutic neutrality. Currently, therapists are more inclined to use brief approaches offering an admixture of limit setting and interpretation, within a theory of object relations personality development.

In our experience, all three approaches may be appropriate, depending on a client's needs and presenting symptoms. The best approach is to try out a strategy and see whether it works. However, the pursuit of insight alone is rarely helpful and almost never leads to meaningful behavioral change. Furthermore, impulsive individuals may often need to rely on limit setting provided by the therapist and often benefit from a therapeutically neutral object allowing them to cathect. In cases of extreme impulsiveness, the therapist begins to assume some of the ego-executive functions of the client, acting as a surrogate parent. When this occurs, it may be difficult to "wean" clients away from treatment, and problems with excessive dependence may soon emerge.

Brief Therapies

The previous discussion highlights the use of a time-limited initial treatment contract. However, a more radical scheduling proposal is possible: time-limited or brief therapy with the limit specified to the client in advance. Brief therapies are gaining in popularity, especially with "resistant" clients. Wells and Gianetti (1993) present several useful frameworks for helping clients quickly, and our clinical experience with brief therapy is that it may have some impact on a client's impulsiveness. The effects, however, may not be immediate or long-lasting. It is not uncommon for very brief therapies to simply "plant the seed" for future change, as suggested by various chaos theorists. Brief therapy becomes problematic when it is the only treatment option and clients who would benefit from longer term intervention are denied this possibility. In this case, such a therapy is probably unethical.

However, brief therapies with impulsive individuals may have a number of advantages over longer term methods (McCown & Johnson, 1993) and may be worth consideration. They allow the impulsive client to focus on a single issue, thereby maximizing the possibility for success. They provide a "breathing space" so that neither the client nor the therapist becomes exhausted by treatment efforts. Brief therapies also provide a way of ethically limiting the tendency of impulsive individuals to acquire large financial debt to the therapist. Whether brief therapies are any more efficacious with impulsive people than longer term therapies is a question for future research.

Family and Group Therapies

Family therapies have frequently been suggested for impulsive or crisis-prone people or systems (Kaplan, 1986; Pittman, 1987). Family intervention is also commonly suggested for children of impulsive individuals (Levy & Rutter, 1991), either as secondary or tertiary prevention. Additionally, families with children diagnosed with attention-deficit hyperactivity disorder can also benefit from systemic interventions. Family therapies clearly appear to be useful for discrete disorders of impulse control, such as child abuse (Asen, George, Piper, & Stevens, 1989), incest (Cornille, 1989), juvenile delinquency (Atkinson & McKenzie, 1987), and substance abuse.

McCown and Johnson (1993) have expanded on Pittman's (1987) work and developed a specific method of intervention for family systems that are treatment resistant and crisis prone, which are common characteristics of families with several impulsive members. An advantage of such an intervention is its economy. It is probably cheaper and more efficient to treat a family conjointly than it is to treat several of its members separately. This is true even when the family is seen more frequently than once a week. However, more detailed research is needed to determine the impact of diverse models of family therapy on impulsive behaviors, and, until such time, individual therapy remains the treatment of choice.

Group therapies have become somewhat less popular in recent years despite the fact they have a long and successful tradition in the field of psychotherapy (Edelwich & Brodsky, 1992). There is little data regarding the efficacy of group psychotherapy for impulsive individuals, although our clinical experience suggests that group therapies may be at least as useful as individual therapy. Again, more detailed studies are urgently needed.

Multifocal Therapeutic Intervention: The Treatment of Choice?

In our experience, perhaps the most important intervention into impulsive behavior is a personological explanation offering a cognitive *and* dynamic interpretation of the behavior in question. In other words, we tell the client that he or she has a trait of impulsivity and that the purposes of our therapy are to find the cause for this and to enable the client to learn ways of modifying, coping, or otherwise working around it. Clinical experience suggests that the client is often both surprised and relieved when his or her problems of impulse control are directly identified. Once the client feels that he or she is no longer battling with "insanity" but instead has a dysfunctional trait, he or she is empowered with the beginnings of an understanding that, the therapist hopes, will result in greater mastery of impulsive behavior.

A multifocal approach toward the treatment of impulsive disorders recognizes that the optimal therapeutic approach may involve a variety of techniques. The goal is not necessarily a "cure" but more of a rehabilitation of a troubling personality characteristic. Within this conceptual framework, any technique in the therapeutic armory may be selected to achieve a specific goal.

In a multifocal therapy approach to impulsivity, neither insight, behavioral change, nor cognitive change is given priority. Instead, all are viewed as being important to the process of adjusting to the environment. Any of these three areas can lead to improvements in functioning or to changes in levels of the other two areas. For example, increased insight may precipitate cognitive change and behavioral variation, whereas the behavioral methods of relaxation and desensitization may promote reduced anxiety levels that ultimately foster development of insight. Treatments are targeted to reduce the most distressing and dangerous symptoms first, so that the client is able to continue treatment. The role of social and biological factors is clearly acknowledged, as is the capacity of the individual to overcome these predisposing influences.

Perhaps some of our failure in working with impulsive clients involves our own theoretical narrowness. Clinicians with the least success and greatest "burnout" seem to be those who are rigidly welded to a particular set of theories and techniques that they apply uniformly. No other type of client that the impulsive one demands such technical and theoretical eclecticism or is as demanding and taxing. A multifocal approach may lessen some of the burnout while simultaneously providing the best chance that at least one intervention will be successful (Kottler

& Blau, 1989). Future empirical studies designed to determine the comparative efficacies of specific treatments will probably involve comparison with a multifocal intervention because such interventions are potentially the most promising.

CONCLUSION

Treating the impulsive adult client is one of the most demanding clinical tasks. It is also a common one, inasmuch as impulsiveness is a problem associated with a variety of clinical conditions. Despite its frequency, clinicians are often ill-prepared to assess impulsive behaviors. One of the premises of this chapter is that such assessment should become more standardized and routine and may serve as the basis for more rapid and comprehensive treatment.

Regarding the treatment of impulsiveness, there is no clear empirical support for one class of psychotherapeutic interventions over another. Generally of more concern than questions regarding what type of treatment are pragmatic issues such as how to deal with clients who fail to show up for sessions or refuse to pay fees. These issues are complex and do not have clear answers. It is important, however, for clinicians to acknowledge that impulsive clients are often difficult and frustrating and that they create a number of countertransferential issues for the therapist. The treatment of such clients frequently involves a substantial crisis intervention component, and the competent therapist needs to be able to negotiate a balance between meeting the client's needs for crisis resolution and avoiding excessive dependency.

There are few data regarding what type of therapy works best for impulsive clients. In our experience, almost every approach has useful features and theories that may be relevant to specific problems. This includes the judicious use of medication in some circumstances. Controlled studies are needed to assess the efficacy of family, group, and brief therapies with impulsive people. Perhaps the most promising type of intervention is a multifocal eclecticism with the flexibility to examine behaviors, connotations, biological predispositions, and historical factors. Regardless, it is usually a mistake to view impulsiveness as a set of behaviors that can be cured or extinguished. The ultimate goal of therapy with impulsive people is to increase their adjustment to life, enhancing their satisfaction with others and with themselves.

REFERENCES

American Psychiatric Association. (1987). *Diagnostic and statistical manual of mental disorders* (3rd ed., rev.). Washington, DC: Author.

Aronson, T. (1989). A critical review of psychotherapeutic treatments of the borderline personality: Historical trends and future directions. *Journal of Nervous and Mental Disease, 177,* 511–528.

Asen, K., George, E., Piper, R., & Stevens, A. (1989). A systems approach to child abuse: Management and treatment issues. *Child Abuse and Neglect, 13,* 45–57.

Atkinson, B., & McKenzie, P. (1987). Family therapy with adolescent offenders: A collaborative treatment strategy. *American Journal of Family Therapy, 15,* 316–325.

Auld, F., & Hyman, M. (1991). *Resolution of inner conflict: An introduction to psychoanalytic theory.* Washington, DC: American Psychological Association.

Austin, S., & McCown, W. (in press). Inability of clinicians to assess trait impulsiveness from clinical interviews. *Contemporary Psychodynamics.*

Barratt, E. S., & Patton, J. H. (1983). Impulsivity: Cognitive, behavioral, and psychophysiological correlates. In M. Zuckerman (Ed.), *Biological bases of sensation seeking, impulsivity, and anxiety* (pp. 77–116). Hillsdale, NJ: Erlbaum.

Bellack, A. S. (Ed.). (1990). *International handbook of behavior therapy and modification.* New York: Plenum Press.

Butcher, J. (1991). *User's guide for the MMPI-2 Minnesota report: Alcohol and drug abuse systems.* Minneapolis, MN: National Computer Systems.

Butler, J. (1992). Of kindred minds: The ties that bind. In M. Orlandi, R. Weston, & L. Epstein (Eds.), *Cultural competence for evaluators: A guide for alcohol and other drug abuse prevention practitioners working with ethnic/ racial communities* (pp. 23–54). Rockville, MD: U.S. Department of Health and Human Services.

Bütz, M. (1993, August). *Medication: Is it psychology's Jurassic Park?* Paper presented at the Third International Conference of the Society for the Study of Chaos Theory in Psychology, Lake Geneva, Ontario.

Cornille, T. (1989). Family therapy and social control with incestuous families. *Contemporary Family Therapy, 11,* 101–118.

Cowdry, R., & Gardner, D. (1988). Pharmacotherapy of borderline personality disorder. *Archives of General Psychiatry, 45,* 111–119.

Craighead, L., Craighead, W. E., Kazdin, A., & Mahoney, M. (1993), *Cognitive and behavioral interventions: An empirical approach to mental health problems.* Des Moines, IA: Longwood Division, Allyn & Bacon.

Dobson, K. S. (1988). *Handbook of cognitive–behavioral therapies.* New York: Guilford Press.

Edelwich, J., & Brodsky, A. (1992). *Group counseling for the resistant client: A practical guide to group process.* Lexington, MA: Lexington Books.

Eysenck, H., & Eysenck, M. (1985). *Personality and individual differences.* New York: Plenum Press.

Eysenck, S. B. G., Pearson, P. R., Easting, G., & Allsopp, J. F. (1985). Age norms for impulsiveness, venturesomeness and empathy in adults. *Personality and Individual Differences, 6,* 613–619.

Forrest, G. (1992). *Chemical dependency and antisocial personality disorder: Psychotherapy and assessment strategies.* Binghamton, NY: Haworth Press.

Giles, T. (Ed.). (1993). *Handbook of effective psychotherapy.* New York: Plenum Press.

Hartmann, D., Roper, B., & Bradford, D. (1979). Some relationships between behavioral and traditional assessment. *Journal of Behavioral Assessment, 1,* 3–21.

Herron, W., & Welt, S. R. (1992). *Money matters: The fee in psychotherapy and psychoanalysis.* New York: Guilford Press.

Johnson, J., & McCown, W. (in press). Family therapy of neurobehavioral disorders. Binghamton, NY: Haworth.

Kaplan, L. (1986). *Working with multiproblem families.* Lexington, MA: Lexington Books.

Kottler, J. (1992). *Compassionate therapy: Working with difficult clients.* San Francisco: Jossey-Bass.

Kottler, J., & Blau, D. (1989). *The imperfect therapist.* San Francisco: Jossey-Bass.

Layden, M. A., Newman, C. F., Freeman, A., & Morse, S. B. (1993). *Cognitive therapy of personality disorder.* Des Moines, IA: Longwood Division, Allyn & Bacon.

Lanyon, R. I., & Goodstein, L. D. (1982). *Personality assessment.* New York: Wiley.

Levy, S., & Rutter, E. (1991). *Children of drug abusers.* Lexington, MA: Lexington Books.

Linehan, M. (1993a). *Cognitive–behavioral treatment of borderline personality disorder.* New York: Plenum Press.

Linehan, M. (1993b). *Skills training manual for treating borderline personality disorder.* New York: Plenum Press.

Links, P., Steiner, M., Boiago, I., & Irwin, D. (1990). Lithium therapy for borderline patients: Preliminary findings. *Journal of Personality Disorders, 4,* 173–181.

Marcus, E. (1992). *Psychosis and near psychosis: Ego function, symbol structure, and treatment.* New York: Springer-Verlag.

McCown, W. (1993, August). *The ideodynamics of impulsive families.* Paper presented at the 101st Annual Convention of the American Psychological Association, Toronto, Ontario, Canada.

McCown, W. G., & Johnson, J. L. (1993). *Therapy with treatment resistant families: A consultation-crisis intervention model.* Binghamton, NY: Haworth Press.

Meehl, P. (1960). The cognitive activity of the clinician. *American Psychologist, 15,* 19–27.

Mischel, W. (1961). Delay of gratification, need for achievement, and acquiescence in another culture. *Journal of Abnormal and Social Psychology, 62*, 543–552.

Mischel, W. (1983). Delay of gratification as process and as person variable in development. In D. Magnusson & V. P. Allen (Eds.), *Interaction in human development* (pp. 149–165). San Diego, CA: Academic Press.

Mischel, W. (1986). *Introduction to personality* (4th ed.). New York: Holt, Rinehart & Winston.

O'Leary, K. D., & Wilson, G. T. (1987). *Behavior therapy: Application and outcome.* Englewood Cliffs, NJ: Prentice Hall.

Pittman, F. (1987). *Turning points: Treating families in transition and crisis.* New York: Norton.

Romney, D., & Bynner, J. (1992). A simple model of five DSM-III personality disorders. *Journal of Personality Disorders, 6*, 34–39.

Rutter, M. (1987). Temperament, personality and personality disorder. *British Journal of Psychiatry, 150*, 443–458.

Sinclair, D. (1993). *Horrific traumata.* Binghamton, NY: Haworth Press.

Svanum, S., & Ehrmann, L. (1992). Alcoholic subtypes and the MacAndrew Alcoholism Scale. *Journal of Personality Assessment, 58*, 411–422.

Wallen, J. (1992). *Addiction in human development: Developmental perspectives on addiction and recovery.* Binghamton, NY: Haworth Press.

Wells, R., & Gianetti, V. (1993). *Casebook of the brief psychotherapies.* New York: Plenum Press.

Wolman, B. (1992). *Personality dynamics.* New York: Plenum Press.

Wolpe, J. (1990). *The practice of behavior therapy* (4th ed.). Elmsford, NY: Pergamon Press.

Zuckerman, M. (1991). *Psychobiology of personality.* Cambridge, England: Cambridge University Press.

17

LEGAL ISSUES IN TREATING THE IMPULSIVE CLIENT

ERIC Y. DROGIN and LAURIEANN Y. DROGIN

The first question raised when addressing legal issues in treating the impulsive client is whether these issues are specific to impulsive clients or simply a natural extension of those inherent in any psychotherapeutic relationship. To provide some parameters in answering this question, it is helpful to keep in mind three basic assumptions:

1. In any form of clinical intervention, be it therapy, diagnostic assessment, consultation, or even referral, legal issues are always present and must be considered.
2. In situations in which a client's impulsive behavior or behavioral tendencies have been identified, certain specific legal issues and the specter of legal complications in general are thrust into the forefront of the conceptualization and day-to-day handling of each case.
3. However, no matter how grave or imminent these legal issues become, the clinician must never allow them to subsume the therapeutic agenda and transform the caregiving relationship into an adversarial one.

By this last assumption we do not mean to suggest that clinical situations never reach the point at which legal intervention becomes necessary. Instead, we suggest that the clinician maintain the therapeutic relationship, to whatever extent possible, in order to avoid crossing the line between clinician and litigant or, at the extreme, prosecutor. There will be times when it seems as if the legal pressures inherent in some treatment paradigms (or paradoxes) are forcing the caregiver into an adversarial role. Despite this, it is the clinician's responsibility to remain sensitive to these pressures and to maintain the clinical depth and flexibility necessary either to sidestep these issues appropriately, co-opt them for the sake of treatment, or summon help from another quarter in resolving the issue outside of the context of the therapeutic relationship.

In this chapter we explore a number of different legal issues that may arise in treatment of the impulsive client. These issues include the so-called *duty to warn*, unwarranted litigation, provision of testimony, inappropriate or extratherapeutic relationships, management of violence in private and institutional settings, and the potential for self-harming behaviors. Although many of these are properly identified as clinical issues, they carry with them legal considerations that can unduly complicate treatment and attach additional responsibilities to the already weighty charge of the clinician or supervisor.

VIOLENT IMPULSIVE BEHAVIOR

Perhaps the most important legal issue in the treatment of the impulsive client involves what is commonly referred to as the duty to warn, the obligation to inform potential victims of the dangers they may face from a client who presents a serious danger to other people. Examination of a few cases in this area demonstrates the basic legal theories underlying the doctrine and also provides some real-life examples of the kinds of situations that can arise in the practice of psychotherapy with potentially impulsively dangerous clients.

The leading case in this area is, of course, *Tarasoff v. Regents of the University of California* (1976). It is from this decision that various jurisdictions have derived the reasoning and persuasive authority necessary to craft their own duty to warn requirements. The *Tarasoff* case actually involved two hearings: an initial, subsequently vacated ruling by the Supreme Court of California in 1974 and a rarely granted rehearing at the petition of the defendants, which yielded an expanded discourse on the duty to warn that remains a model more than 15 years after its adoption.

The facts in this case were as follows: Prosenjit Poddar was a naval architecture student at the University of California, Berkeley. A voluntary outpatient with Lawrence Moore, a psychologist at the university's student health facility, Poddar informed Moore that he was going to kill a woman named Tatiana Tarasoff, a fellow student who had spurned Poddar's affections. Poddar also told Moore that he had purchased a gun.

After consulting his supervisors, Moore concluded that Poddar should be placed under observation in an institution, and notified campus police that he would seek Poddar's involuntary civil commitment. Moore then wrote a letter to the chief of the local police department, requesting his assistance in having Poddar confined.

Campus police apprehended, questioned, and then released Poddar after determining that he was rational, having secured his promise that he would stay away from Tarasoff. One of Moore's supervisors, psychiatrist Harvey Powelson, then requested that Moore's letter to the local police chief be returned, had all copies of the letter and Moore's therapy notes destroyed, and ordered that plans for involuntary civil commitment of Poddar be abandoned. Two months later, Poddar, who understandably never returned for further psychotherapy, shot and stabbed Tarasoff to death.

Tarasoff's parents, as plaintiffs, set forth three distinct causes of action, alleging that (a) the various doctors negligently failed to detain a dangerous patient who subsequently committed a murder, (b) the various doctors negligently failed to warn the victim's family that she was in grave danger from the patient, and (c) the supervisor maliciously and oppressively abandoned a dangerous patient.

A fourth cause of action, including specific references to a broad duty both to the patient and to the public, was legally indistinguishable from the first.

The court dismissed the first and third causes on the bases of governmental immunity (given that the doctors were state employees) and the lack of availability of punitive damages in a wrongful death action, respectively.

This left the issue of whether a cause of action could be stated against the doctors for their "negligent failure to protect Tatiana" (Tarasoff v. Regents of the University of California, 1976, p. 22).

The court acknowledged the general rule that an individual has no duty to control the conduct of another, nor does he or she have a duty to provide any warning to those threatened by that conduct. An exception was, however, identified for individuals who stood in some "special relationship" (Tarasoff v. Regents of the University of California, 1976, p. 23) either to the person threatening such conduct or to that conduct's foreseeable victim. The cited authority for this proposition was the Restatement (Second) of Torts § 315 (1965), which states that a duty may spring from

> (a) a special relation . . . between the actor and the third person which imposes a duty upon the actor to control the third person's

conduct, or (b) a special relation ... between the actor and the other which gives to the other a right of protection.

The court went on to note that, although no special relationship existed between the doctors and Tarasoff, a special therapist–patient relationship existed between the doctors and Poddar that might support a duty to third parties who might be affected by Poddar's actions. In support of this theory, persuasive authority from other jurisdictions was cited, including cases in which doctors or their employers had been held liable to third parties for failing to diagnose contagious diseases (*Hoffman v. Blackmon*, 1970); failing to warn of a properly diagnosed contagious disease (*Wojcik v. Aluminum Co. of America*, 1959); and failing, after arranging a work placement, to warn an employer of a mental patient's psychiatric background (*Merchants National Bank & Trust Co. of Fargo v. United States*, 1967).

The court concluded that a cause of action might indeed be against the doctors in this case under the following new rule:

> When a therapist determines, or pursuant to the standards of his or her profession should determine, that his or her patient presents a serious danger of violence to another, he or she incurs an obligation to use reasonable care to protect the intended victim against such danger.
>
> The discharge of this duty may require the therapist to take one or more of various steps, depending upon the nature of the case. Thus it may call for him to warn the intended victim or others likely to apprise the victim of the danger, to notify the police, or to take whatever other steps are reasonably necessary under the circumstances. (*Tarasoff v. Regents of the University of California*, 1976, p. 20)

Because it is not often noted in scholarly treatments of his case, it may be of some interest to note the ultimate fate of Prosenjit Poddar. After his original conviction for second-degree murder was thrown out on a technicality, Poddar was convicted of voluntary manslaughter and briefly confined in a California medical facility before returning to India, where he was soon, by his own report, happily married (Keith-Spiegel & Koocher, 1985).

As the *Tarasoff* doctrine began to spread to other jurisdictions, numerous attempts were made to argue against its imposition on various legal and policy grounds. *McIntosh v. Milano* (1979) is a classic example of how state courts have managed to dispense with these objections. In this case, Michael Milano, a New Jersey psychiatrist, had been treating a patient, Lee Morgenstein, for approximately 2 years when Morgenstein shot and killed Kimberly McIntosh, Morgenstein's erstwhile neighbor and alleged former girlfriend.

Morgenstein went into psychiatric treatment with Milano at the age of 15, when his school psychologist referred him for weekly treatment of what was initially diagnosed as an "adjustment reaction to adolescence" (*McIntosh v. Milano*, 1979, p. 503). Over the course of the next 2 years of therapy, Morgenstein revealed an ongoing drug problem and various fantasies involving fear of others, personal heroism and villainy, and threatening or retaliating against others with knives.

Morgenstein also claimed to have become involved sexually with McIntosh, who was living next door to him with her parents at that time. Milano eventually came to accept these assertions as truthful, despite initial reservations that McIntosh was, at that time, 20 years of age and thus 5 years Morgenstein's senior. From the start, Morgenstein expressed possessive feelings toward McIntosh and eventually claimed, according to Milano, to be "overwhelmed" (*McIntosh v. Milano*, 1979, p. 503) by the relationship.

Upon discovering that McIntosh was dating other men, Morgenstein admitted in therapy that he wished McIntosh would "suffer" (*McIntosh v. Milano*, 1979, p. 503) as he did, that he was jealous of these other men, and that he was very angry at being unable to obtain McIntosh's telephone number when she moved from her parents' home. Morgenstein also told Milano that he had fired a BB gun at a car that belonged to McIntosh or to one of her boyfriends and that he had shown Milano a knife that he had purchased in order to menace anyone who tried to intimidate him.

During one session, Milano briefly left the room, at which point Morgenstein stole a prescription blank from the doctor's desk. When Morgenstein tried later the same day to obtain 30 Seconal tablets from a local pharmacy, the suspicious druggist telephoned the doctor. Milano told him to send the boy home and later tried unsuccessfully to reach Morgenstein by telephone. That evening, Morgenstein waylaid McIntosh as she attempted to visit her parents' home, took her to a local park, and fatally shot her in the back.

Milano advanced four separate arguments in asserting that the *Tarasoff* doctrine should not be applied to his case, namely, that such a duty to warn would (a) impose an unworkable duty on therapists, given that a patient's dangerousness cannot be predicted in a reliable fashion; (b) eliminate confidentiality and thus interfere with effective treatment; (c) discourage therapists from treating potentially violent individuals because of a fear of malpractice claims; and (d) burden the state by increasing the number of involuntary commitments, civil and otherwise.

Addressing the issue of the unreliability of mental health professionals' predictions of dangerousness, the court stated the following:

> It may be true that there cannot be 100% accurate prediction of dangerousness in all cases. However, a therapist does have a basis for giving an opinion and a prognosis based on the history of the patient and course of treatment. Where reasonable men might differ

and a fact issue exists, the therapist is only held to the standard for a therapist in the particular field in the particular community. Unless therapists clearly state when called upon to treat patients or to testify that they have no ability to predict or even determine whether their treatment will be efficacious or may even be necessary with any degree of certainty, there is no basis for a legal conclusion negating any and all duty with respect to a particular class of professionals. This is not to say that isolated or vague threats will of necessity give rise in all circumstances and cases to a duty. (*McIntosh v. Milano*, 1979, p. 508)

Milano's assertion that making confidentiality a casualty of this duty would impair the ability of mental health professionals to provide effective treatment was dismissed as follows:

The need for confidentiality cannot be considered either absolute or decisive in this setting. A patient is entitled to freely disclose his symptoms and condition to his physician "except where the public interest or the private interest of the patient so demands." A patient therefore, possesses a "limited right" to confidentiality in extra-judicial disclosures, "subject to exceptions prompted by the supervening interest of society" (*Hague v. Williams*, 1962, p. 336), just as a lawyer has no privilege in the lawyer–client relationship to protect or conceal intent to commit a crime. . . . Thus, there is no out-and-out professional prohibition against certain types of disclosure by therapists. (*McIntosh v. Milano*, 1979, pp. 512–513)

The court went on to cite various treatises on psychiatric and other ethical systems before concluding that "considerations of confidentiality have no over-riding influence here" (*McIntosh v. Milano*, 1979, p. 513).

Less effort was required to obviate Milano's remaining two arguments. The court pointed out the inconsistency of claiming that a duty to warn would decrease services to patients who might act out violently in the future, wondering aloud, "If the psychiatrist claims inability to predict dangerousness or detect a dangerous person, how will he make the determination to weed out 'potentially violent patients'?" (*McIntosh v. Milano*, 1979, p. 514). Even if reasonable, it was asserted, this threat of diminished service provision should not prevent victims of a breach of duty from obtaining any remedy whatsoever. Finally, it was held that there was "no reliable statistical support or backing" for a claim that imposition of a duty to warn would result in increased commitments, "even if relevant to the issue of providing a remedy for a possible wrong" (*McIntosh v. Milano*, 1979, p. 515).

One federal case is particulary instructive in demonstrating the great lengths to which courts have gone to infer a duty to warn where none had existed previously. In *Lipari v. Sears* (1980), Ulysses L. Cribbs, Jr., bought a shotgun from a Sears store in Bellevue, Nebraska. Cribbs had recently undergone inpatient treatment at a Veterans Administration (VA) hospital, and, subsequent to buying the gun, entered psychiatric day-care treatment, again with the VA. Cribbs was seen for approximately 3½ weeks before terminating treatment against medical advice. A little more than a month after termination, Cribbs walked into an Omaha nightclub and fired his shotgun directly into the crowded dining room, seriously wounding Ruth Lipari and killing her husband, Dennis.

Lipari and the coadministrators of her husband's estate sued for the wrongful death of her husband and for her own personal injuries, alleging that it was the negligence of Sears, in selling a gun to someone they knew or should have known was a psychiatric inpatient, that had caused the incident.

Sears was understandably disinclined to face these charges alone. They filed a third-party complaint against the United States, under the Federal Tort Claims Act, claiming that the VA had been negligent in failing "to take those steps, and to initiate those measures and procedures customarily taken or initiated for the care and treatment of mentally ill and dangerous persons by mental health professionals practicing in the community" (*Lipari v. Sears*, 1980, p. 187), even though they knew or should have known that Cribbs was dangerous to himself or others. Once this step was taken by Sears, Lipari and the estate coadministrators stepped in and sued the VA as well, claiming in a more straightforward fashion that the VA had been negligent in failing either to detain Cribbs or to have him involuntarily civilly committed.

The VA attempted to have these complaints dismissed, claiming, among other issues, that there was no law in Nebraska imposing a duty on a psychotherapist to protect the potential victims of a patient who is or should be understood to present a danger to others. The Nebraska Division of the United States District Court agreed that the Nebraska Supreme Court had never addressed this issue directly but assumed the duty of ascertaining what legal rule the Supreme Court would have adopted, on the basis of any analogous Nebraska authorities and even on "the case law of other jurisdictions, to the extent that it suggests the rule of law which [the court] would be likely to adopt" (*Lipari v. Sears*, 1980, p. 188).

Having defined this as a case of first impression, the district court went on to discuss in detail the facts and holdings in *Tarasoff* and in *McIntosh*, citing specifically the reliance in those decisions on the "special relationship" theory of the Restatement (Second) of Torts § 315 (1965). Actually, although no Nebraska case could be found that expressly adopted § 315, two cases were unearthed that "implicitly" (*Lipari v. Sears*, 1980, p. 190) recognized a special relationship theory, including one that relied on a duty for jailers to protect prisoners from the violent acts of others.

With a special relationship rule now established, the district court retrieved a 1920 Nebraska case (*Simonsen v. Swenson*) that held that a doctor would not be held liable for disclosing information given by a patient in confidence when that disclosure was necessary to prevent an outbreak of contagious disease. One passage in *Simonsen* referred to "a duty [that] may be owing to the public and, in some cases, to other individuals" (*Simonsen v. Swenson*, 1920, p. 832). Although this particular case did not necessarily imply a duty to warn, it did, in effect, establish that such a concept indeed necessitated an exception to the usual rule of clinical confidentiality.

Thus, in a state where no case law or statutory reference to a psychotherapist's duty to warn had previously existed, a federal court went to considerable lengths to impose such a duty and claim its basis in state law. In *Lipari v. Sears* (1980), it was held that an affirmative duty existed on the part of psychotherapists to third parties for those parties' benefit and that the VA was therefore liable to suit for negligence in that regard. The warning to be taken from this case is that, simply because a particular jurisdiction may not have addressed a duty to warn in the past, is not to assume that no such duty shall be inferred.

On the other hand, this is not to create the impression that all jurisdictions routinely infer a duty to warn in cases in which it is raised as the basis of a negligence theory in tort. In *Brady v. Hopper* (1983), the psychiatrist who treated John Hinckley, Jr., prior to his attempt to assassinate President Reagan was found not to have failed in any duty to warn, despite knowing or having access to information that Hinckley had identified with assassins, collected articles and books on political assassination and on Ronald Reagan, and owned and practiced with guns and ammunition. Similarly, VA hospital personnel in *Leedy v. Hartnett* (1981), who knew that a soon-to-be discharged patient had a tendency to be violent while drinking, and who knew the people with whom the patient would stay when released, were found to have no duty to warn those individuals. In *White v. the United States* (1986), a psychiatrist was found to have no duty to warn of his patient's fantasy of assaulting his wife, although the patient subsequently walked off hospital grounds and stabbed his wife 55 times with a pair of scissors.

Given that all psychotherapists treating the impulsive client are accountable to some duty to warn, whether on the basis of ethical, statutory, or case law formulations, they need to understand that the parameters of this duty stand to vary in each jurisdiction in which they practice and in each clinical situation they may face. Nor is the issue always as clear-cut as it may appear in the cases just described.

It would seem that therapists practicing in jurisdictions with no specific policy regarding a duty to warn must rely on their own ethical principles as related to clinical situations. As the professional party in what has been identified as a "special relationship," it is understandable that a therapist may adhere to an ethical standard of confidentiality that militates against any duty to warn.

If a client is not perceived to be dangerous enough to require commitment, when does this client become threatening enough to warrant the warning of potential victims of impulsive behavior, overriding considerations of client confidentiality? In the case of *Lipari v. Sears* (1980), for example, the question may be raised as to how, in lieu of involuntary civil commitment, any duty to warn might have been carried out without interfering with the civil rights of the client. The means of warning necessary to prevent a client from purchasing a weapon, be it firearms or any other legal weapon, would be tantamount to branding any client with the potential for an impulsive act of violence.

One can easily grasp the complexity of such an issue when considering, for example, a sexually impulsive client known to be infected with the human immunodeficiency virus (HIV). To whom does a therapist owe a duty to warn if indeed such a duty exists in this case? Perhaps the reasoning in *Simonsen v. Swenson* (1920) and similar cases could be used as a guideline in exempting such a situation from normal confidentiality restrictions. On the other hand, a public "warning" in the case of an HIV-infected client may be so socially stigmatizing as to prevent the client from engaging in gainful employment, fair access to housing, and a general pursuit of happiness.

There are, of course, numerous instances of potentially dangerous impulsive behavior in domestic situations involving family relationships. A client who is an abusive spouse, prone to committing impulsive violent acts, presents a physical danger to the spouse, as well as an emotional danger to any other people, particularly children, living in the home. If a child is the direct object of physically violent outbursts, laws in virtually every jurisdiction address this situation directly, making the therapist's duty clear. On the other hand, if the client–parent is impulsive to the detriment of the minor child in other nonphysical ways, this clarity soon disappears and standards become murky.

NONVIOLENT IMPULSIVE BEHAVIOR

Impulsive behavior that is not viewed as violent is a parent impulsively spending money to the extent of endangering the child's future well-being. When does this type of impulsive behavior warrant therapist intervention on behalf of a minor, or, to the contrary, would such an intervention, solely on the basis of the therapist's opinion, be overstepping the carefully drawn and legally protected line between therapeutic and adversarial relationships?

There may be other situations in which people at risk because of a client's impulsive behavior are in positions dependent on that client's behaving in a thoughtful, responsible manner. When, for whatever reason, a client is unable to control his or her impulses to act irresponsibly, what is the clinician's duty to warn, for example, that client's employees?

The question of nonviolent impulsive behavior on the part of a client is not only an issue when the detrimental result of such behavior affects others but also when the behavior is self-harming. If, returning to the impulsive spending example for the purposes of illustration, in the course of therapy a clinician becomes aware of a client's impulsive spending, can that clinician be held liable when the client is no longer able to afford basic living expenses? Furthermore, if a family member subsequently institutes guardianship proceedings, are the therapist's records protected by standards of confidentiality and privilege, or can they be subpoenaed by the courts for use in such a proceeding?

There are, of course, specific mental disorders that predispose a client to impulsive actions. Compulsive gambling is one of these particular diagnoses, as is attention-deficit hyperactivity disorder (ADHD). Compulsive thrill seeking and risk taking may be harbingers of danger. Although no direct physical danger may exist, a clinician may wonder whether he or she could be held liable for "warning" the family of a compulsive gambler that their life savings are in danger of depletion.

Assuming that many, if not most, voluntary psychotherapy clients perceive some danger to themselves or to others, be it physical, emotional, or some combination thereof, nearly all patients could then conceivably be the subject of some type of warning. Ludicrous as this idea may seem, it points out the need for specific guidelines within the psychotherapeutic community, establishing a professional standard regarding a duty to warn. Similarly, the importance of detailed written documentation, particularly in relation to clients diagnosed with a disorder commonly associated with impulsive behavior, must be impressed on all mental health professionals.

Only when the profession establishes such a standard can the mental health community adequately and fairly inform clients of the clinician's duty prior to engaging in a therapeutic special relationship. Given the variety of laws and the breadth of their interpretaion in each of 51 jurisdictions, it is incumbent on the therapist to become familiar with his or her state's duty-to-warn statutes, case law, or both and to seek legal consultation when the answers to these and other questions appear unclear.

CLINICIAN SELF-PROTECTION AND TREATMENT OF THE IMPULSIVE CLIENT

All impulsive behavior exhibited by a client does not take place outside of the therapist's office or institution. Clinicians are not only subject to legal responsibilities, but they may also claim some legal rights. For example, situations may occur in which a client becomes hostile or even violent toward the therapist. What are the legal rights of clinicians and clinical staff to protect themselves, as well as other clients, from one client's impulsive behavior? In the case of a client impulsively becoming enraged and attempting to leave the therapist's office, execution of the duty to warn is colored by the more immediate need for self-protection.

In an outpatient setting, assuming that the clinician has done everything possible to contain events with the scope and standards of professional treatment, the clinician and staff retain their usual right to self-protection. An institutional setting, on the other hand, lends itself to the further complication of controlling the impulsive inpatient for the client's own protection, as well as for the protection of others within that setting.

Once the impulsive client's propensity toward dangerous behavior has been established and involuntary civil commitment has been instituted, legal issues arise concerning the way in which such detainees are to be treated. Several relatively recent cases within the past decade (*Rennie v. Klein*, 1982; *Rogers v. Okin*, 1980; *Youngberg v. Romeo*, 1982) have held that civilly committed patients maintain the right to refuse to participate in various forms of treatment that are seen as being exceptionally arduous or intrusive.

In *Youngberg v. Romeo* (1982), 33-year-old Nicholas Romeo was assessed as having the mental capacity of an 18-month-old child with an IQ between 8 and 10. When Romeo's father died, his mother arranged for him to be admitted to Pennhurst State School and Hospital. Romeo was allegedly injured on 63 occasions, by himself and by others in reaction to his impulsive behavior. A suit was brought over inappropriate preventive procedures that included placing Romeo in restraints in alleged violation of his Eighth Amendment and Fourteenth Amendment rights against cruel and unusual punishment. An amended complaint charged failure to provide appropriate treatment or programs that adequately addressed Romeo's mental retardation.

The United States Supreme Court held the following:

1. If it defied the Eighth Amendment to hold convicts in unsafe conditions, it certainly did so to hold the involuntarily committed, who were not to be punished at all.
2. Liberty from arbitrary bodily restraint was a basic aspect of the Fourteenth Amendment due process clause.
3. The state bore no constitutional duty to provide substantive services for those within its borders, but it was compelled to provide minimally adequate or reasonable training to ensure safety and freedom from undue restraint.

4. The standard for determining these issues was that courts make sure that some form of professional judgment was, in fact, applied to each case, with the understanding that the court was not to second-guess which professional conclusion would be the best out of all acceptable choices offered.
5. The state continued to be held to its conceded duty to provide adequate food, shelter, clothing, and medical care.

In contrast to the circumstances of Nicholas Romeo's confinement, *Rennie v. Klein* (1982) concerned a highly intelligent ex-pilot and flight instructor whose incipient mental illness was exacerbated by his twin brother's death in an airplane crash. Rennie evidenced symptoms of depression and suicidality and was eventually diagnosed as a paranoid schizophrenic. In and out of mental hospitals on numerous occasions, Rennie expressed delusions involving his belief that he was Jesus Christ and that he planned to kill then-President Gerald Ford.

The United States District Court, District of New Jersey produced a series of holdings in this case as follows:

1. In the specific case of forced Prolixin, side effects did not appear to be unnecessarily harsh, so that no Eighth Amendment cruel and unusual punishment was found.
2. The temporary disorientation or diminution of mental capacity occasioned by forced medication did not amount to a violation of Rennie's First Amendment right to freedom of expression.
3. The right of privacy, however, as inferred from various amendments constituting the Bill of Rights, did include the right to protect mental processes from state interference, thus requiring a strong state interest to justify such interference by way of forced medication.
4. The state's police power allowed the power to confine but did not provide for involuntary treatment unless risk of harm to other patients was involved.
5. If a parens patrie theory of provision of care for Rennie's case was invoked, a competency hearing had to be held, involving consideration of least restrictive alternatives and a full and reasonable treatment plan.
6. Regarding Fourteenth Amendment due process issues, the requisite hearing had to include informed participation in decisions regarding treatment, assistance of counsel (provided, if necessary, by the state), and review of treatment decisions by independent, state-elected psychiatrists.

Rogers v. Okin (1980) involved a class action suit against Boston State Hospital for perpetuating a policy of involuntary seclusion and forced medication in other than emergency situations. It was alleged that, in addition to violating medical practice standards, such a policy interfered with patients' constitutional rights as addressed in the following holdings by the United States District Court, District of Massachusetts:

1. Committed patients were to be presumed, barring evidence to the contrary, to be capable of making decisions about their own treatment in nonemergency situations.
2. If patients were incompetent, duly appointed guardians could exercise any decision-making rights on behalf of patients.
3. A substantial likelihood of physical harm to patients or others was required to forcibly apply medication.
4. Arbitrary patient seclusion violated Fourteenth Amendment due process liberty interests, so that seclusion required either the occurrence or a serious threat of attempted suicide, personal injury, or extreme violence on the part of a patient.

Thus, it can be seen that various jurisdictions have addressed the right of treatment refusal on the part of impulsive mental health detainees in a variety of ways, including specific reference not only to the circumstances under which forcible treatment might be applied but to the standards necessary for review of that treatment's use in institutional settings. The cases of *Romeo*, *Rennie*, and *Rogers* lead to consideration of legal guardianship for inpatients. Mere compliance with local standards of treatment is not enough. There must be, particularly in the forcible control of impulsive behavior, an advocate for the best interests of the patient. Before engaging in any control-oriented remediation, the guardian for the client who is unable to make reasonable treatment compliance decisions must be consulted regarding the risks and benefits of any treatment to be administered for purposes of behavior control. In some way, the patient's right of informed consent to or refusal of treatment must always be protected.

Once again, the importance of maintaining accurate and up-to-date records of treatment cannot be overstated. It is the responsibility of the attending clinician to show that the intervention of choice is the least restrictive treatment modality that protects the best interests of the patient, as well as the safety of others.

The clinician with a professional or academic interest in such issues should be aware that, in addition to a jurisdiction-specific analysis of this issue, it is necessary to review standards pertaining to different settings, such as correctional facilities. For example, in *Vitek v. Jones* (1980), the United States Supreme Court mandated a hearing for transfer from prisons to mental hospitals, during which the inmate–patient retained the right to be heard.

SEXUALLY IMPULSIVE CLIENTS

A controversial situation involving impulsive behavior within the office or institution is that of the sexually impulsive client who becomes sexually involved with the therapist. The degree to which the impulsive client may contribute to incidences of therapist–patient sexual intimacy (TPSI) remains a hotly debated topic among mental health professionals of all orientations (Davis & Drogin, 1992). Perhaps the best formulation of this issue is that offered by Gutheil (1991), who stated two axioms for examination of medicolegal aspects of client responsibility in these cases: (a) Only the clinician is culpable, liable, or criminal, and (b) both therapist and client are competent adults, except in cases of infancy, intoxication, or competence-barring mental illness.

Therefore, both are responsible because both contribute interactively within the context of the psychotherapeutic dyad, but only the clinician is blameworthy.

Legislative responses to TPSI are growing and usually involve criminalization of this conduct. Iowa, Tennessee, and Arizona are among the jurisdictions whose legislations are considering bills that would, in some cases, make engagement in such extratherapeutic relationships a felonious offense by the psychotherapist ("Patient–Therapist Sex," 1991).

CONCLUSION

As evidenced by the preceding examination, legal issues in the treatment of impulsive clients are not simple, nor has the evolution of these issues reached any static conclusion. Any clinician engaging in the covenant of psychotherapy is undertaking a serious responsibility. In the special instance of treating the impulsive client, this responsibility may be compounded by a duty to warn, involuntary civil commitment proceedings, the client's right to refuse treatment, and his or her right to the least restrictive measures necessary to prevent harm to self and society. When undertaking such responsibility, and engaging in the special relationship of therapist to client, it is incumbent on the professional clinician to know and understand the legal rights, restrictions, and responsibilities in the jurisdiction and therapeutic community in which psychotherapy is to be practiced.

In order to ensure continued competence in treating the impulsive client within the scope of the law, it is advisable to receive ongoing supervision from an experienced psychotherapist in the area served. For the legal protection of the clinician, as well as to preserve the best therapeutic interests of the client, thorough documentation and record keeping that cites potential for, incidence, and therapeutic remediation of impulsive behavior is essential. Perhaps most important, the clinician must know, understand, and maintain the role of therapist at all times, seeking the counsel of one specifically trained and experienced in psycholegal matters when necessary, so that the therapist can continue about the business of treating clients, impulsive or otherwise.

REFERENCES

Brady v. Hopper, 570 F. Supp. 1333 (D. Colo. 1983).

Davis, M. H., & Drogin, E. Y. (1992). *Redrawn boundaries and mended fences: Treatment in the aftermath of therapist–patient sexual intimacy.* Manuscript submitted for publication.

Gutheil, T. (1991). Patients involved in sexual misconduct with therapists: Is a victim profile possible? *Psychiatric Annals, 21,* 661–667.

Hague v. Williams, 37 N.J. 328, 181 A.2d 345 (1962).

Hoffman v. Blackmon, 241 So.2d 752 (Fla. App. 1970).

Keith–Spiegel, P., & Koocher, G. (1985). *Ethics in psychology: Professional standards and cases.* New York: Random House.

Leedy v. Hartnett, 310 F. Supp. 1125 (M.D. Pa. 1981).

Lipari v. Sears, 497 F. Supp. 185 (D. Neb. 1980).

McIntosh v. Milano, 168 N.J. Sup. 466, 403 A.2d 500 (1979).

Merchants National Bank & Trust Co. of Fargo v. United States, 272 F. Supp. 409 (D.N.D. 1967).

Patient–therapist sex, civil commitment getting states' legislative attention. (1991, December 20). *Psychiatric News,* p. 4.

Rennie v. Klein, 476 F. Supp. 1294 (D. N.J. 1979), aff'd in part and rev'd in part, 653 F.2d 836 (3d Cir. 1981), remanded, 458 U.S. 1119 (1982).

Restatement (Second) of Torts § 315 (1965).

Rogers v. Okin, 478 F. Supp. 1342 (D. Mass. 1979), aff'd in part, rev'd in part, 634 F.2d 650 (1st Cir. 1980).

Simonsen v. Swenson, 104 Neb. 224, 177 N.W. 831 (1920).

Tarasoff v. Regents of the University of California, 13 Cal.3d 177, 118 Cal. Rptr. 129, 529 P.2d 553 (1974), vacated, 17 Cal.3d 425, 131 Cal.Rptr. 14, 551 P.2d 334 (1976).

Vitek v. Jones, 445 U.S. 480 (1980).

White v. United States, 780 F.2d 97 (D.C. Cir. 1986).

Wojcik v. Aluminum Co. of America, 18 Misc.2d 740, 183 N.Y.S.2d 351 (1959).

Youngberg v. Romeo, 457 U.S. 307 (1982).

18

THE IMPULSIVE WOMAN AS CLIENT: TREATING THE LEGACY OF SHAME

JUDITH L. JOHNSON and KATHERINE BISHOP

Many clinicians have noted that a principal issue in psychotherapy with women is the exaggerated shame often experienced and uncovered during treatment (e.g., Jones, 1990; Rawlings & Carter, 1977; Resneck-Sannes, 1991; Scheff & Retzinger, 1991). Research also suggests that women may be more likely to feel a more profound humiliation regarding the commission of impulsive actions than do men (Johnson, McCown, & Booker, 1986). These feelings are associated with a variety of behaviors that men may find less stigmatizing, including violence (Scheff & Retzinger, 1991), sexual behavior (Reiss, 1990), and alcohol and other drug abuse (Miller, 1980). Women are also likely to experience serious humiliation and stigmatization when they are victims of the impulsive behavior of others, as is indicated by the negative social response to sexual abuse victims (Blume, 1990; Lowery, 1987).

In this chapter we present several new studies reiterating that impulsive actions committed by women are judged more harshly than are

We wish to thank Deborah Barkhausen for her contribution to this chapter.

similar behaviors performed by men. We discuss some of the potential reasons for this phenomenon and the implications of these findings for therapy with women. We then briefly suggest a general therapeutic orientation that will assist clinicians in successfully treating impulsive women and discuss specific techniques that we have found to be useful in treating of impulsive women. Finally, we present a case study of a successful eclectically oriented psychotherapy involving a woman with a history of multiple impulsive behaviors, illustrating how therapists can empower women to change through combatting the shame associated with impulsiveness.

IMPULSIVENESS AND GENDER DIFFERENCES IN BEHAVIORAL STIGMA

In American culture there is a mystique associated with being reckless, daring, and impulsive. This is evident in a cursory view of American literature: Tom Sawyer, Martin Eden, Dean Moriarty, and even Billy the Kid are prototypic literary heroes. All of these figures form part of the charm, enchantment, and even charisma of American culture. All of these characters are also male. In the European–American literature tradition, male impulsiveness is often celebrated as a positive attribute, whereas the same behaviors in women are usually labeled much more pejoratively or are even subject to moral scorn.

This double standard is also prevalent in popular culture. Johnson (in press) investigated gender bias in writers and historians of popular culture. She found that authors tend to ignore the negative aspects of impulsive behavior in men while criticizing the same behaviors in women. For example, historical accounts of the great jazz singer Billie Holiday afford significantly more content to discussing her drug and alcohol problems than they do the similarly severe difficulties of Charlie Parker, an equally great—and equally drug-dependent—jazz figure. Similarly, the popular blues–rock singer Janis Joplin is more frequently discussed in the context of having a drug problem than is the rock guitarist Jimi Hendrix, who, like Joplin, died from a drug overdose (and within days of Joplin). Certainly, there are a few exceptions to this bias, especially in circles noted for their avant-garde standards and liberalism (e.g., dance and Isador Duncan). However, in general, the media and popular culture stigmatize women who behave impulsively while tolerating or even lauding the same behaviors in men (Saakvitne & Pearlman, 1993).

In a related line of research, Johnson and her associates assessed how women feel about behaving impulsively or being labeled as being "impulsive personalities." One study (Johnson, in press) is representative of the approach of this group and will be described here in some detail.

Subjects were 100 male and 100 female subjects of various ages who were solicited to participate via random-dialing telephone contact. Subjects were then mailed a series of hypothetical vignettes and a questionnaire measuring their reaction to the vignettes. The vignettes involved a series of 30 situations consisting of the description of a variety of impulsive behaviors. The content of the vignettes was designed to reflect a slight modification of the theory of S. B. G. Eysenck and associates (see chapter 8, this volume). Three principal components of impulsiveness were postulated: (a) impulsive pleasure seeking (extraversion-based impulsiveness); (b) impulsive action for harm avoidance (neuroticism-based impulsiveness); and (c) disinhibitory impulsive action (psychoticism-based impulsiveness).

Ten vignettes each were designed to illustrate three classes of situations reflecting the hypothetical cause of impulsiveness postulated by Eysenck. For example, 1 of the 10 vignettes that described a disinhibitory impulsive action featured a protagonist who acted cruelly to a figure who had harmed her. One of the vignettes involving avoidance impulsiveness described a woman who ran from the scene of an automobile accident when she realized she did not have her driver's license in her wallet. Vignettes were sex matched, with both sexes receiving stories altered to involve only people of their own sex. Otherwise, the vignettes were identical. Each vignette was approximately 250 words, and all were constructed to require an eighth-grade reading level.

Subjects rated each of the 30 vignettes on a 7-point, Likert-type scale for answers to the following questions: "How badly would you feel if you had done what the person in this story did?" and "How ashamed would you be if this had been you in the story?" Because preliminary data indicated that responses to these questions correlated very highly (.92) within vignettes, these scores were summed to produce a single scale. This produced a possible score of 0 (*no shame and no bad feelings*) to 12 (*extreme shame*) for each vignette. Aggregate "shame scores" were then derived for each of the three types of impulsiveness by summing up scores across vignette categories.

For all three types of impulsiveness included in this study, women had significantly higher shame scores than did men. Surprisingly, this included impulsive behavior designed to avoid harm, pleasure-seeking impulsiveness, and sensation-seeking impulsiveness. The major conclusion of this study was that women felt greater guilt for behaving impulsively compared with men regardless of the implicit reasons for their behavior. Another finding of this study was that women were more likely to experience shame for impulsive disinhibition than for pleasure seeking or harm avoidance. These within-groups differences were not found in the group of men. Age effects were also noted. Older subjects of both sexes were more likely to feel greater shame for pleasure-seeking impulsive behavior but less shame for harm-avoidant impulsiveness.

A third study by Johnson (in press) further underscores the issues of sensitivity to shame, impulsiveness, and sex. Johnson examined variables associated with recovery from nonpharmacological addictions in women and men. These included behaviors such as "compulsive" shopping, subjectively labeled excessive sexual behavior, and chronic and inappropriate gambling. People who completed inpatient treatment for nonpharmacological addictions were surveyed at 6 months posttreatment regarding factors associated with a self-perceived positive outcome. For women, the most consistent factor associated with self-rated treatment success was the belief that a treatment program accepted their clinical condition or "disease" in a nonjudgmental fashion. For men, the greatest predictor of positive treatment outcome was acceptance of the disease model per se, followed by a personal belief that the individual in question had reached a subjective life low point (i.e., "bottomed out"). Current levels of social contrition for previous addictive behaviors were positively correlated with self-rated treatment success in men ($.37, p < .01; n = 134$), but negatively correlated with self-rated treatment success in women ($-.44, p < .01; n = 96$).

In a fourth study, Johnson (in press) further examined sex differences in the issue of sensitivity to impulsiveness. College-age women ($n = 100$) and men ($n = 100$) were administered a written version of the Thematic Apperception Test (TAT; Murray, 1943). Responses were scored for impulsiveness on the basis of a method developed by McCown and Barkhausen (1993). As predicted, women had lower overall TAT impulsiveness scores, indicating that they were less likely to generate scenarios involving impulsive behavior. Moreover, and also as predicted, women had a significant tendency to attach negative outcomes to stories in which impulsive behaviors were committed by female protagonists. This tendency was not found for women's thematic content regarding male figures in thematic narratives.

Men also viewed the behavior of impulsive women in a disapproving fashion, as indicated by the unpleasant outcomes impulsive female protagonists encountered. Interestingly, there were no significant differences between how "harshly" men viewed impulsive female protagonists as compared with women. Both sexes viewed this behavior negatively. On the other hand, men were significantly more likely to attach favorable outcomes to male TAT protagonists who "behaved" impulsively. Whereas women saw impulsive behavior in men as being neutral, men were more likely to end thematically generated stories in which male protagonists behaved impulsively with a fortuitous or otherwise positive outcome.

Johnson and her associates have also found that legal authorities and other agents of social control were apt to judge impulsiveness more severely when it was done by women than by men. For example, in a survey of presentencing probationary personnel, McCown and Johnson

(1985) found that men who committed unplanned and spontaneous crimes against property were more likely to be recommended for court-supervised psychotherapy or pharmacotherapy compared with women who committed similar offenses and had corresponding mitigating circumstances and criminal records. Furthermore, probationary personnel of both sexes were more likely to view impulsive behaviors of men as being "situational" (e.g., "He just had too much to drink") while ascribing characterological etiology to the similar behavior of women (e.g., "evidence of her criminal tendencies").

Regrettably, there is evidence that this sex-based bias regarding situational versus characterological attribution of impulsiveness also exists in the work of psychologists. Johnson (in press) examined psychological testing reports of female and male clients with a history or presenting problem of impulsive behaviors. These included individuals who had a history of violence, substance abuse, undesired sexual promiscuity, or "compulsive" gambling or shoplifting. Reports prepared both by female and male psychologists were examined by raters who were unaware of the tester's sex. Regardless of the sex of the psychologist preparing the report, test reports were significantly more likely to suggest environmental contributions to the behavior of male clients (e.g., "client is under financial stress") and characterological etiologies of the behaviors of women (e.g., "clear evidence of long-standing borderline disorder"). Furthermore, shame or contrition on the part of women was more apt to be labeled as feigned than it was when noticed in men. In other words, psychologists were much more likely to ascribe a characterological explanation to the behavior of women and to discount the shame that their behaviors caused.

ETIOLOGY OF HEIGHTENED STIGMATIZATION IN IMPULSIVE WOMEN

The findings just discussed should not be surprising to clinicians who work with women or listen to their stories. They suggest that society, including women themselves, judge impulsiveness in women very harshly. Unfortunately, legal authorities and even psychologists fall into this same pattern. Even among professionals, impulsive behavior is apparently more stigmatizing when its source is a woman.

The reasons for heightened stigmatization regarding impulsive behaviors committed by women are probably complex and determined by a number of interrelated causes. We will now briefly discuss some hypothetical causes.

The Base-Rate Issue

One reason for the greater stigmatization among women may relate to the relative infrequency of the occurrence of impulsive behaviors in the female population. Statistically, women as a group are less impulsive than are men (H. J. Eysenck & Eysenck, 1985; S. B. G. Eysenck, Pearson, Easting, & Allsopp, 1985). No replicated studies in the behavioral science literature have indicated that mean scores of any kind of measure of trait impulsiveness are as elevated in a comparable group of women as they are in men. (However, individual behaviors associated with impulsiveness, such as bulimia or self-destructive wrist-slashing, may have higher prevalence rates in women than men. The data are still not clear regarding the prevalence of these disorders.)

With these sex differences in mind, some authors, such as H. J. Eysenck and Eysenck (1985) and H. J. Eysenck and Gudjonsson (1989), have suggested a connection between impulsiveness and gonadal hormones. However, this remains speculative, especially given the strong link between serotonin and impulsiveness suggested by data from a variety of sources (see chapter 5 by Zuckerman, this volume). The manner in which gonadal hormones hypothetically might interact with the serotonergic system has never been specified and remains difficult for the neuroscientist to pinpoint. Until such a link is found, caution is indicated in postulating a causal link between gonadal hormones and impulsiveness.

Regardless, it is inappropriate to extrapolate on the causation of stigmatization from frequency or epidemiology alone. Rates of prevalence do not in themselves determine social stigmatization. Two examples can readily dispel any notion of causality of stigmatization from sex-related base rates. Women have a lower prevalence rate of learning disabilities than do men, yet girls and women do not encounter a substantial stigma when they manifest this syndrome (Johnson & McCown, in press). Schizophrenia is also perceived to be more common in men, especially younger men (Gottesman & Shields, 1982), yet this does not differentially stigmatize the female schizophrenic. Clearly, factors other than simple prevalence are operating to provide the social and self-imposed stigmatization of impulsive women.

Sociobiological Explanations

Many differences in gender behaviors have been analyzed fruitfully from a sociobiological framework (Wilson, 1975), the popular perspective that attempts to explain complex social behaviors by evolutionary causation. A sociobiological explanation of the phenomenon of heightened shame associated with impulsiveness in women might argue that impulsive women are much less likely to take care of their children adequately (Rushton, 1988). Therefore, proponents of this theory contend that societies that do not discourage impulsiveness in women are at an evolutionary disadvantage and are less likely to endure. Consequently, the stigma associated with impulsiveness in women might be genetically based and acquired through natural selection because of its survival value.

On closer scrutiny, this theory is less tenable than it initially seems. From a purely biological vantage point, impulsivity in women might be a positive adaptation because it would encourage larger genetic interaction (i.e., more matings outside of kin or social groups). This would most likely result in healthier progeny who were more disease resistant and genetically more adaptable. Tolerance of high societal rates of impulsiveness in women would also result in larger family sizes because this trait would encourage more frequent sexual behavior (Sherfey, 1972). It is hard to see how these reproductive advantages would work against a group's survival. Regardless, a sociobiological explanation of stigmatization associated with impulsiveness is essentially untestable and highly speculative. Researchers will probably have to look beyond biology to account for the dissimilarity in attitude between men and women regarding impulsive behavior.

Myths Regarding Optimal Mental Health

Another source of explanation for the sex-related difference in stigmatization may be in theories regarding optimal mental health. These begin with Freud, whose influence still remains the strongest on clinical practice. Freud (1905/1953) held that repression is a cornerstone of socialization, particularly for women. Furthermore, to Freud, these differences are a consequence of the anatomical distinction between men and women (Freud, 1925/1961). Therefore, a heightened sense of shame for impulsive behaviors would be considered natural in women, at least by many psychoanalytic thinkers (J. Brown, 1940).

Many of the contemporary biases regarding mental health are derived either overtly or subtly from the psychoanalytic perspective. Even when it has been misinterpreted, the Freudian perspective has had a profound impact on how women view themselves (Brennan, 1992; Gilligan, 1982), and very often this impact has been mostly negative (Fine & Gordon, 1989). Consequently, it is not hard to imagine that at least some of women's negative views of impulsiveness are absorbed from the prevailing mental health ideologies rooted in the Freudian legacy that emphasize a fundamental difference between women and men (Chodorow, 1989).

Socialization

In contrast with the Freudian position, which argues for an inherent dissimilarity between men and women, most cross-cultural research suggests that there are few or no absolute personality or cognitive ability differences between the sexes (Chodorow, 1989; Feingold, 1988; Hyde, 1990). Rather, perceptions of sex differences are largely subjective and vary between cultures. This was noted much earlier in this century by Margaret Mead (1935, 1949), who observed that many characteristics assumed to be gender related, such as sensitivity, stoicism, and endurance, are not universally gender specific, as they are often thought to be in Western societies. Although we are unaware of any cross-cultural studies of sex and impulsiveness, these observations regarding other "male" and "female" behaviors suggest that increased stigma attached to impulsiveness in women is socially learned rather than immutable.

An obvious potential source for these differences is the societal process of childhood and adolescent socialization. Gilligan (1982) has argued that women in Western society are often socialized to be "others" oriented and to make decisions from a nurturing perspective. Men, on the other hand, are more often socialized to take a "justice-based" viewpoint. This propensity for considering others, and particularly the judgments and feelings of others, may play a large role in the shame associated with women's impulsivity because impulsive behavior typically has a substantial "self-oriented" focus. This tendency may also determine women's excessive responses to behaviors that make them less physically attractive, such as excessive eating (Barth, 1988; Ogden, 1992).

A more feminist interpretation of social stigma and impulsive behavior declares that the sense of shame women feel for behaving impulsively is deliberately induced in the socialization process to support the interests of a patriarchical culture. Numerous feminist authors have asserted that aspects of the socialization process serve to foster the interests of men over women (Brown, 1992; Hare-Mustin & Maracek, 1990; Lips, 1988). It is conceivable to these and other feminist scholars that women are socialized to feel greater shame for behaviors that are not under the control of men (Rawlings & Carter, 1977). Consequently, behaviors contrary to the gender interests or desires of men may be paired with excessive stigma (Frayser, 1985). This labeling is especially likely to be true for sexual behavior (Foucault, 1981) because the socialization of sexual behavior in women is a significant aspect of the developmental process (Parsons & Bales, 1955). Frequently, religion may also be used as the vehicle for this socialization (Ranke-Heinamnn, 1990).

The socialization of women concerning sexuality and impulsivity was evident in the interviews of adolescents conducted by L. M. Brown and Gilligan (1992). One 13-year-old interviewed referred to the behavior of a friend who "goes out with guys" and "goes farther than most people would" as "disgusting." She also noted that if she had done the same things as her friend, she would feel like "total dirt" and "totally worthless." Clinical experience suggests that such shame is not developmentally limited but may be endemic in a larger population of women who feel guilt and self-reproachment for behaviors that men consider appropriate or even laudatory.

However, given the tenets of feminist theorists, it is likely that any evidence of female autonomy is likely to be problematic for men, especially if this autonomy results in behavior that is uncontrollable and unpredictable (Foucault, 1981). As traditional sociological theory has long noted, an effective manner in which to reduce the frequency of a behavior is to label it negatively, preferably as being "deviant" or "immoral" (Merton, 1968). It is not surprising, feminists argue, that women therefore become socialized to believe that any behavior that is outside of the control of their family or male partner is "wrong." Because impulsive behavior is by definition under less control of a partner than more predictable behaviors, impulsiveness quickly becomes excessively stigmatized (Meadow & Weiss, 1992). Essentially, according to the feminist perspective, impulsiveness in women is especially bad because it allows women to be less under the control of men.

FEMINIST PSYCHOTHERAPY AND TREATMENT FOR IMPULSIVE WOMEN

Perhaps the most scientifically justified solution to the problem of determining the etiology of stigma associated with impulsive behavior by women is to realize that there are not yet any satisfactory answers. Although a feminist critique of stigma and impulsiveness emphasizing socialization is probably difficult to empirically test, it may have substantial heuristic value. Clearly, the socialization and societal role of women is far different from that of men (Hyde, 1991; J. B. Miller, 1988). These differences could easily be related to increased feelings of shame and guilt associated with impulsivity and other behaviors. Whether this socialization can be altered without substantial social change is beyond the scope of this chapter. However, a natural starting point for therapy with impulsive women is suggested by the implications of feminist theory. Women themselves should work to redefine what is the appropriate range of feminine behavior, much as many women are doing regarding the aging process (Thone, 1992), sexuality (Reiss, 1990), and other behaviors that were previously prescribed for them by patriarchically oriented others.

For many potential reasons, women are much more stigmatized by behaving impulsively. This shame often causes them to seek psychotherapy. However, it is important to realize that the traditional model of therapy may reflect a style that is ill-equipped to help women cope with this shame. This is because traditional therapies may be geared more toward men than women because they often emphasize objective, unemotional, and impersonal attitudes in a manner that may be alien to the experience of women (Ballou & Gabalac, 1985). This traditional model is built on the assumptions that the therapist must be relatively dispassionate and not care "too much" for the patient (Okun, 1990), an orthodoxy common in almost every psychotherapy textbook. Consequently, both female and male therapists are socialized toward a professional standard of care perhaps more appropriate for the developmental experiences of men than of women, whose socialization often reflects opposing values.

As an example, consider the treatment of substance abuse. A usual method of therapy involves methods that increase contrition, stigma, or remorse (e.g., admitting powerlessness and helplessness, public confession of shortcomings, and semipublic retribution; cf. Brantley & Sutker, 1984; McKenna, 1979). These methods usually result in increased humiliation and disempowerment to the substance abusers. The implicit theory applied by such treatment practitioners or programs is usually that substance abusers lack sufficient social sensitivity to their impulsive behaviors or that they are perhaps sociopathic (Alterman & Cacciola, 1991; Craig, 1986). However, adding shame and anxiety to disempowered individuals who already are exquisitely sensitive regarding the social ramifications of their behaviors is clearly countertherapeutic and may serve to increase the prevalence of the behavior in question. This suggests that many of the therapy methods that are effective with men may not necessarily be effective with women and that therapeutic techniques that take into account women's particular needs may be the most useful (Bepko, 1992; Meadow & Weiss, 1992; Way, 1992).

Feminist critiques of psychotherapy note that conceptions of what is "male" and what is "female" are based on subjective cultural ideals first identified by Mead (1949). In turn, they argue that psychotherapy must be designed to take into account the effect that these ideals may have on the individual (e.g., the shame a woman may feel for certain actions over which a man would be much less disturbed). The purpose of this brief chapter is not to provide a comprehensive review of feminist psychotherapy or feminist theories of development. There are many primary sources and even dedicated journals that are now available for this objective. Instead, our goal is to highlight the potential role of a feminist-based therapy in treating the impulsive problems of women.

GUIDELINES FOR A FEMINIST THERAPY FOR IMPULSIVE WOMEN

The following guidelines for working with impulsive women are tentative, flexible, and useful suggestions derived from clinical experience. To date, they have received no substantial empirical research and are based more on observation and our own biases than we would wish. We hope that someday funding can be found for a careful and controlled study regarding the efficacy of feminist therapy techniques. Unfortunately, this has not been the case. Until then, we believe that these guidelines will be useful for practitioners working with women who display a variety of impulse-related problems. These include criminality, substance abuse, child abuse, and nonpharmacological addictions, all areas in which substantial orthodox treatment methodologies are already in place. Theorists committed to only one narrow school of intervention will find dissatisfaction with our recommendations because they combine aspects of humanistic, psychodynamic, behavioral, and cognitive–behavioral therapies. For these critics, as well as others wedded to traditional treatment ideologies, we encourage an open mind and urge practitioners to conduct some honest personal experimentation with their clinical styles and attitudes.

Therapy Should Emphasize Self-Respect

It has been suggested that the feminist therapist's role is not to "change" the woman who suffers from a disorder such as bulimia but to accept and understand who the woman is and to respect how she came to have her particular behavioral pattern or symptom constellation (Barth, 1988). In this manner, self-respect is promoted and modeled. Although we believe that behavioral change is the final goal of any therapy, in our clinical experience we have found that the one uncompromisable aspect of any form of psychotherapy with impulsive women is that it must foster a more positive self-outlook as a major treatment goal. Women who are impulsive or have been labeled impulsive need a self-respecting identity (Way, 1992). Clinically, this means that impulsiveness in women should be treated in as nonjudgmental a manner as possible. Unconditional respect, if not complete unconditional positive regard by the therapist, is the strongest tool in one's therapeutic arsenal. Rogerian principles are usually necessary, and sometimes sufficient, for helping impulsive women attain satisfactory behavioral change.

Although our preference is for therapy that involves more immediate behavioral change, we have noticed that the therapist's respect is occasionally sufficient by itself to spark a personal examination of behavior by impulsive women that may eventually result in a positive change. However, this process may not begin until long after the formal therapy has ceased. Consequently, it is difficult to quantify with "linear" models because the change may not occur in easily analyzable sequences. For example, a client may abruptly terminate a stormy treatment and apparently revert to former behaviors. By almost all outcome criteria, she would be considered a therapeutic "failure." However, many months or even years later, she reestablishes contact with the therapist with a call or brief letter. She reports that her life is proceeding relatively well and is fairly stable. She reports significant behavioral changes, such as substance abstinence, and credits the therapist with instigating the change process. When the therapist somewhat dumbfoundedly denies being of substantial assistance, the client respectfully disagrees and credits the initiation of behavioral change to the fact that the therapist treated the client with dignity and respect.

Similarly, overt or even subtle disrespect may have long-term deleterious effects, given that they are apt to be internalized by women clients (Crawford, Kippax, Onyx, Gault, & Benton, 1992). Therapists frequently display a judgmental reaction to their client's impulsive behaviors. This reaction can be conveyed either verbally or nonverbally. When it is conveyed verbally, it is often caused by the therapist's clumsiness, a convenient term to highlight the technical lapses that occur when one's skills are not as proficient as necessary. Regarding verbal clumsiness, most clinicians have sufficient training to avoid extraordinary mistakes. However, well-meaning therapists often convey a hostile or judgmental attitude while trying to follow up questions of clinical interest (e.g., "It's rare to see a woman who has been married as many times as you have. I'm wondering if you could tell me more about this"). Much more effective with the impulsive woman is an empathetic inquiry designed to elicit further information in a nonjudgmental manner that is client focused rather than event focused (e.g., "Married five times. Hmmm. That must have been very hard on you"). This clinical subtlety is especially important in psychological assessment (L. S. Brown, 1990), where it is paramount that the client not feel even more stigmatization.

We believe that it is equally important to avoid nonverbal stigmatization because the experience of women makes them very likely to react negatively to such stigma (Surrey, 1985). Most therapists try to monitor their own nonverbal communications and know enough to avoid displaying expressions of shock or scorn. However, therapists that attempt a "blank slate" may be likely to communicate subtle rejection or judgment. Consequently, we believe a warm, empathetic, but realistic, approach is the best demeanor in working with impulsive women.

Therapy Should Reflect the Realities of Women's Disempowerment

Therapy with impulsive women cannot escape the double standard that exists regarding impulsiveness in American society. Our own clinical experience suggests that this social issue should be tackled gingerly yet firmly. Often, clients will seek therapy for what are obviously minor issues, such as "excessive spending" confined to one or two sessions per year, a single sexual episode, or a few bouts of excessive drinking. Although the importance of the symptom to the client should not be discounted, the behavior should also be placed in its appropriate cultural context

(Espin & Gawelek, 1992; Way, 1992). This can usually be done by asking the client, "If you were a man, would this bother you?" Very often, such questions spark profound insight. When answered in the affirmative, we are usually able to work more effectively with a client knowing that she has violated her own personal code rather than the social mores that are meaningless to her (Meadow & Weiss, 1992).

In American culture, it is much too common for women to be sent to therapists as a means of social control. Although this is not uncommon for a variety of disempowered groups (Ramirez, 1991), clinicians often seem to forget this fact regarding impulsive women, frequently because their presenting problems are often told in a compelling manner and are believed to be their only reason for treatment. It is of foremost importance that therapy not revictimize women. Women who attend therapy because their boyfriends or husbands perceive a problem in their behavior that they themselves do not see are not likely to form a meaningful therapeutic alliance. They will often make superficial progress on discrete behaviors yet will apparently sabotage prospects for long-term change. In our clinical experience, we have found it useful to examine apparent cases of therapeutic failure by women and ask ourselves whether resolution of the target symptoms would cause the woman in unsuccessful treatment to lose autonomy and empowerment. In retrospect, the answer has too often been yes.

As an example, women frequently enter treatment because of involvement with social or criminal justice agencies. Treatment may be required because placement of the woman's children outside the home is imminent. In our experience, therapy with these women is rarely effective until such women embrace their own personal goals as being important in the therapy process. This often means that women need to hear that their lives are important independently of their children. One of the problems that we have noted in the rash of recently federally funded "child-centered programs" is that they forget the fact that children have mothers who are also people, people with aspirations, dreams, hope, and dignity.

Therapy Must Recognize Social Realities

As much as possible, our therapy of impulsive women attempts to embrace feminist treatment goals. On the other hand, therapy must not ignore the cultural realities in which clients live (Fox, 1992). It is easy for the educated, affluent therapist to tell a working-class woman that she should reject the rigid restraints placed on her by her sociocultural status (e.g., to rebuff a patriarchical religious tradition). It is quite another thing for the therapist to be able to replace the deficits that the client faces when she turns her back on her restrictive peers, family, social institutions, and other relationships. In our opinion, it is bad therapy and ultimately cruel to encourage women to embrace new attitudes and behaviors that have little likelihood of being maintained in their home and community environments. Consequently, therapy must consider the realities in impulsive women's lives.

Therapy with impulsive women is not unlike therapy with any disenfranchised person or group. The practitioner must carefully weigh the role of intervention against the realities in the client's world. Didactic or consciousness-raising approaches, popular in the late 1970s and still used in some circles, are probably of limited value. This is because the process of therapy involves negotiating a new identity and determining "what fits" for a person and in which environment (Bynum, 1992; Fox, 1992). This highly personal task of therapeutic process may more resemble a voyage into unchartered territory than an informational seminar or spiritual revival. Psychotherapy must be tailored to take into account the social constructs that serve as a background for women's actions and their socialization from a very young age. Women's feelings of guilt regarding impulsive behaviors should be legitimized, and the particular individual's experiences need to be taken into account (L. S. Brown, 1992).

The (Un)Importance of Theoretical Orientation and the Usefulness of Eclecticism

So far, we are not convinced that the psychotherapy literature indicates that one type of therapy is more effective than any other for impulsive women. We believe that more important than specific treatment modalities are the metaframeworks endorsed by the therapist, as well as the cultural awareness that the therapist demonstrates for each client's life space. Our group of associates, supervisees, colleagues, and trainees has treated impulsive women with a variety of therapeutic techniques and from a number of theoretical perspectives. These include psychodynamic, behavioral, and cognitive–behavioral modalities, which we use as deemed appropriate and generally in a symptom-focused manner. As in any good therapy, problems are triaged so that behavioral management of life-threatening and more serious difficulties is attempted first. However, although there is no clear scientific literature to suggest that one type of modality may be the most effective, our clinical experience has enabled us to endorse several techniques that we think are the most useful.

We believe that group therapy may be especially effective with impulsive women. Many authors have written extensively regarding the efficacy of group therapy, and we do not try to duplicate these discussions. However, for impulsive women, perhaps the foremost advantage of groups is the manner in which they facilitate growth through the removal of stigma (L. S. Brown, 1987). The obvious manner in which women model social skills, behavioral restraint, and a therapeutic alliance in the group setting is another cluster of advantages that individual therapy does not have. Structured, task-oriented groups may be as effective as traditional psychotherapy groups. For example, the group approach suggested by McAuliffe and Albert (1992) for cocaine recovery can be modified to include a variety of specific impulse-related problems.

Self-help groups, especially Twelve Step, are a useful treatment modality for most impulse control disorders (Bepko, 1992). This is true despite the fact that impulsivity is a major predictor for self-help attrition (McCown, 1989, 1990). In our clinical experience, impulsive women are more likely to attend self-help groups when at least some of the groups that they attend are composed only of women. Attendance at women's groups also minimizes the likelihood that the impulsive woman will be "thirteenth stepped," Twelve Step group vernacular for being sexually exploited by someone connected with the self-help group. To work through the shame associated with impulsiveness, it may be useful to prescribe bibliotherapy or various paratherapeutic modalities, including structured writing tasks. L'Abate and Cox (1993) have assembled a series of structured therapeutic writing assignments that we have found very useful in reducing impulsive behaviors. Therapeutic writing may also be helpful because it is relatively anonymous. Clients are able to disclose their feelings at their own pace and thus regulate their expressions of shame to a tolerable level.

Use of Metaphors: Chaos as a Construct of Change

Chaos theory is one of the most exciting developments in the physical sciences since Heisenberg's uncertainty principle (Gleick, 1987). Chaos theory has also found useful and growing applications in a variety of areas of the behavioral sciences (Kaplan & Kaplan, 1991; Kelso & Fogel, 1987), including psychotherapy (Chamberlain, 1989; McCown & Johnson, 1992). An excellent introduction to chaos theory for psychologists is presented by Marks-Tarlow (chapter 7, this volume). Although the implications of chaos theory for psychotherapy are vast, the notion of chaos as a metaphor for behavioral change may be very appropriate for changing impulsive behavior, especially in women.

Perhaps the central principle of chaos theory is that behavior of a complex system that appears random and unpredictable is actually causally determined (Abraham, Abraham, & Shaw, 1991). Systems undergoing change frequently oscillate wildly around a point or plane called an *attractor*. However, because of what is known as "sensitivity to initial conditions," it is impossible to make accurate predictions regarding this pattern of behavior. Often, however, given enough data (sometimes several thousand data points), it is possible to portray, graphically or mathematically, the behavior of the system in question. Seemingly random behaviors are frequently found to have a complex pattern called a *strange attractor*. As Marks-Tarlow indicates (chapter 7, this volume), even if one knows the shape of the strange attractor, one's capacity to make predictions about the system is limited.

Another interesting phenomenon of complex systems and chaos is the capacity of patterns to spontaneously organize out of chaotic conditions (Garfinkel, 1987). Chaotic behavior is frequently predictive of a new and more finely structured organization that emerges from within the system. This is often referred to as "self-organization" because the new structure occurs independently of outside influence. An apparent prerequisite for the emergence of self-organization in any complex system is a prior interval of aperiodic (i.e., chaotic) behavior. The ancient Chinese proverb that chaos is a time for potential growth seems to have a mathematical and empirical basis.

Both the notions of chaos and self-organization can be useful metaphors for impulsive women in treatment. We emphasize to our clients the notion of impulsive behavior being chaotic (i.e., determined, but with no discernable pattern unless viewed from a distance and over time). This metaphor implies two useful concepts. First, it stresses that the impulsive behavioral history of the client was caused by an outside force: biochemistry, role incongruence, sexual abuse, socialization, or whatever. In other words, "Your behavior is impulsive because something external is acting on you." Some of the shame from past actions is often mitigated when causes are seen as external.

Second, this metaphor suggests that insight (e.g., "Why did I behave that way?") may emerge only after a time, when the impulsive behavior is able to be viewed historically. Insight developed while the problem is current may be only transitory or may reflect an inaccurate formulation of causality. The clearest view of complex systemic behavior occurs some distance from it (Abraham, Abraham, & Shaw, 1991). As a client told us recently, "For a long time I was concerned about the *why* of what I was doing. Now I know that you can't tell the why until you've been away from it for a while."

Chaos theory also suggests a second metaphor, namely, that of self-organization. Chaos allows the client the freedom for creating the self anew. In nature and perhaps also in the human psyche, disorder allows the creation of new structures, new behaviors, and new rules (Kaplan & Kaplan, 1991; Loye & Eisler, 1987). In this manner, the treatment of impulsive women reiterates the useful ideology from disease-model treatment of substance abuse, which argues that an individual is not responsible for his or her own chaos. The person is, however, free to change it and is therefore responsible for his or her recovery. Self-organization occurs when a chaotic system spontaneously develops a higher, more intricate organization than before the chaos occurred.

The Sex of the Therapist

Although there have been few published studies regarding this topic, we have found that impulsive women work best with female therapists. This is true for a number of reasons. Women may find it easier to trust other women (Matlin, 1987), and this trust may assist in the development of a therapeutic alliance. Women may also serve as gender-appropriate role models for other women. Finally, the therapist must consider the very unpleasant topic of sexual exploitation of the impulsive client by the therapist. Although we do not have substantial data, our clinical experience suggests that impulsive women in treatment are more vulnerable to the initiation of sexual contact by therapists than are other clients. Sexual contact between the therapist and client in any form is now regarded as sexual abuse (Bates & Brodsky, 1989). Male therapist–female client sexual abuse is generally the most common source of concern because it appears to be the most frequent type (Pope, 1990; Rutter, 1989), although other types of abuse may be on the rise or underreported. Hence, well-trained female therapists may offer yet another advantage over their male counterparts.

Men are often—but not always—less sensitive regarding what constitutes inappropriate sexual behavior (Stakes & Oliver, 1991). Consequently, they may not be able to assist the impulsive woman in defining her own sexual behavior in a manner that is both empowering and also culturally appropriate. Men, and often women too, may become inappropriately concerned about the sexual histories of their clients. This may at times become a form of voyeurism and is certainly exploitative. Regardless, such interest reduces time spent pursuing more productive aspects of therapy.

A CASE STUDY: WORKING THROUGH THE SHAME

We close this chapter with a case history of an impulsive woman treated by one of us. The example we chose illustrates the manner in which shame regarding impulsive behaviors interferes with the process of therapeutic change in women. This case also illustrates some of the therapeutic techniques outlined in the previous discussion and how they may be implemented with difficult clients that have proved resistant to psychotherapeutic change. Unlike some other feminist therapists, the approach we advocate is subtle and flexible and recognizes that the client may reject our interpretation of her problems. We chose this case study in part to illustrate how a feminist-oriented therapist can gently use her knowledge regarding impulsiveness and women, even when the client specifically denies the importance of gender in her life experience.

"Pam" was a high-functioning professional, a physician in her late 30s who was the junior partner in a large private practice. At the time that she was seen by one of us, she had been involved in five different types of therapy with seven therapists. Most recently, this included psychoanalysis three to four times a week. Despite the amount of therapy Pam had received, she stated that she felt "chronically suicidal" and that "I'm a worthless piece of shit."

Pam stated that she was in therapy to "work on issues regarding my alcoholic parents." Her father, a well-respected surgeon and prominent Republican, died of complications related to alcohol abuse. Her mother, "a closet drinker," had lost a number of nursing jobs because of her drinking. Although there was no evidence that her parents had sexually abused her, Pam felt persistent vague memories involving inappropriate sexual contact "with someone in the home."

The principal issues of therapy with her previous therapist, the analyst, had been her "guilt and excessive rumination about my past sexual behavior." Her analyst noted that Pam was making strides toward accepting her parents' dysfunctionality and the impact that it had on her life. The analyst also noted that in sessions, "Pam does most of the work . . . she's already decided what issues are important and almost conducts the session herself."

What the analyst did not notice was that despite her therapeutic glibness, Pam's personal life was not proceeding at all well. Pam displayed numerous behaviors congruent with what may be labeled *multi-impulsiveness*. She spent money that she did not have and was always in serious debt. She had brief periods of bulimia, usually following major life stress. She also had multiple and poorly thought-out romantic and sexual relationships, many of which evolved into emotional abuse. Furthermore, Pam had periods of cocaine binges, usually over long weekends. Finally, she had difficulty getting along with others and actually had no female confidants outside of her office. Needless to say, Pam was depressed about her life and her behavior.

Pam's formal diagnosis was cyclothymic personality, although her psychiatrist noted that "she seems to fall between the cracks of the *DSM* [*Diagnostic and Statistical Manual of Mental Disorders*]." Psychological testing had been performed by a psychology intern when Pam entered treatment. (There was no evidence that the information from the testing was used by the treating physician.) The 4–7 profile on the Minnesota Multiphasic Personality Inventory ($Pd = 70$, $Pt = 67$) suggested a person who cycled between periods of impulsiveness and periods of regretful guilt (Dahlstrom, Welsh, & Dahlstrom, 1972).

Pam began seeing one of us when her analyst left on maternity leave. She and her new therapist decided to make her impulsiveness a central issue of treatment. The new therapist was able to help Pam identify several specific behaviors that she wished to change, such as avoiding cocaine use, refraining from sex with a partner until she knew him well, and restraining her verbally hostile and impulsive comments toward co-workers and potential friends. These behaviors were modified using common cognitive–behavioral techniques, which were implemented with little difficulty despite Pam's insistence that she needed insight prior to any meaningful behavioral change.

Throughout the therapy the new therapist's attitude was a warm, positive regard for Pam despite whatever behavioral shortcomings she seemed to demonstrate. The therapist even (gingerly) disclosed stories from her own life regarding impulsive behavior, primarily to let Pam realize that such behaviors were not unchangeable. Pam responded to this warmth and sharing almost immediately. She noted that although she used to look forward to her appointments for therapy "about like you'd look forward to the dentist," the warmth and respect the therapist showed "makes me feel a lot better about my problems and who I am. It definitely gives me more hope."

Pam was introduced to some of the concepts of feminist psychotherapy but had difficulty accepting most of them. For example, she stated that "I really don't care about those theories (about patriarchical domination and the stigmatization of impulsive behavior in women). I've done pretty well in this culture, so I can't knock it. Besides, I'm too conservative to believe that stuff." Despite the potential importance the therapist attached to feminist interventions, she did not argue with Pam or interpret her behavior as being resistant. Instead, the therapist realized that her interpretation of Pam's situation was not congruent with Pam's life experiences. It would be counterproductive and disempowering to force Pam to accept a definition of her life situation that she did not believe. The therapist simply decided to drop this line of intervention for the time being and introduce other treatment strategies.

At this point, metaphors regarding chaos theory and its applicability to human behavior were introduced and explained. The therapist advanced the notion that much of Pam's behavior had been determined by unknown factors that future therapy could explore if Pam wished. The conflicting demands that these determinants made on her resulted in apparently impulsive and goalless periods in her life, analogous to the chaotic behavior of complex systems. Her oscillation between personal extremes, for example, sexual indulgence one week and commitment to abstinence the next, was nothing more than a pattern of bifurcating commonly observed in systems undergoing change. Once Pam was presented with this metaphor, she began to feel less humiliation and shame.

She also felt herself to be more empowered to take charge of her life. The therapist then discussed the notion with her that she could probably gain insight into these causes only once she was removed from them. Pam immediately smiled because she had been thinking the same thing.

During the next therapy session, Pam suddenly announced that she had reached "a major conclusion about life: Most of what I've been upset about is no one's damn business but mine. I really cause all my own depression by being sorry for no good reason. . . . I really don't have to give account to anyone about myself but me." She offered a few examples: "I've done a lot of drugs in the past, and my mother wants me to feel guilty over that. Sometimes people at work try to make you feel guilty because you didn't work harder, put in the extra hours. Or sometimes I might be dating someone who tries to make me feel guilty about sleeping with someone or something like that in the past. I've just got to learn to ignore them."

She concluded the session with an impressive analysis of her life: "All of my life, there have been these forces trying to dictate to me what I should be. They make me feel guilty when I don't conform. It's just like that chaos stuff, being pulled in all those different directions. . . . Mostly they pulled me by making me feel guilty. That's got to stop before I can become myself."

At this point the therapist and Pam agreed to work on modifying Pam's feelings of self-punishment. A number of cognitive interventions were tried, with substantial success. Within just a few months Pam was able to reach a personally satisfactory level of adjustment. As her guilt decreased, she also stopped using cocaine and reported fewer incidences of bulimia. She next decided on a "personal moratorium" on sexual relationships "until things clear out with me." Eventually, Pam began dating again, happier with the slower pace that occurred when she set limits on herself. Within 5 months of treatment with her new therapist, Pam felt confident enough to take a break from therapy, something she had not done for several years.

Pam terminated therapy with the knowledge that she might want to continue in the future to gain additional insight or if any behavioral problems reemerged. A year later, Pam sought treatment again, following the depression accompanying a breakup of a relationship and apparent flashbacks of sexual abuse from childhood. Within four sessions Pam was able to identify several incidences of sexual abuse by an uncle that had occurred when her parents were drunk. She joined a self-help group for support but soon found it unnecessary. After a few more months of therapy, Pam stated that she felt her life was "going along better than it ever has" and decided to temporarily suspend formal treatment. At the time of this writing, Pam has been symptom-free for 2-1/2 years. She has also become politically active (in conservative organizations, much to the dismay of the therapist's liberal leanings) and occasionally calls her former therapist to rib her about being "too politically correct."

CONCLUSION

For a number of reasons, women frequently react to their own impulsive behaviors with intense shame. This shame is usually counterproductive, interfering with a woman's self-esteem. It is often necessary to treat this shame directly before women can experience therapeutic change. The treatment of women who behave impulsively is not limited to the techniques described in this chapter and, indeed, a variety of eclectic methods are useful with such clients. Generally, however, a feminist orientation is the best for treating impulsive women. It is important, however, that the therapist's interventions are congruent with the client's worldview and the social realities in her life.

It is often the practice to end research articles or chapters with a plea for future studies and for more empirical research. It is rarely the practice to do so in an "applied" clinically oriented chapter such as this one. However, we must take exception and call strongly for more empirical efforts from feminist-oriented psychologists. It is doubtful that additional theorization without verifying data will do much to advance knowledge. Although substantial research is needed on the treatment of impulsive individuals, even more efforts are needed on the successful treatment of impulsive women. Feminist psychologists and other mental health practitioners need to discover what works and how. This will undoubtedly involve a major research effort and substantial collaboration and cost. However, unless researchers realize that women can and do behave impulsively, this line of inquiry is apt to be a low priority.

REFERENCES

Abraham, F. D., Abraham, R. H., & Shaw, C. D. (1991). *A visual introduction to dynamical systems theory for psychology*. Santa Cruz, CA: Aerial Press.

Alterman, A. I. & Cacciola, J. S. (1991). The antisocial personality disorder diagnosis in substance abusers: Problems and issues. *The Journal of Nervous and Mental Disease, 179*, 401–409.

Ballou, M., & Gabalac, N. W. (1985). A feminist position on mental health. Springfield, IL: Charles C Thomas.

Barth, F. D. (1988). The treatment of bulimia from a self-psychological perspective. *Clinical Social Work Journal, 16*, 270–289.

Bates, C. M., & Brodsky, A. M. (1989). *Sex in the therapy hour: A case of professional incest*. New York: Guilford.

Bepko, C. (Ed.). (1992). *Feminism and addiction*. Binghamton, NY: Haworth.

Blume, S. (1990). *Secret survivors: Uncovering incest and its after effects in women*. New York: Wiley.

Brantley, P. J. & Sutker, P. B. (1984). Antisocial behavior disorders. In H. E. Adams & P. B. Sutker (Eds.), *Comprehensive handbook of psychopathology*, (pp. 439–478). New York: Plenum Press.

Brennan, T. (1992). *The interpretation of the flesh: Freud and femininity*. New York: Routledge.

Brown, J. (1940). *The psychodynamics of abnormal behavior*. New York: McGraw-Hill.

Brown, L. M. & Gilligan, C. (1992). *Meeting at the crossroads: Women's psychology and girls' development*. Cambridge, MA: Harvard University Press.

Brown, L. S. (1987). From alienation to connection: Feminist therapy with post traumatic stress disorder. *Women and Therapy, 5*, 13–26.

Brown, L. S. (1990). Gender and assessment. *Professional Psychology: Research and Practice, 21*, 12–17.

Brown, L. S. (1992). A feminist critique of the personality disorders. In L. Brown & M. Ballou (Eds.), *Personality and psychopathology: Feminist reappraisals* (pp. 206–228). New York: Guilford Press.

Bynum, B. (1992). *Transcending psychoneurotic disturbances: New approaches in psychospirituality and personality development*. New York: Haworth.

Chamberlain, L. (1989). *Chaos and change in systems: A model for understanding suicidal behavior in families*. Unpublished manuscript, University of Denver, School of Professional Psychology.

Chodorow, N. (1989). *Feminism and psychoanalytic theory*. New Haven, CT: Yale University Press.

Craig, R. J. (1986). The personality structure of heroin addicts. In S. I. Sazara (Ed.), *Neurobiology of behavioral control in drug abuse* (NIDA Research Monograph 74, pp. 253–266). Washington, DC: U.S. Department of Health and Human Services.

Crawford, J., Kippax, S., Onyx, J., Gault, U., & Benton, P. (1992). *Emotion and gender: Constructing meaning from memory*. London: Sage.

Dahlstrom, W., Welsh, G., & Dahlstrom, L. (1972). *An MMPI handbook; Vol. 1: Clinical interpretation* (2nd ed.). Minneapolis: University of Minnesota Press.

Espin, O., & Gawelek, M. A. (1992). Women's diversity: Ethnicity. race, class, and gender in theories of feminist psychology. In L. Brown & M. Ballou (Eds.), *Personality and psychopathology: Feminist reappraisals* (pp. 88–110). New York: Guilford Press.

Eysenck, H. J., & Eysenck, M. W. (1985). *Personality and individual differences: A natural science approach*. New York: Plenum Press.

Eysenck, H. J., & Gudjonsson, G. (1989). *The causes and cures of criminality*. New York: Plenum Press.

Eysenck, S. B. G., Pearson, P. R., Easting, G., & Allsopp, J. F. (1985). Age norms for impulsiveness, venturesomeness and empathy in adults. *Personality and Individual Differences, 6,* 613–619.

Feingold, A. (1988). Cognitive gender differences are disappearing. *American Psychologist, 43,* 95–103.

Fine, M., & Gordon, S. M. (1989). Feminist transformation of/despite psychology. In M. Crawford & M. Gentry (Eds.), *Gender and thought: Psychological perspectives* (pp. 146–174). New York: Springer.

Foucault, M. (1981). *The history of sexuality: Vol 1. An introduction.* Harmondsworth, England: Viking.

Fox, R. (1992). *Elements of the helping process: A guide for clinicians.* Binghamton, NY: Haworth.

Frayser, S. (1985). *Varieties of sexual experience: An anthropological perspective on human sexuality.* New Haven, CT: HRAF Press.

Freud, S. (1953). Three essays on sexuality. In J. Strachey (Ed. and Trans.), *The standard edition of the complete psychological works of Sigmund Freud* (Vol. 7). London: Hogarth Press. (Original work published in 1905)

Freud, S. (1961). Some physical consequences of the anatomical distinction between the sexes. In J. Strachey (Ed. and Trans.), *The standard edition of the complete psychological works of Sigmund Freud* (Vol. 19, pp. 248–258). London: Hogarth Press. (Original work published 1925)

Garfinkel, A. (1987). The virtues of chaos. *Behavioral and Brain Sciences, 10,* 178-179.

Gilligan, C. (1982). *In a different voice.* Cambridge, MA: Harvard University Press.

Gleick, J. (1987). *Chaos: Making a new science.* New York: Viking Penguin.

Gottesman, I., & Shields, J. (1982). *Schizophrenia: The epigenetic puzzle.* Cambridge, England: Cambridge University Press.

Hare-Mustin, R., & Maracek, J. (1990). *Making a difference: Psychology and the construction of gender.* New Haven, CT: Yale University Press.

Hyde, J. (1990). Meta-analysis and the psychology of gender differences. *Signs: Journal of Women in Culture and Society, 16,* 55–73.

Hyde, J. (1991). *Half the human experience: The psychology of women* (4th ed.). Lexington, MA: Heath.

Johnson, J., (in press). Shame and sex bias in the impulsive behaviors of women and men: Four empirical studies. *Contemporary Psychodynamics: Theory, Research and Application.*

Johnson, J., & McCown, W. (in press). *Family therapy of neurobehavioral disorders.* New York: Haworth Press.

Johnson, J., McCown, W., & Booker, M. (1986, April). *MMPI profiles of multiply abused and sheltered women.* Paper presented at the Meeting of the Midwestern Psychological Association, Chicago, Illinois.

Jones, D. (1990). Social analysis in the clinical setting. *Clinical Social Work Journal, 18,* 393–406.

Kaplan, M. L., & Kaplan, N. R. (1991). The self-organization of human psychological functioning. *Behavioral Science, 36*, 161–178.

Kelso, J. A., & Fogel, A. (1987). Self-organizing systems and infant motor development. *Developmental Review, 7*, 39–65.

L'Abate, L., & Cox, J. (1993). *Therapeutic writing.* New York: Brunner-Mazel.

Lips, H. (1988). *Sex and gender: An introduction.* Mountain View, CA: Mayfield.

Lowery, M. (1987). Adult survivors of childhood incest. *Journal of Psychosocial Nursing, 25*, 27–31.

Loye, D., & Eisler, R. (1987). Chaos and transformation: Implications of nonequilibrium theory for social science and society. *Behavioral Science, 32*, 53–65.

Matlin, M. (1987). *Psychology of women.* New York: Holt, Rinehart, & Winston.

McAuliffe, W., & Albert, J. (1992). *Clean start: An outpatient program for initiating cocaine recovery.* New York: Guilford Press.

McCown, W. (1989). Impulsivity and twelve step self-help "success." *British Journal of Addictions, 84*, 91–93.

McCown, W. (1990). Impulsivity and twelve step self-help "success": A prospective study. *British Journal of Addictions, 85*, 635–637.

McCown, W., & Barkhausen, D. (1993). *A Rasch scaled measure of impulsiveness suitable for use with Thematic Apperception tests.* Unpublished manuscript, Nathan Kline Institute, Orangeburg, New York.

McCown, W., & Johnson, J. (1985, March). *Factors related to the decision to recommend court ordered psychotherapy by probationary personnel.* Paper presented at the Annual Meeting, Society of Criminal Justice Sciences, Las Vegas, Nevada.

McCown, W., & Johnson, J. (1993). *The treatment resistant family: A consultation/crisis intervention model.* New York: Haworth Press.

McKenna, G. J. (1979). Fitting different treatment modes to patterns of drug use. In H. A. Wishnie & J. Nevie-Olsen (Eds.), *Working with the impulsive person* (pp. 113–123). New York: Plenum Press.

Mead, M. (1935). *Sex and temperament in three primitive societies.* New York: Dell.

Mead, M. (1949). *Male and female.* New York: William Morrow.

Meadow, R., & Weiss, L. (1992). *Women's conflicts about eating and sexuality: The relationship between food and sex.* Binghamton, NY: Haworth Press.

Merton, R. (1968). *Social theory and social structure.* Glencoe, IL: Free Press.

Miller, D. (1980). Women in pain: Substance abuse/self-medication. In M. P. Kirkin (Ed.), *The social and political contexts of family therapy* (pp. 179–192). Boston, MA: Allyn & Bacon.

Miller, J. B. (1988). *Toward a new psychology of women* (2nd ed.). Boston: Beacon.

Murray, H. (1943). *The Thematic Apperception Test.* Cambridge, MA: Harvard University Press.

Ogden, J. (1992). *Fat chance: The myth of dieting explained.* New York: Routledge.

Okun, B. (1990). *Seeking connections in psychotherapy.* San Francisco: Jossey-Bass.

Parsons, T., & Bales, R. (1955). *Family, socialization and interaction process.* New York: Free Press.

Pope, K. (1990). Therapist–patient sex as sex abuse: Professional and practical dilemmas in addressing victimization and rehabilitation. *Professional Psychology: Research and Practice, 21,* 227–239.

Ramirez, M. (1991). *Psychotherapy and counseling with minorities: A cognitive approach to individual and cultural differences.* Needham, MA: Allyn & Bacon.

Ranke-Heinamann, U. (1990). *Eunuch for the kingdom of heaven: Women, sexuality, and the Catholic church.* New York: Penguin Books.

Rawlings, E., & Carter, D. (Eds). (1977). *Psychotherapy for women: Treatment towards equality.* Springfield, IL: Charles C Thomas.

Reiss, I. (1990). *An end to shame: Shaping our next sexual revolution.* Buffalo, NY: Prometheus.

Resneck-Sannes, H. (1991). Shame, sexuality, and vulnerability. *Women and Therapy, 11,* 111–125.

Rushton, J. P. (1988). Genetic similarity, mate choice, and fecundity in humans. *Ethological Sociobiology, 9,* 329–335.

Rutter, P. (1989). *Sex in the forbidden zone.* Los Angeles: Tarcher.

Saakvitne, K., & Pearlman, L. (1993). The impact of internalized misogyny and violence against women on feminine identity. In E. P. Cook (Ed.), *Women, relationships, and power: Implications for counseling* (pp. 247–274). Alexandria, VA: American Counseling Association.

Scheff, T., & Retzinger, S. (1991). *Emotions and violence: Shame and rage in destructive conflicts.* Lexington, MA: Lexington/Heath.

Sherfey, M. J. (1972). *Nature and evolution of female sexuality.* New York: Stratford Press.

Stakes, J., & Oliver, J. (1991). Sexual contact and touching between therapists and clients: A survey of psychologists' attitudes and behaviors. *Professional Psychology: Research and Practice, 22,* 297–307.

Surrey, J. L. (1985). *Self-in-relation: A theory of women's development.* Wellesley, MA: Stone Center of Developmental Studies and Services.

Thone, R. R. (1992). *Women and aging: Celebrating ourselves.* Binghamton, NY: Haworth.

Way, K. (1992). *Anorexia nervosa and recovery.* Binghamton, NY: Haworth.

Wilson, T. (1975). *Sociobiology.* Cambridge, MA: Harvard University Press.

19

INTERPERSONAL COGNITIVE PROBLEM SOLVING AS PREVENTION AND TREATMENT OF IMPULSIVE BEHAVIORS

MARIA E. TOUCHET, MYRNA B. SHURE,
and WILLIAM G. McCOWN

A consistent body of research has long shown that children who exhibit dysfunctional behaviors at an early age, including the prevalence of impulsive behaviors, have a disproportionately higher rate of later psychopathology (Achenbach, 1988; Cowen, Pederson, Babigian, Izzo, & Trost, 1973; Eisenberg, 1988; Felner, Jason, Moritsugu, & Farber, 1983). For example, Loeber (1982) demonstrated that children exhibiting high rates of impulsive antisocial behavior continued to display these and other dysfunctional behaviors later in life at much higher rates than did children who exhibited low rates of these problems. Attempts to explore possible mediating factors for this phenomenon include the work of Spivack, Platt, and Shure (1976), who demonstrated that the cognitive steps people use to solve everyday problems are crucial elements of personal adjustment and mental health. Impulsive or aggressive responses to interpersonal problems can be an indication of deficient cognitive skills, and these deficits are repeatedly associated with further dysfunctional behaviors (Shure & Spivack, 1988). Shure and Spivack have postulated that improvement in specific problem-solving abilities will decrease the occurrence of dysfunctional behaviors and subsequently lessen the likelihood of later psychopathology. In this chapter we briefly review some of the research program and treatment methodologies of Shure, Spivack, and their associates, suggesting that impulsive behaviors can be reduced by psychosocial interventions designed to teach specific interpersonal problem-solving skills.

COGNITIVE PROCESSES AND IMPULSIVENESS

The prevalent view of many psychologists prior to the 1970s was that childhood socialization primarily involved mastery of a number of discrete rules and situation-specific rules (Scarr, Weinberg, & Levine, 1990). An emphasis on rule-governed behavior was an appropriate and obvious starting point for the study of childhood development (Thomas, 1985). Adults often teach children the "correct" behavior for different situations inasmuch as children are told to perform or not perform certain acts because of a rationale provided by adults (Collins & Kuczaj, 1991). Hence, children are provided with varying degrees of external adult structure and guidance, with possible rewards and punishment being contingent on displayed behavior elicited in specific environments (Skinner, 1957).

However, a frequently overlooked aspect of development is that the use of external techniques to modify desired or unwanted behavior limits the child's ability to internally generate effective ways of dealing with personal and interpersonal problems (Spivack & Shure, 1974). In this situation, the child is unable to create solutions to his or her own problems and is consequently restricted in the competence to develop solutions in the absence of assistance from others. As a result, such children—and later the adults they become—behave impulsively and often aggressively in the absence of external structures or in novel situations.

An individual's proficiency in the cognitive process associated with solving interpersonal problems has been found to correlate with a variety of measures of personal and social adjustment (Spivack et al., 1976). The reasons for this are not hard to understand. Durlak (1983) noted the following:

> Good problem solvers . . . are flexible and adaptable in different social circumstances, able to deal effectively with stress, and able to develop suitable methods to attain personal goals and satisfy their needs. Moreover, repeated success in problem-solving would be expected to

heighten self-confidence, motivation, and perseverance, thus facilitating future task performance. (p. 31)

Indeed, nearly 25 years ago, Shure and Spivack (1970) found that inner-city school students' abilities to simply list several different solutions to a specific problem positively correlated with indexes of behavioral adjustment. Children who were rated as less well adjusted generated fewer alternate solutions than did their better adjusted counterparts. Later, Richard and Dodge (1982) found that aggressive and isolated second-through fifth-grade boys were less likely to develop varied solutions to a given problem than were the peer-rated "popular" boys. Similarly, Asarnow and Callan (1985) found a difference in fourth- and sixth-grade boys who had positive and negative peer status in their ability to generate alternate solutions. The results indicated that boys with negative peer status "generated fewer alternative solutions, proposed fewer assertive and mature solutions, showed less adaptive planning, and evaluated physically aggressive responses more positively and positive responses more negatively than did P [positive peer status] boys" (Asarnow & Callan, 1985, p. 80).

Hence, to facilitate social adjustment and mental health better, we believe that adults must teach children *how* to think, not *what* to think; emphasis must be placed on the process of problem solving and not merely on the content of the solution (Shure & Spivack, 1988). This is certainly not a new concept. Baron and Brown (1991) noted that its roots extend at least as far back as Socrates. More specifically, the tradition of cognitive problem solving in this century can be traced to John Dewey and his intellectual descendants. Nevertheless, only during the past 25 years have psychologists paid extensive attention to the development of theories and technologies designed to teach problem-solving skills.

Consistent with this trend in research, the *interpersonal cognitive problem solving (ICPS) skills training program* was developed by Shure (reviewed by Spivack & Shure, 1985). Although the ICPS is only one approach to the general schema of teaching interpersonal problem-solving skills (see Baron & Brown, 1991), it is among the most systematic and empirically researched problem-solving interventions (Pelligrini & Urbain, 1985). The ICPS approach is based on several theoretical concepts supported by previous research. These include the notion that people who are unable to envision (or who have no concern for) the consequences of their actions, those who place importance on the end rather than the means, and those who are unable to develop alternate routes toward their goals are more susceptible to impulsivity, aggression, or the avoidance of problems through withdrawal (Shure & Spivack, 1988). The assumption is that intervention with ICPS training can greatly reduce the need for impulsive and aggressive actions and consequently restrain other dysfunctional behaviors.

CHILDREN AND PROBLEM SOLVING

Children can be differentiated according to their ability to solve social interpersonal problems as early as age 4 (Shure & Spivack, 1970). The major differences that can be demonstrated are (a) the ability to develop alternative solutions to an interpersonal problem and (b) the ability to foresee the consequences to an interpersonal action. For example, suppose a child wishes to obtain a toy from another child. If this child responds impulsively to the obstacle by hitting the second child, this action could be related to one of two possible deficiencies in the child's ICPS skills. First, the child could be unable to visualize the consequences of impulsive behavior; thus, the possibility of such consequences are not accessible as a deterrent. Alternatively, the child could be able to conceive the repercussions of his or her actions but hit the child anyway because he or she is unable to develop an alternate plan of action for meeting his or her goals. The emphasis of the ICPS is to enhance the child's ability to generate novel solutions that, because of their more positive outcome, become self-reinforcing and are eventually integrated into the child's natural behavioral repertoire.

A principal belief of ICPS proponents is that the intervention skills training is the most productive and advantageous if applied at an early age. ICPS-deficient children have been found to experience more psychological and social abnormalities than ICPS-competent ones (Spivack et al., 1976). Therefore, early intervention (age 4 and beyond) is logically more likely to prevent the development of later dysfunctional behaviors. ICPS deficiencies exist independently of IQ scores (Spivack & Shure, 1974) and can be found across all socioeconomic and racial populations. However, impoverished urban children were chosen for the initial implementation of ICPS training and its subsequent development. Because poor urban children are at a greater risk for social maladjustment and dysfunctions caused by the effects of greater levels of environmental stress, intervention may appear to be the most beneficial in this milieu and is probably maximally cost-efficient.

Implementing ICPS Training

Although variants of ICPS training have been used in clinical settings for a number of discrete problems (e.g., D'Zurilla, 1986; Platt, 1989; Shure & Spivack, 1978), the ICPS is rooted more firmly in community than in clinical psychology. Early in the history of problem-solving training, it was decided that interventions would be targeted at an entire classroom within a school-based program, as opposed to testing children for ICPS competence and then conducting training with only the ICPS-deficient students (Shure & Spivack, 1970). Administering the program to the entire class of students at once eliminates many potential problems. For example, children initially identified as ICPS competent might actually be deficient and thus be overlooked. Furthermore, ICPS-competent children help to avoid initial silence in small training groups. Perhaps equally important, larger groups of children can be reached in a shorter period of time and no child feels left out (Shure & Spivack, 1988). Also, an improvement in ICPS skills is beneficial for everyone, so by targeting the entire classroom, no student would be harmed.

Structured training manuals have been developed and recently revised for teaching problem-solving skills to different age groups (Shure, 1992a, 1992b, 1992c). These "scripted" texts make it relatively easy for teachers or even paraprofessionals to master the philosophy of interpersonal social problem-solving approaches and to consistently model, teach, and reward problem-solving ability in children of any age. The programs for preschool and for kindergarten and the primary grades developed by Shure (1992a, 1992b) take 4 months to execute when conducted for approximately 20 minutes per day. The teachers work in small mixed groups (about 6-10 students) of ICPS-competent and ICPS-deficient children. Although the formal implementation is a script, informal use of the principles everyday while the children are in school appears to supplement learning consolidation (Spivack & Shure, 1974).

ICPS, administered at any age, begins by assuming little about a child's interpersonal cognitive problem-solving abilities. For example, an error often made by adults is to believe that children always have a full understanding of the vocabulary necessary for problem solving. A familiar statement made by a parent is something similar to, "You can't hit Jimmy because you might hurt him." What is not understood, however, is that children frequently cannot grasp words such as *can't* because they have no context to place them in. The first lessons of the program follow the assumption that children cannot comprehend the vocabulary and must be taught certain basic language concepts essential for problem solving. Children are thus brought through a series of games to strengthen their language skills in order to better facilitate clear communication with others.

Although children use basic words such as *or*, *and*, and *not*, some educators have found that children frequently do not possess a clear understanding of these words (Bereiter & Englemann, 1966). By teaching children the precise meaning and use of these linguistic concepts, an understanding of thinking in terms of alternative options ensues. For example, the word *or* is essential if a child is to be able to later develop alternate solutions. Children at this age are able to distinguish between boys and girls; therefore, by using the context of gender (a concept that the child is able to grasp), the teacher might ask, "Are you a boy *or* are you a girl?"

Children need to comprehend the principle of negation in order to be effective problem solvers, so the word *not* is taught as an aid to this understanding. To illustrate two opposing notions, the teacher might say "Johnny is a boy. Johnny is *not* a girl." This can then be posed as a question to the children, "Johnny is a boy. Johnny is *not* a _____ ?" The responses could then be humorous to further illustrate the negation, "Johnny is a boy. Johnny is not a piano." Letting children develop the many things that "Johnny is not," assists in their learning the linguistic nuances of the word.

The word *and* is important if children are going to be able to think in terms of more than one aspect of a situation at a time. Teachers may say, "Sally *and* Mary are girls." A combination of the terms may then be applied: "Am I pointing to Richard *and* Sally *or* am I pointing to Peter *and* Mary? [children respond] Good, I am pointing to Peter *and* Mary. I am *not* pointing to _____ *and* _____ ."

Additional language skills taught during this prerequisite program are essential for later problem solving. The words *some*, *same*, and *different* are helpful for teaching children that others may possess preferences different from their own; for example, "*different* people like *different* things; *some* people like apples and *some* do not; not everybody likes the *same* things" (Spivack & Shure, 1974, p. 24). The understanding that not all children want the same things will later help a child avoid arriving at erroneous conclusions in interpersonal relationships. These words can be taught through word games such as "A hat is *different* from a flower. A hat is *not* the *same* as a flower."

The concept of *if–then* is introduced by the teacher in order to strengthen the child's understanding of cause and effect and help them later to perceive the consequences of his or her actions. A tenet of the ICPS is that the child must discover that there is an aftermath or consequence for every action he or she performs. To familiarize children with this concept, the teacher might say, "*If* Billy is a boy, *then* Billy is *not* a girl" or "*If* your name is Susan, *then* you *can* jump." This class of intervention eventually leads to the understanding of "*If* I hit Peter, *then* he will cry."

Beyond the Basic: Becoming Aware of Others

The next section of the ICPS program consists of lessons aimed at teaching children how to become aware of other people's emotions. Sensitivity to others' emotions is closely related to understanding that people have different preferences. In ICPS, children learn that things that make one person happy might make another person sad and that something that makes someone happy does not necessarily always evoke that same emotion (Shure, 1992a, 1992b, 1992c). Another important point taught is that there is always more than one way to determine how a person is feeling in a certain situation.

It is first critical to teach children how to identify emotions. Research indicates that high-risk children may have deficits in the ability to recognize the emotions of others (McCown, Johnson, & Austin, 1986). Using pictures that depict different situations, the teacher would say, "Here is a picture of a girl. What is she doing? [group responds] Yes, she is crying. How do you think she feels? Yes, *sad*. What makes you think she feels *sad*? [group responds] Yes, she is crying" (Shure, 1992a, p. 67).

Once the children have learned the names for and expressions of emotions, they must learn how to identify the emotions of other people in different situations. Through this learning to distinguish the emotions of others, children begin to internalize the feelings and become more able to view the situation through the eyes of another. The ICPS stresses that there are many ways to discover how a person is feeling (Shure, 1992a, 1992b, 1992c). A sample lesson is the following: "I am laughing. [demonstrate] Am I *happy* or am I *sad*? [group responds] How can you tell I'm *happy*? [If response is 'you're laughing': How can you tell I am laughing?] Did you see me with your eyes? Did you hear me with your ears?" (Shure, 1992a, p. 72). An additional lesson that is taught is that if none of the senses seem adequate enough to determine how a person is feeling, the child can always directly ask how the other person is feeling.

Children are also taught that cause–effect relationships are directly related to consequences and are guided toward identifying the causal connection between an act and its consequences. The children are directed toward learning that behaviors have causes (*why–because* relationships), and they learn to avoid making rash assumptions about others (*might–maybe* relationships). The teacher could say, "I like birthday parties. Can you guess *why* I like birthday parties? [children respond] Very good. I *might* like birthday parties *because* [repeat response]. Now let's think of a *different because*. I like birthday parties *because* _____" (Shure, 1992a, p. 106).

The final section of the program consists of three parts: alternative solutions, alternative consequences, and solution and consequence pairing. Once the language skills required for problem solving are mastered, children can proceed to games and dialogues focused on instilling the program's major interpersonal problem-solving skills.

The first subgroup in this section, alternative solutions, prompts children to generate as many solutions as possible when presented with a typical interpersonal problem. The lessons are designed to use all that the child has learned in the preceding lessons, stimulating the child to ask, "What else can I do?" The importance of this section is to have the children develop as many *different* solutions as possible, not to evaluate which solutions are right and which are wrong. Hence, the solution of "I'd bite her," is just as valid as, "I'd ask her to share." No value judgments about the solutions can be made by the trainers. The children must be encouraged only to produce as many different solutions to the problem as possible.

Once the children have learned to form several solutions to a given problem, the next step is to have them identify what would happen if their solutions to the problem were actually implemented in the given situation. They are asked to develop as many different consequences of their actions as possible, and then they are asked to make a judgment about which consequence would be the most feasible or realistic. It is in this section that judgments about responses are made. No value judgments are made by the teachers; instead, the evaluations are made by the children themselves. The children must learn to decide what is a good idea and what is not; they must learn to think for themselves and not just passively follow the directions of others.

A sample script narration for consequential thinking is as follows:

Show children illustration of a boy and a girl.

> *Teacher:* Let's pretend the problem here is that this boy [point] wants this girl [point] to let him feed the fish. What can the boy do or say so the girl will let him feed the fish? I'm going to put *all* your ideas on this side [left side] of the chalkboard.

[Elicit three or four solutions. Write each one as given.]

> *Teacher:* Now listen very carefully. We're going to change the question. *If* the boy [pushes the girl out of the way], *then* what *might* happen next? I'm going to write what you think *might* happen next over here [point to right side of the board].

[Children respond. Teacher writes the consequence on the right side of the board and very dramatically draws an arrow from the solution to the consequence.]

Teacher: OK, she *might* [hit him]. What *might* happen next?

[Teacher elicits additional consequences, adding them to the list and drawing arrows from the solution. Teacher should remember to watch for and classify enumerations, for example "telling her father" and "telling her mother" and place them with the like consequence. Teacher then classifies by saying, "Telling her mother and telling her father are kind of the *same because* they are both telling someone. What *might* happen next that is *different* from telling someone?"]

[When a number of consequences have been generated]

Teacher: Look at *all* the things that can happen just from this one act, [pushing the girl out of the way]. Maybe *some* of you think [pushing her out of the way] *is* a good idea. If you think [pushing her out of the way] *is* a good idea, raise your hand. *Why* do you think that *is* a good idea? OK maybe it *is* a good idea *because* (repeat child's reason).

[Teacher asks each child who raises his or her hand to state a reason.]

Teacher: Maybe *some* of you think [pushing her out of the way] is *not* a good idea. If you think [pushing her out of the way] is *not* a good idea, raise your hand. *Why* do you think that is *not* a good idea? OK maybe that's *not* a good idea *because* [repeat child's response].

[Teacher asks each child who raises his or her hand to state a reason.]
(Shure, 1992a, pp. 233–235)

The final section, solutions and consequences pairing, is an incorporation of all that has been taught previously. The children are asked to develop a solution for a problem, determine the consequence of this solution, and evaluate the solution given by weighing the pros and cons of the consequence that could possibly ensue. The children are then asked to return to the same problem and develop another solution and consequence pair. This cycle continues in order to help the child determine which of the solutions would elicit the most favorable consequence before the child takes action; the child therefore learns to think through the situation thoroughly before reacting impulsively.

RESEARCH EVALUATIONS

Previously, we cited studies suggesting that ICPS-competent children are more socially and behaviorally adjusted than their ICPS-deficient counterparts. A second pertinent question is how effective ICPS training actually is in modifying maladaptive and impulsive behavior and fostering emotional health.

To begin evaluating the program's impact, certain questions had to be addressed. First, the extent to which a 4-month intervention program was sufficient to successfully improve a child's ICPS skills and improve behavioral adjustment needed to be established. Additionally, it was necessary to demonstrate whether behavioral gains were largely attributable to the ICPS training or to extraneous sources of variance. Another issue concerned longitudinal maintenance of treatment gains and cross-situational effects of the training. Finally, an important question concerned whether ICPS training was effective only at lessening dysfunctional behaviors in children already exhibiting maladaptive behavior or if it could also prevent the later appearance of maladaptive behaviors in otherwise adjusted children.

To address many of these issues, Shure and Spivack (1982) evaluated an ICPS intervention program over a 2-year period. The subjects were African-American urban preschool and kindergarten children. In the first year, 113 children took part in the training program and 106 were not trained. One hundred thirty-one children in the kindergarten class were divided into four groups: once trained in preschool, once trained in kindergarten, twice trained, and never-trained controls.

ICPS training was found to have a major impact on the behavioral adjustment of preschool and kindergarten children. It was especially effective at reducing maladaptive behaviors such as impulsiveness and inhibited behavior. Of the children who entered the training as impulsive—on the basis of a multimethod of assessment—50% became better adjusted as compared with only 21% of the control children. It was also found that these improvements were stable over time and lessened the chances of a later reoccurrence. At 1- and 2-year follow-ups, the behavioral gains were still intact.

Spivack and Swift (1977) found that the likelihood of dysfunctional behaviors increased as low socioeconomic children progressed through the early grades of elementary school. Shure and Spivack (1982) found that children classified as adjusted before implementation of ICPS skills training were less likely than the untrained adjusted children to develop maladaptive behaviors when they entered the more rigorous higher grades. Hence, it was found that ICPS training not only reversed already-present maladaptive behaviors but prevented the development of new ones. This also addresses the durability of ICPS skills training, the effects of which were found to remain stable as the children progressed through the lower grades.

The group with the most amelioration in alternative solution skills was the group that was exposed to two 3-month-long training programs. However, it was found that one intervention alone was sufficient to foster healthier social adjustment and interpersonal competence. The data indicated that for those children who did not receive training in preschool, kindergarten was not too late. ICPS training was most effective at terminating maladaptive behaviors, but it was also found to enhance the children's social competencies. ICPS-trained children were found to be more concerned with the feelings of other children and better liked when compared with the control children. One important aspect of the training was discovered to be the need to implement informal dialogue with the children. The formal scripts provide the children with a context through which to view the problem, but the informal dialogues then aid the children in progressing from the hypothetical to the actual. Along these lines, R. J. Allen (1978) found that the formal ICPS script combined with informal problem-solving dialogue techniques were able to notably alter dysfunctional behaviors. Indeed, Durlak and Sherman (1979) found that the use of the formal scripts alone might not be sufficient to modify behavior.

In addition to the school-based program, many inner-city mothers were trained to implement this training at home. Children trained at home by their mothers were compared with controls, and the at-home trained children were rated better on social competency scales by teachers who had no knowledge of the training (Shure & Spivack, 1978). This research also showed that the benefits of the training were the most dramatic when the inner-city mothers (most of whom were ICPS deficient themselves to begin with) were taught how to effectively problem solve. The mothers became more adept at dealing with interpersonal problems that arose each day with their children, and this subsequently fostered a healthier home environment for both the children and mothers. This at-home application also shows that the behavioral advancements were not limited to the classroom.

The research cited thus far has established that an implementation of ICPS training is beneficial for a variety of adjustment factors encompassing self-concept, academic adjustment, psychiatric and behavioral difficulties, delinquency, and interpersonal functioning (Feis & Simons, 1985). ICPS training has been demonstrated to be effective not only in modifying existing maladaptive behaviors but also in fostering mental health in high-risk but otherwise normal children. This is quite promising for a primary preventive measure because the training can be instituted in a relatively short period of time and does not require high-level professionals for implementation.

Meta-analyses of the ICPS literature have been conducted by Denham and Almeida (1987). These authors examined reported relations between children's interpersonal cognitive problem-solving skills and adjustment. They also attempted to specify the specific effects of ICPS training. In general, the relation between ICPS and adjustment appeared to be robust. Furthermore, interventions yielded clear increases in ICPS skills. Intervention effects on specific behavioral adjustment scores were found to be somewhat more equivocal; meta-analytic results differed depending on whether behavioral ratings or observations was the dependent variable. Teacher–child dialogues on ICPS real-life situations, the expertise of the investigator, the source and quality of the publication, and the length of interventions mediated the magnitude of certain intervention effects. In general, however, results were highly favorable to the assumptions that have guided the research of Shure and her associates.

However, it should be noted that meta-analysis also suggests that the impact that ICPS training has on behavioral adjustment in older children has not been as firmly established or apparently as positive as the school-based programs for younger children. Similar findings were reported by Shure and Spivack (1982). Research on such groups indicates that either dysfunctional behaviors do not change substantially (G. Allen, Chinsky, Larcen, Lochman, & Selinger, 1976) or that behavior slightly improves but that the connection between behavioral gains and ICPS training is not as substantial as that observed in younger children (Elias, 1980; Weissberg et al., 1981). An example of some of the research relevant to the question of maintenance of change in older students is a study by Shure (1986), who implemented a 4-month ICPS intervention program with urban fifth graders. The program was shown to have an initial positive effect on behavioral adjustment, but the longevity of effects was not as pronounced as with younger children. A program reexposure was needed in the sixth grade in order to counteract the impulsive and inhibited behaviors that had spontaneously reappeared or developed within that time. Although Shure's (1986) research suggests that earlier intervention may have more immediate impact on behavior than later, new evidence suggests that training in middle school and junior high can have a profound impact on behavior as well (Elias & Clabby, 1989, 1992; Jackson, Fontana, & Weissberg, 1992; Weissberg, personal communication, 1993; Weissberg & Gesten, 1982).

In our clinical experience the major problem that has been encountered with implementation is that ICPS trainers often expect results too quickly and may become discouraged if behavioral gains are not automatically apparent. Teachers need to be reminded that if continued for the full scripted time, change is likely to occur but that it will not occur overnight. Also, the informal dialogue required of the program is difficult to learn. Parents and teachers are used to telling children what to do and then providing a rationale to the children for that decision; it is difficult to learn to let the child think through the situation themselves. Trainers are mistaken in the belief that children will become able to talk them into something they do not want the children to do. It is important to learn that one must guide the child through the thought process but not determine what the solution should be. These are minor problems, however, that if trained properly, the trainer can overcome.

OTHER USES OF ICPS TRAINING

The ICPS approach was designed to be applicable across racial, ethnic, and age groups. ICPS-inspired interventions have been designed for application from the preschool level through adulthood (see Spivack et al., 1976, for a review), as well as across socioeconomic groups (Camp & Bash, 1981; Elardo & Cooper, 1977; Weissberg, Gesten, Liebenstein,

Schmid, & Hutton, 1979). Various scripts that are based on a social problem-solving approach have also been modified for use with abused children and their families (Nesbitt et al., 1980), for emotionally disturbed children and their parents (Yu, Harris, Solovitz, & Franklin, 1986), for developmentally disabled children at high risk for human immunodeficiency virus disease (McCown, 1993), for elderly people, (McCown & DeSimone, 1993), and for learning-disabled children (Weiner, 1978). The diversity of the training groups exemplifies the impact and acceptance that ICPS training has had on many different fields, including mental health, education, and social welfare. Johnson and McCown (in press) have used an ICPS-based approach for persons with mild-to-moderate neurobehavioral disorders. Shure and McCown (1993) have even directly adopted ICPS training for use with a radio format, with listeners calling in to receive script-based solutions to problems associated with child-rearing, substance abuse, and relationship issues.

Along with the previous applications of ICPS training, this intervention has been used in the treatment of other dysfunctional behaviors, including clinical syndromes closely related to impulsiveness. The cognitive processes used to problem solve have been shown to directly affect behavioral adjustment. It has been theorized that these cognitive processes play a large role in the development of drug addiction (Platt, 1989). This suggests that a possible reason addicts are unable to achieve their goals is that they are deficient in their ability to think through the specific steps involved in reaching their goal. Indeed, empirical research suggests that this is the case. Platt, Scura, and Hannon (1973) found that compared with an equivalent group of nonaddicts, addicts were much less capable of conceptualizing the steps through which one achieves a goal in actual interpersonal problem situations. Platt (1989) has subsequently proposed that the individual who is at the greatest risk for drug addiction is one who has higher achievement potential than his or her peers and who sets very high personal goals. When combined with deficient problem-solving skills, such a person is likely to divert some of this achievement motivation to the use of drugs.

The efficacy of ICPS training for substance abusers has been demonstrated in research by Platt and his associates. A 3-year study (Platt, Morell, Flaherty, & Metzger, 1982) was conducted in order to assess the feasibility of providing ICPS training to heroin addicts within methadone treatment programs. The results showed that significant acquisition of ICPS skills, primarily those related to means–ends thinking, occurred in the experimental group. A follow-up test showed that these skills still endured 12 months later. The teaching of ICPS skills has proved to be so successful in the recovery process that a treatment regimen has been designed specifically for this group of addicts. ICPS skills have been shown to be good predictors of both retention in treatment and successful

discharge from treatment. This training is also important in the rehabilitation process because it can be taught in small-group settings and
does not require the employment of higher level professionals in the
treatment process. The reason that ICPS is apparently successful with
adult drug addicts but not as successful with older children is an empirical
question for future research.

CONCLUSION

Research has shown that ICPS training promotes interpersonal social competence and reduces impulsiveness. This series of interventions
appears to be therapeutic by instilling the ability to more successfully
cope with interpersonal problems. This is accomplished by helping children and others understand and verbalize certain interactions, by encouraging them to visualize and verbalize the numerous possibilities available to them when dealing with others, and by having them open their
minds to the feelings of others. Personal adjustment of young children
has been found to be enhanced because the children learn to appreciate
different ways of handling problems and become sensitive to the possible
consequences of their actions.

The ICPS is not a panacea. However, the intervention program is
apparently successful because it actively engages children with language
and situations they understand and shows them how to move from the
hypothetical to the actual. It is easy to learn and implement, and the
formal script extends for only a relatively short period of time but produces
concrete results. Indeed, when one considers the relatively inexpensive
costs of implementing the ICPS or similar cognitive-based programs to
reduce impulsive behaviors and other developmentally related psychopathologies, the need for their more extensive use for prevention becomes
clear. A broader question concerns whether society will make the rational
choice to invest now for the future dividends associated with its children's
well-being. Unfortunately, this is not a question that can be answered at
this time.

REFERENCES

Achenbach, T. (1988). Developmental psychopathology. In M. Borstein & M.
    Lamb (Eds.), *Developmental psychology: An advanced textbook* (2nd ed., pp.
    549–595). Hillsdale, NJ: Erlbaum.

Allen, G., Chinsky, J., Larcen, S., Lochman, J., & Selinger, H. (1976). *Com-
    munity psychology and the schools: A behaviorally oriented multilevel preventive
    approach.* Hillsdale, NJ: Erlbaum.

Allen, R. J. (1978). *An investigatory study of the effects of cognitive approach to interpersonal problem solving on the behavior of emotionally upset psychosocially deprived preschool children.* Unpublished doctoral dissertation, Union Graduate School, Washington, DC.

Asarnow, J. R., & Callan, J. W. (1985). Boys with peer adjustment problems. *Journal of Consulting and Clinical Psychology, 53,* 80–87.

Baron, J., & Brown, R. (Eds.). (1991). *Teaching decision making to adolescents.* Hillsdale, NJ: Erlbaum.

Bereiter, C., & Englemann, S. (1966). *Teaching disadvantaged children in the preschool.* Englewood Cliffs, NJ: Prentice Hall.

Camp, B. W., & Bash, M. A. (1981). *Think Aloud: Increasing social and cognitive skills—A problem solving program for children, primary level.* Champaign, IL: Research Press.

Collins, W. A., & Kuczaj, S. A. (1991). *Developmental psychology: Childhood and adolescents.* New York: Macmillan.

Cowen, E. L., Pederson, A., Babigian, H., Izzo, L. D., & Trost, M. A. (1973). Long-term follow-up of early detected vulnerable children. *Journal of Consulting and Clinical Psychology, 41,* 438–446.

Denham, S,. & Almeida, M. C. (1987). Children's social problem-solving skills, behavioral adjustment, and interventions: A meta-analysis evaluating theory and practice. *Journal of Applied Developmental Psychology, 8,* 391–409.

Durlak, J. A. (1983). Social problem–solvings as primary prevention strategy. In R. Felner, L. Jason, J. Moritsugu, & S. Farber (Eds.), *Preventive psychology: Theory, researching, and practice* (pp. 31–48). Elmsford, NY: Pergamon Press.

Durlak, J. A., & Sherman, D. (1979, September). Primary prevention of school maladjustment. In J. A. Durlak (Chair), *Behavioral approaches to primary prevention: Programs, outcomes and issues.* Symposium conducted at the 87th Annual Convention of the American Psychological Association, New York.

D'Zurilla, T. J. (1986). *Problem solving therapy.* New York: Springer.

Eisenberg, N. (1988). *The development of prosocial and aggressive behavior.* In M. Borstein & M. Lamb (Eds.), *Developmental psychology: An advanced textbook* (2nd ed., pp. 461–496). Hillsdale, NJ: Erlbaum.

Elardo, P. T., & Cooper, M. (1977). *Aware: Activities for social development.* Reading, MA: Addison-Wesley.

Elias, M. J. (1980). *Developing instructional strategies for television-based preventive mental health curricula in elementary school settings.* Unpublished doctoral dissertation, University of Connecticut, Storrs.

Elias, M. J., & Clabby, J. F. (1989). *Social decision-making skills: A curriculum guide for the elementary grades.* Rockville, MD: Aspen.

Elias, M. J., & Clabby, J. F. (1992). *Building social problem solving skills: Guidelines from a school-based program.* San Francisco: Jossey-Bass.

Feis, C., & Simons, C. (1985). Training preschool children in interpersonal cognitive problem-solving skills: A replication. *Prevention in Human Services, 3,* 59–70.

Felner, B., Jason, L., Moritsugu, J., & Farber, S. (Eds.). (1983). *Preventive psychology: Theory, research, and practice*. Elmsford, NY: Pergamon Press.

Jackson, A. S., Fontana, T., & Weissberg, R. P. (1992). *Crossroads: A program to promote responsible decision making* [The New Haven Social Development Program, ninth-grade module (pilot edition)]. New Haven, CT: Yale University Press.

Johnson, J., & McCown, W. (in press). *Family therapy and neurobehavioral disorders*. Binghamton, NY: Haworth.

Loeber, R. (1982). The stability of antisocial and delinquent child-behavior: A review. *Child Development, 53*, 1431–1446.

McCown, W. (1993). *ICPS curriculum for reducing HIV risk behavior in developmentally disabled adolescents*. Unpublished manuscript, Nathan Kline Institute for Psychiatric Research, Orangeburg, NY.

McCown, W., & DeSimone, P. (1993). *ICPS curriculum for boosting healthy behavior in normal elderly*. Unpublished manuscript, Nathan Kline Institute for Psychiatric Research, Orangeburg, NY.

McCown, W., Johnson, J., & Austin, J. (1986). Inability of delinquents to decode facial expressions of emotion. *Journal of Social Behavior and Personality, 1*, 91–97.

Nesbitt, A., Madren-Braun, J., Bruckner, M., Caldwell, R., Dennis, N., Liddell, T., & McGloin, J. (1980). *A problem-solving approach: Final Evaluation*. Report to Law Enforcement Assistance Administration #77–2A (1)–36–52. Commerce City, CO: Children's Resource Center, Adams County Department of Social Services.

Pellegrini, D., & Urbain, E. (1985). An evaluation of interpersonal cognitive problem solving training with children. *Journal of Child Psychology and Psychiatry and Allied Disciplines, 26*, 17–41.

Platt, J. (1989). *Heroin addiction: Theory, research, and treatment*. Malabar, FL: Kreiger.

Platt, J., Morell, J., Flaherty, E., & Metzger, D. (1982). *Controlled study of methadone rehabilitation process* (Final Rep. No. R01-DA01929). Philadelphia: National Institute on Drug Abuse.

Platt, J., Scura, W., & Hannon, J. R. (1973). Problem-solving thinking of youthful incarcerated heroin addicts. *Journal of Community Psychology, 1*, 278–281.

Richard, B. A., & Dodge, K. A. (1982). Social maladjustment and problem-solving in school-aged children. *Journal of Consulting and Clinical Psychology, 50*, 226–233.

Scarr, S., Weinberg, R. A., & Levine, A. (1990). *Understanding development*. San Diego, CA: Harcourt Brace Jovanovich.

Shure, M. B. (1986). *Problem solving and mental health of ten- to twelve-year olds* (Final Summary Rep. No. MH35989). Washington, DC: National Institute of Mental Health.

Shure, M. B. (1992a). *I Can Problem Solve (ICPS): An Interpersonal Cognitive Problem Solving Program* [preschool]. Champaign, IL: Research Press.

Shure, M. B. (1992b). *I Can Problem Solve (ICPS): An Interpersonal Cognitive Problem Solving Program* [kindergarten and the primary grades]. Champaign, IL: Research Press.

Shure, M. B. (1992c). *I Can Problem Solve (ICPS): An Interpersonal Cognitive Problem Solving Program* [intermediate elementary grades]. Champaign, IL: Research Press.

Shure, M. B., & McCown, W. (1993). *Radio-based ICPS interventions for enhancing community public health.* Unpublished manuscript, WJJZ FM Radio, Philadelphia.

Shure, M. B., & Spivack, G. (1970). *Cognitive problem-solving skills, adjustment and social class* (Research Rep. No. 26). Philadelphia: Department of Mental Health Sciences, Hahnemann Medical College and Hospital.

Shure, M. B., & Spivack, G. (1978). *Problem-solving techniques in childrearing.* San Francisco: Jossey-Bass.

Shure, M. B., & Spivack, G. (1982). Interpersonal problem-solving in young children: A cognitive approach to prevention. *American Journal of Community Psychology, 10,* 341–355.

Shure, M. B., & Spivack, G. (1988). Interpersonal cognitive problem solving. In Price, R. H., Cowen, E. L., Lorion, R. P., & Ramos-McKay, J., *Fourteen ounces of prevention* (pp. 69–82). Washington, DC: American Psychological Association.

Skinner, B. F. (1957). *Verbal behavior.* Englewood Cliffs, NJ: Prentice Hall.

Spivack, G., Platt, J., & Shure, M. B. (1976). *The problem solving approach to adjustment.* San Francisco: Jossey-Bass.

Spivack, G., & Shure, M. B. (1974). *A cognitive approach to solving real-life problems.* San Francisco: Jossey-Bass.

Spivack, G., & Shure, M. B. (1985). ICPS and beyond: Centripetal and centrifugal forces. *American Journal of Community Psychology, 13,* 226–243.

Spivack, G., & Swift, M. (1977). The Hahnemann High School Behavior (HHSB) Rating Scale. *Journal of Abnormal Child Psychology, 5,* 299–308.

Thomas, R. M. (1985). *Comparing theories of child development* (2nd ed.). Belmont, CA: Wadsworth.

Weiner, J. A. (1978). *A theoretical model of the affective and social development of learning-disabled children.* Unpublished doctoral dissertation, University of Michigan, Ann Arbor.

Weissberg, R. P., & Gesten, E. L. (1982). Considerations for developing effective school-based social problem-solving (SPS) training programs. *School Psychology Review, 11,* 56–63.

Weissberg, R. P., Gesten, E. L., Carnrike, C. L., Toro, P. A., Rapkin, B. D., Davidson, E., & Cowen, E. (1981). Social problem solving skills training: A competence-building intervention with second- to fourth-grade children. *American Journal of Community Psychology, 9,* 411–423.

Weissberg, R. P., Gesten, E. L., Liebenstein, N. L., Schmid, K. D., & Hutton, H. (1979). *The Rochester Social Problem Solving (SPS) Program: A training manual for teachers of 2nd–4th grade children.* Rochester, NY: Center for Community Study.

Yu, P., Harris, G. E., Solovitz, B. L., & Franklin, J. L. (1986). A social problem solving intervention for children at high risk for later psychopathology. *Journal of Clinical Child Psychology, 15,* 30–40.

AUTHOR INDEX

Numbers in italics refer to listings in reference sections.

Mather, K., 61, 69
Matlin, M., 377, *384*
Matsushima, J., 285, *306*
Matthews, G., 155–158, 161–170, 172–
175, 177–179, *183*
Maxwell, M. E., 200, *221*
May, R., 122, *137*
McAllister, T. W., 312, *321*
McAuliffe, W. E., 232, *242*, 375, *384*
McBride, W. J., 234, *243*
McCalley-Whitters, M. K., 230, *246*
McCaul, M. E., 229, *239*
McClain, J., 152, *181*
McClough, J. F., 212, *218*
McCombs, A., 112, *115*
McConaghy, N., xviii, *xxix*
McCord, C., 228, *243*
McCord, W., 228, *243*
McCormick, D. A., 27, 29, *36*
McCown, W., xix, *xxix*, 4, 17, 20, 26, 131,
132, 138, 225, 226, 228, 236,
238, *243*, 265, 266, 268, 269,
*276, 280, 285, 289, 290, 292,
296, 297, 304, 306*, 323, 325,
*326, 328, 330, 331, 333, 335,
337, 339, 342, 343, 361, 364–
366, 375, 383–384, 393, 399,
402, 403*
McCrae, R. R., xviii, *xxviii*, 47, 55, 79, 81,
88, 151, *183*, 250, 251, *260*
McCullers, C., 135
McDermott, M., 214, *222*
McDevitt, S. C., 25, *35*
McDiarmid, C. G., 42, *56*
McElroy, S. L., 204, *222*
McGloin, J., 399, *402*
McGree, S. T., *219*
McGurk, B. J., 145, *149*
McIntosh, K., 348–349
McIntyre, M., 82, *89*
McKenna, G. J., 225, 235, *243*, 370, *384*
McKenzie, P., 339, *342*
McLeer, S., 300, *305*
McLellan, A. T., 229, 230, 236, *243, 246*
Mead, M., 368, 370, *384*
Meadow, R., 369, 370, 373, *384*
Meehl, P., 328, *343*
Mehrabian, A., 144, *149*
Meichenbaum, D., 299, *306*
Meisler, A. W., 227, *240*
Melges, F. T., 52, *56*
Menninger, K., 13

Merckelbach, H., 195, *220*
Mercy, J., xvi, *xxix*
Merry, J., 203, 205, *218*
Merton, R., 369, *384*
Mesibov, G. B., 26, *37*
Messer, S., 282, 291, *306*
Messer, S. B., 152, *183*, 226, *243*
Metzger, D., 399, *402*
Meyer, D. E., 152, 166, 172, 177, *181*,
267, *275*
Meyer, R. E., 229, 230, *243*
Meyers, K., 231, 236, *242*
Michael, J. L., 229, 244, *246*
Michaelson, B. S., 230, *245*
Milano, M., 348–350
Milich, R., 279–283, *306*
Millar, K., 154, *183*
Miller, D., 361, *384*
Miller, H., 135
Miller, J. B., 369, *384*
Miller, L., 82, *89*
Miller, L. A., 31, *36*
Miller, N. E., xix, *xxviii*, 42, *56*
Miller, R. L., 102, *117*
Millon, T., 6, *20*
Milner, B., 82, *89*
Milstein, R. M., 192, *222*
Minors, D. S., 154, *183*
Mirin, S. M., 227, 229, 244, *246*
Mischel, W., xix, *xxix*, 17, 20, 24, *36*, 57–
58, 69, 281, 297, *306*, 324, 325,
344
Missildine, W. H., 248, *262*
Mitchell, J. E., 198, 205, 206, 210, 214,
219, *222*
Moldofsky, H., 211, *219*
Molliver, M. E., 235, *244*
Monk, T. H., 154, *182*
Moore, C. A., 247, *260*
Moore, K., xvii, *xxix*
Moore, L., 346–347
Moore, M., 17, *21*, 286, 295, *305*
Morell, J., 399, *402*
Morey, L., xviii, *xxix*
Morgenstein, L., 348–349
Moritsugu, J., 387, *402*
Morrell, W., 201, *224*
Morris, J., 253, *263*
Morrison, G., 283, *302*
Morrison, R. L., 229, *244*
Morrison, V., 230, *241*
Morse, S. B., 337, *343*

SUBJECT INDEX

Autogenesis, 131
Avoidant procrastination, 269, 271–272, 274
Aw (awareness) skills, 97–98
Awareness (Aw) skills, 97–98
Awareness of others' emotions, 392–395

Barratt, Ernest, 15–16, 175–176
Barratt Impulsiveness Scale (BIS), 41, 43–44, 52, 53
Basic Personality Inventory (BPI), 251–253
BDI (Beck Depression Inventory), 255, 257
Beck Depression Inventory (BDI), 255, 257
Beck Hopelessness Scale, 257
Behavior, theories on interpretation of, 99
Behavior Problem Checklist, 284
Behavioral approach, 47, 48
  neurobehavioral disorders, 319
  techniques, 337
Behavioral Inhibition System, 83
Behaviorism, operant, 15
Belvzov-Zhabotinsky (BZ) reactions, 126–127
Bender-Gestalt Test, 285
Bible, 5–7
Bifurcation point, 126–127
Binge eating. *See also* Bulimia
  alcohol abuse and, 203–204
  familial impulsivity and, 201–202
Biological approaches to impulsivity, 15–16, 47, 48, 64–66
BIS (Barratt Impulsiveness Scale), 41, 43–44, 52, 53
Bonet, T., 9
Borderline personality disorder (BPD), 133, 230–231, 337
BPD. *See* Borderline personality disorder
BPI (Basic Personality Inventory), 251–253
Brady v. Hopper, 352
Brain dysfunction, impulsivity from, 42–43, 311–318
  AIDS dementia, 315–316
  closed-head injury, 313–314
  damage and trauma, 65, 82–83, 292, 312–313
  dementia of frontal lobe type, 314–315

organic personality syndromes, 317–318
organization, brain, 24–25, 33–34, 53
spontaneous electrical activity, 29–30
Brief therapies, 338–339
Broadbent, D. E., 271–272
Brown Holtzman Study Habits Inventory, 45
Bulimia, 198–216
  affective disorder and, 210
  alcohol abuse and, 198–200, 202–208
  chemical dependency and, 205–206
  familial impulsivity and, 198–202
  impulsivity measurements and, 207–208, 211–213
  kleptomania and, 204, 206, 208–209, 213
  obesity and, 213–214
  substance abuse and, 207, 212–213
  treatment results, 215–216
Burka, J. B., 266
Buss, Arnold, 16, 294
BZ (Belvzov-Zhabotinsky) reactions, 126–127

C (context) skills, 98
Caffeine, 153, 154, 156, 160–167
Calvin, John, 8
CAPS (Child-Adolescent Perfectionism Scale), 252–253
Caregivers, interactions with, 26
CC (conductive/creative) style (intimate relationships), 103–104, 106
Cerebrospinal fluid (CSF), 27–28
CFQ (Cognitive Failures Questionnaire), 271–272
Chaos, impulsivity and, 120, 129–136
  addictions, 132–133
  age factor, 129–130
  borderline personality disorder, 133
  control, development of, 130–131
  creativity, 134–135
  crisis, role of, 132
  pathological impulsivity, 131–133
  self-psychology theory, 133–134
  testing for relationship of, 129
Chaos theory, 120–127, 375–376
  feminist therapy, relevance to, 376–377

Denial of impulsivity (by client), 327
Depression
    impulsive behavior and, in children and adolescents, 290–291
    perfectionism and, 258
    and socially prescribed perfectionism, 250–251
Descriptive levels, 99
Developmental competence theory, 99
Developmental disabilities
    impulsiveness and, 291
Developmental models of impulsivity, 294–295
Developmental theories, 17–18
Dewey, John, 389
Dexedrine, 27
DFT (demential of frontal lobe type), 314–315
Diagnostic and Statistical Manual of Mental Disorders (DSM), 198
    childhood disorders in, 287
    impulsivity in, 120, 323, 325
    Kraeplin's role in, 12
    personality as defined in, 51
    substance abuse in, 227–230
Dickman, S. J., 58–59, 267–268, 270–271
Dickman Scale, 208
Dieting, impulsive eating and, 193–196
Differential rate of low responding procedure (DRL), 281–282
Differentiation, personality, 102
Dilantin (Phenytoin), 42
Dimensions of Temperament Survey (DOTS), 81
Direct observation, rating impulsive behavior by, 283–284
Disease model of impulsive behavior, xix
Dopamine (DA), 28, 65, 84–86, 293
DOTS (Dimensions of Temperament Survey), 81
Draw-A-Line-Slowly test, 281
DRL (differential rate of low responding procedure), 281–282
Drug abuse. *See* Alcoholism and alcohol abuse; Substance abuse
DSM. *See Diagnostic and Statistical Manual of Mental Disorders*
Durlak, J. A., 399–389
Dysfunctional impulsivity
    eating behavior, 191–193, 195–196

encoding tasks, performance on, 155, 157–160
functional impulsivity vs., 58–59, 152, 267–268, 270–271
procrastination and, 273–274

E (emotionality) skills, 97
E (extraversion) scale, 77, 84, 153
E factor (Extraversion-Introversion), 142
EASI impulsivity scale, 75
Eating disorders, impulsivity and, 185–217
    arousability, role of, 187–189
    binge eating, 201–202
    definition of impulsive eating, 186–187
    dieting, 193–196
    dysfunctional impulsive eating, 191–193, 195–196
    external responsiveness, role of, 189–190
    family history, role of, 197–202
    kleptomania, association with, 204, 206, 208–209
    other impulsive disorders, association with, 202–215
    risk factor, impulsivity as, 197–202
    substance abuse, association with, 203–204
    treatment implications, 215–217
EEG (electroencephalogram), 29–30, 66
EFT (Embedded Figures Test), 157, 158, 175, 295
Ego, 13, 248
Electroencephalogram (EEG), 29–30, 66
Embedded Figures Test (EFT), 157, 158, 175, 295
Emotional expression, effect of neurobehavioral disorders on, 310
Emotionality (E) skills, 97
Emotions, awareness of others', 392–395
Empathy, 143–147
Encoding ability, impulsivity and, 155–158, 174, 177
Enlightenment, 9
Entropy, 126, 127
Environmental factors in impulsivity, 26, 61–64
EPI (Eysenck Personality Inventory), 141, 142, 227
Epilepsy, 65, 82–83
Epistasis, 62

Gordon, M., 281–282
Gordon Diagnostic System, 300
Greece, ancient, 6–7
Gross motoric measures of impulsivity, 280
Group therapies, 339–340
    for impulsive women, 374–375
Grubin, D. H., 231–232

Hallucinogens, 235
Hammurabi, Code of, 6
Haptic Matching Task, 295
Harmavoidance subscale (Personality Research Form), 72–73
Heart rate, fluctuations in, 128
Hendrix, Jimi, 362
Hepatic encephalopathy (portal-systemic encephalopathy), 316–317
Heritability of impulsivity, 61–64
Heymans, G., 14
HIAA (5-hydroxyindoleacetic acid), 28–29, 235
Hinckley, John, Jr., 352
Hippocrates, 6
Hirt, Ehrlheim, 11–12
Holiday, Billie, 362
Homeostasis, 128
Homovanillic acid (HVA), 28
5-HT. See Serotonin
Hull, C. L., 40–41
Human Figures Drawing Test, 285
Human physiology, fractal structures in, 125–126
Huntington's chorea (Huntington's disease), 315
HVA. See Homovanillic acid
5-hydroxyindoleacetic acid. See HIAA
Hyperthymic personality, 12
Hysterical personality, 13

I_5 questionnaire, 143–146
I_7 questionnaire, 146–148
ICCI, 298
ICPS. See Intrapersonal cognitive problem solving
Id, 12–13
Ideographic approach, 11–13
Imipramine, 28
Immediate, emphasis on, 99–101
Importance, attribution of, 105–106
ImpSS factor, 77–78
    eating behavior and, 191–193

Impulse Control Categorization Instrument, 285
Impulses (term), 4
Impulsion (term), 7
Impulsive (term), 8
Impulsive Unsocialized Sensation Seeking (ImpUSS), 75–78, 85
Impulsiveness/impulsivity (terms), xv, 4–5, 23, 33, 120, 279, 325. *See also under individual headings*
ImpulsivenessN (ImpN), 143
Impulsivity subscale (Personality Research Form), 72–73
Impulsivity (term). *See* Impulsiveness/impulsivity
ImpUSS (Impulsive Unsocialized Sensation Seeking), 75–78, 85
Infants
    chaotic behavior in, 129
    impulsivity in, 23
    later years, impulsive behavior in, 25–26
    of overweight parents, 192
    self-organization in, 131–132
    sensation seeking in, 80
Information processing, impulsivity and, 151–180
    arousal, 153–155
    attention, role of, 176–180
    Barratt's theory, 175–176
    complex problem solving, 165–166
    encoding ability, 155–158
    Eysenck's theory, 168–170
    individual differences, role of, 152–153
    memory, 162–165
    motor response, 166–167
    Revelle's theory, 171–175
    spatial ability, 160
    visual comparison tasks, 158–160
    visual search tasks, 160–162
Inhibition of common motor responses, impulsiveness and, 281–283
Insulin release, eating behavior and, 190
Internal-external hypothesis of obesity, 186, 189
Interpretation of behavior, 99
Interventional relevance, 96
Interview sensitization techniques, 45
Interviews, limitations of clinical, 327–328
Intimacy, 106–107

Intimate relationships, styles in, 102–104
Intraclass correlation, 61
Intrapersonal cognitive problem solving
    (ICPS), 387–400
    awareness of others, 392–395
    implementing training for, 390–392
    non-impulsivity applications of, 398–
        400
    research evaluations, 395–398

Johnson, J., 269–270, 362–365
Joplin, Janis, 362
Juvenile delinquency, 45–46, 287–288
    eating disorders and, 204

Kagan, Jerome, 17
Kagan Matching Familiar Figures Test, 46
Kahn, E., 12
Kant, Immanuel, 10
Kennedy, H. G., 231–232
Kleptomania, eating disorders and, 204,
        206, 208–209, 213
Knox, John, 8
Kraeplin, Emil, 12
Kretschmer, Ernst, 14

Lacey, J. H., 209–210, 231
Language comprehension, children's prob-
        lem-solving abilities and, 391–392
Lazare, A., 48
Learning disabilities, 289, 366
Leedy v. Hartnett, 352
Legal issues in client treatment, 345–359
    nonviolent impulsive behavior, 353–
        354
    self-protection of clinician, 354–357
    sexually impulsive clients, 358
    violent impulsive behavior, 346–353
Leonard, Linda, 135
Limit cycle, 123
Lipari, Dennis, 351
Lipari, Ruth, 351
Lipari v. Sears, 350–352, 353
Lithium carbonate, 52, 336
Liveliness factor, 143
Long-term memory (LTM), 163–164, 168–
        169
Lorenz, Edward, 121–122
Lorenz attractor, 123
Lowe, M. R., 193–194
LTM (long-term memory), 163–163, 168–
        169

Luther, Martin, 8

Maddi, Salvatore, 14
Mann, L., 270
MAO. *See* Monoamine oxidase
Matching Familiar Figures Test (MFFT),
        25
    epileptics, 82–83
    impulsivity, 159–160
    MAO levels, 84
    reflection-impulsivity, 152, 282–283
    thrill seeking, 74
    treatment, effects of, 298, 300
McCown, W., 236, 238, 269–270
McCrae, R. R., 250–251
McIntosh, Kimberly, 348–349
McIntosh v. Milano, 348–350
Mead, Margaret, 368
Measurement of impulsivity, 24. *See also*
        Assessment of impulsivity
    artificial motoric tests, 282–283
    in children and adolescents, 280–
        287
    direct observation, 283–284
    everyday behavior, correlations with,
        60
    gross motoric measures, 280
    inhibition of motor responses, 281
    integration of data, 44–47
    physiological indexes, 286
    projective measures, 285–286
    ratings by other individuals, 284–
        285
    reinforcement schedules, 281–282
    self-reports, 285
    and superfactors of personality, 141–
        148
Medical students, 44–45
Melancholic temperament, 14
Memory, impulsivity and, 162–165. *See
        also* Long-term memory; Short-
        term memory
Mental disorders, co-occurrence of sub-
        stance abuse with, 227–232
    Axis I disorders, 228–229
    Axis II disorders, 229–232
    and psychopathology in childhood
        and adolescence, 228
Mental health, impulsiveness in women
        and myths about, 367
Mental retardation, impulsiveness and,
        291

Mesomorphy, 14
Metabolic neurobehavioral disorders, 316–317
3-methoxy-4 hydroxy-phenethyleneglycol (MHPG), 28
Methylphenidate, 27, 299, 300
MFFT. *See* Matching Familiar Figures Test
MHPG (3-methoxy-4 hydroxy- phenethyleneglycol), 28
Michigan Alcoholism Screening Test, 207
Middle Ages, 7–8
Milano, Michael, 348–350
Minnesota Multiphasic Personality Inventory (MMPI), 285, 329
    advantages of, 331
    eating-disordered individuals, 200–202, 208, 209, 211–213
    interview, using data for, 332
    perfectionism, 254–256
    in Social Training program, 113
    weaknesses of, 58
Mischel, Walter, 16
MMPI. *See* Minnesota Multiphasic Personality Inventory
Modification of impulsivity, 26–27
Monoamine oxidase (MAO)
    impulsivity and levels of, 27–29, 32–33, 65–66
    sensation seeking and levels of, 71–72, 83–85
Moore, Lawrence, 346–347
Moral model of impulsive behavior, xix
Morgenstein, Lee, 348–349
Mother, role of, 105–107, 109–112
Motor response, impulsivity and
    Barratt's theory, predictions of, 175
    Eysenck's theory, predictions of, 169
    measurements, 280–281
    motor control, 166–167
    repetitive motor responses, 167
    Revelle's theory, predictions of, 172
Mountain climbers, 71–72
MPS. *See* Multidimensional Perfectionism Scale
Multi-impulsive personality disorder, 231–232
Multidimensional Perfectionism Scale (MPS), 251, 253–255, 257, 258
Multidimensional Personality Inventory, 212

Multifocal therapeutic intervention, 340–341

N factor (Neuroticism-Stability), 142
Narcissistic personality, 13
Narrow heritability, 62
Narrow Impulsivity, 73–74
NE (norepinephrine), 28, 293
Negotiation
    within families, 107–109
    in resource exchange theory, 101
NEO Personality Inventory, 81, 250, 251
Neurobehavioral disorders, impulsivity and, 309–320
    children and adolescents, 292–293
    degenerative diseases, 314–317
    frontal lobe syndromes, 312–314
    organic personality syndromes, 317–318
    sensation seeking, 80–83
    treatment strategies, 318–320
Neurodevelopment, impulsivity and, 23–34
    brain electrical activity, 29–30
    early influences, 25–26
    genetic factors, 24–25
    interventions, 26–27
    monoamine metabolism, 27–29, 32–33
    representational capacity and imbalances, 30–32
    threshold/timing factors, 32
Neurosis, 128, 250–251
Neuroticism-Stability (N) factor, 142
New Testament, 7
Nineteenth Century, theories of impulsivity in, 9–11
Nomothetic approach, 11, 13–16
Nonplanning, 73
Norepinephrine (NE), 28, 293

Obesity
    and impulse control problems among bulimics, 213–214
    theories of, 186
Observation, rating impulsive behavior by direct, 283–284
Obsessive behavior, 128–129
Old Testament, 5–6
Olfactory system, chaos and, 128
Ontogeny, 127–128
Open systems, 127

Openness to experience, 81
Operant behaviorism, 15
Orderliness, 59
Organic personality syndromes, 317–318
Other-oriented perfectionism, 249–250

P factor (Psychoticism-Conformity), 142
P (psychoticism) scale, 75, 76, 81, 82
Parker, Charlie, 362
Pemoline, 27
Perfectionism, impulsivity and, 247–259
 conscientiousness, 250–251
 direct investigations, 252–254
 problem solving, 251
 suicidal impulses, 254–258
 types of perfectionism, 249–250
Personality. *See also* Personality disorders;
  Temperament(s)
 contextuality of development of, 94
 differentiation of, 102
 general systems model of, 46–53
 ideographic theories of, 11–13
 modalities of, 101–102
 nomothetic theories of, 14–16
 traits, 57–58
Personality disorders
 among eating-disordered individuals,
  209–210
 among substance abusers, 229–230
 impulsiveness and, in children and
  adolescents, 291
Personality Inventory for Children (PIC),
 284–285
Personality Research Form, 72–73, 250
Pharmacological treatments, 26, 27, 300,
 336–337
Phase space, 122
Phenotypical descriptive sublevel, 99
Phenytoin (Dilantin), 42
Phlegmatic temperament, 14
Phrenology, 10
PIC (Personality Inventory for Children),
 284–285
Pick's disease, 292
Pinel, P., 9–10
Plasma cortisol levels, impulsivity and,
 234
Plomin, Robert, 16, 294
Poddar, Prosenjit, 346–347
Portal-systemic encephalopathy (hepatic
 encephalopathy), 316–317
Porteus Maze Test, 41, 282, 283, 298

Posttraumatic stress disorder (PTSD), 290
Powelson, Harvey, 347
Prediction, scientific, 119, 121
Prenatal substance abuse, impulsive behavior and, 296
Prichard, J. C., 10
Prigogine, Ilya, 126–127
Principia Theologica, 8
Problem solving. *See also* Intrapersonal
 cognitive problem solving
 by children, 390
 impulsivity and, 165–166
 perfectionism and orientation to, 251
Process/structure dichotomy, 133–134
Procrastination, 265–274
 avoidant, 269, 271–272, 274
 dysfunctional impulsiveness and,
  273–274
 empirical data, 269–273
 motives for, 268–269
 as opposite of impulsiveness, 267–
  268
Projective measures, as indexes of impul-
 siveness in children, 285–286
Pseudopsychopathic personality disorder,
 312, 318
Psychodynamic/psychoanalytic approaches,
 296, 300–301, 337–338
Psychometric assessments, 284, 330–332
Psychopaths, 86
Psychophysiological measures of impulsiveness, 286
Psychosomatic theory of obesity, 186
Psychosurgery, 82
Psychoticism (P) scale, 75, 76, 81, 82
PTSD (posttraumatic stress disorder),
 290
Purging, 210–211

Questionnaires, 24. *See also specific ques-*
 tionnaires
Queyart, F., 11

R (rationality) skills, 97
Rape by adolescents, 287–288
Rapport with impulsive client, 330
Rationality (R) skills, 97
Reactive/repetitive (RR) style (of relat-
 ing), 103–104, 106
Reagan, Ronald, 352
Reducibility (of theories of impulsivity),
 96

Social Training program, 113–114
Socialization, and negative view of impulsiveness in women, 367–369
Socrates, 6–7
Somatonia personality type, 14
Space, 104–105
Spatial ability, impulsivity and, 160
Spitzer, L., 187–189
Spivack, G., 388–392, 396–398
Spontaneity, 59
SPSI (Social Problem Solving Inventory), 251
SSS (Sensation Seeking Scale), 72–73, 75, 81
Stealing. *See* Kleptomania
Stevens, S. S., 14
STI (Strelau Temperament Inventory), 81
Stimulus encoding ability, impulsivity and, 155–158
STM. *See* Short-term memory
Strange attractors, 123, 376
Strelau Temperament Inventory (STI), 81
Stress, eating behavior and, 193
Stroop Color and Word test, 166, 281
Structured Clinical Interview, 230
Substance abuse, 225–239. *See also* Alcoholism and alcohol abuse
  children and adolescents, impulsivity in, 290
  eating disorders and, 203–204, 207, 212–213
  ICPS training as treatment for, 399–400
  mental disorders, co-occurrence with other, 227–232
  prenatal abuse, 296
  and prevalence of impulsiveness, 226–227
  psychobiological links with impulsiveness, 233–235
  treatment for impulsivity and, 235–237
Suicide and suicidal behavior
  bulimics, 208
  perfectionism and, 254–258
Superego, 13, 296
Syndromes of disinhibition, 153
Syphilis, 315
Systems model of personality, 46–53

Tarasoff, Tatiana, 347

Tarasoff v. Regents of the University of California, 346–348
TAS (Thrill and Adventure Seeking), 59–60, 71, 72
TAT (Thematic Apperception Test), 285, 364
Temperament(s)
  as explanation of impulsiveness, 293–294
  Hippocratic concept of, 6
  Hirt's concept of, 11–12
  sensation seeking and, 79–81
Termination of therapy/treatment, 326, 333–335
Thematic Apperception Test (TAT), 285, 364
Therapist, sex of, 377
Therapist-patient sexual intimacy (TPSI), 358
Think Aloud program, 299
Thrill and Adventure Seeking (TAS), 59–60, 71, 72
Time, 105
Time intervals, underestimation of, 52
Time zone anchors, 46
Toddlers, impulsivity in, 23
TPSI. *See* Therapist-patient sexual intimacy
Training programs (for treatment of impulsivity), 26
Traits, as explanation of impulsiveness, 293
Trajectory, 122–123
Tramer, M., 12
Treatment of impulsivity, 332–341
  behavioral and cognitive techniques, 337
  brief therapies, 338–339
  children and adolescents, 297–301
  eating disorders, 215–217
  family and group therapies, 339–340
  fees, payment of, 334–335
  legal issues. *See* Legal issues in client treatment
  medication, 300, 336–337
  multifocal therapeutic intervention, 340–341
  neurobehavioral disorders, 318–320
  psychodynamic techniques, 337–338
  scheduling problems, 332–334
  substance abusers, 235–237

ABOUT THE EDITORS

William G. McCown received his master's and doctorate in psychology from Loyola University of Chicago. Presently, he is a research scientist in neuropharmacology at the Nathan Kline Institute (NKI) in Orangeburg, New York. NKI is an affiliate of the World Health Organization and of the New York University School of Medicine, where Dr. McCown also holds an appointment. Previously, Dr. McCown was an associate professor of mental health sciences at Hahnemann University and a clinical assistant professor at Tulane University School of Medicine.

Judith L. Johnson received her master's degree from George Mason University and her doctorate in Counseling Psychology from Loyola University of Chicago. Presently, she is Assistant Professor, Department of Psychology, Villanova University, and Director of Research, Bucks/Upper Montgomery County Neurologic Group, in North Wales, Pennsylvania. Her postdoctoral training in research neuropsychology was completed at the New Orleans Veterans Administration Medical Center.

Myrna B. Shure is a professor in the Department of Mental Health Sciences at Hahnemann University. Over the past 25 years, she has developed I Can Problem Solve (ICPS): An Interpersonal Cognitive Problem Solving program, which helps children learn to think and to solve problems for themselves. Her programs, which help decrease or prevent impulsive and other behavior problems at school and at home, have been successfully integrated into a number of school and community settings across the country. Dr. Shure received her doctorate in child development and family relationships from Cornell University.